Studies in Applied Philosophy, Epistemology and Rational Ethics

Volume 17

About this Series

Studies in Applied Philosophy, Epistemology and Rational Ethics (SAPERE) publishes new developments and advances in all the fields of philosophy, epistemology, and ethics, bringing them together with a cluster of scientific disciplines and technological outcomes: from computer science to life sciences, from economics, law, and education to engineering, logic, and mathematics, from medicine to physics, human sciences, and politics. It aims at covering all the challenging philosophical and ethical themes of contemporary society, making them appropriately applicable to contemporary theoretical, methodological, and practical problems, impasses, controversies, and conflicts. The series includes monographs, lecture notes, selected contributions from specialized conferences and workshops as well as selected PhD theses.

More information about this series at http://www.springer.com/series/10087

Chiu-Shui Chan

Style and Creativity in Design

 Springer

Chiu-Shui Chan
Architecture Department
Iowa State University
Ames, Iowa, USA

Every effort has been made to contact the copyright holders of the figures and tables which have been reproduced from other sources. Anyone who has not been properly credited is requested to contact the publishers, so that due acknowledgment may be made in subsequent editions.

ISSN 2192-6255   ISSN 2192-6263 (electronic)
Studies in Applied Philosophy, Epistemology and Rational Ethics
ISBN 978-3-319-14016-2   ISBN 978-3-319-14017-9 (eBook)
DOI 10.1007/978-3-319-14017-9

Library of Congress Control Number: 2014958566

Springer Cham Heidelberg New York Dordrecht London
© Springer International Publishing Switzerland 2015

Printed on acid-free paper

Springer International Publishing AG Switzerland is part of Springer Science+Business Media (www.springer.com)

To Hungching, Dexter, Virginia

Preface

Design can be explained from two perspectives. One is from the action point of view that means to do something deliberately; the second can be seen as a product created intentionally by human beings to fit certain purposes. The first definition is the sequence and the last one is the result. Both are actually mental steps or cognitive results of going through sequences of manipulating various kinds of external information from sensory input and internal knowledge from memory, blending them together into an integrated one to create an artifact that satisfies certain requirements or constraints. The various sequences of these mental activities as they unfold in the designers' minds are called *design thinking*.

Studies in design thinking have been approached from areas of problem solving, human cognition, and information processing to explain how a design is cultivated, what methods are used by designers to create a design result, and how to precisely describe the formation of a design from its starting point to its end product. Many studies have been conducted and reported. However, there also are patterns of cognitive processes that trigger some phenomena in the processes and products. These phenomena, namely style and creativity, have been recognized as important factors coming from thinking and used to differentiate design stylistic patterns, classify typologies, and verify. In fact, these phenomena of style and creativity do correlate to each other to some extent, and it is crucial to know how the mind functions creatively and stylistically so that designers can improve design productivity. Yet, notions of the character of style and creativity, and their correlations, have not been systematically discussed in literature from the problem solving, cognition, and information processing points of views. This book serves these purposes.

How did this book come into being? It started in the years I was at Carnegie Mellon University. At that time, I devoted my efforts on the study of design methodologies used by master designers to strategically generate a design. I discovered that design is actually a problem solving sequence, which is operated by cognition. Thus, I took courses in cognitive psychology and tried to explore methods that could be used to study thinking. At that time, notions on association in memory network by John Anderson, parallel distributed processing in mental operation by David Rumelhart, and integrated cognition in cognitive architecture (SOAR) by Alan

Newell inspired me, thus, a number of psychological experiments were conducted to explore psychology in design.

The first experiment I did in 1985 was to study mental image and its internal representation applied by novice and expert designers. Results of the videotaped study showed that expert designers have a richer set of image repertoire stored more hierarchically in memory than novice designers. The rich set of knowledge schemata provides more opportunities for expert designers to quickly and easily combine images together, which explain why expert designers are more creative than novice designers.

The second experiment, conducted in 1986 through the use of protocol analysis, was to study how architects think while they design. In that study, a number of tasks had been carefully developed and applied including: formulated procedures of running the experiment for collecting protocol data, defined methods of coding and decoding verbal data; plus generated systematic techniques of transforming verbal data into a problem behavior graph. In fact, the exact methods of protocol analysis used for studying cognitive processes in design thinking had not been clearly published in detail around 1986. It took me awhile, with much trial and error, to develop the methods on data analysis. Fortunately, I had the opportunities to discuss the technical methods and procedures with the pioneers and forerunners in this field at Carnegie Mellon. Of course, that experiment fascinated me and I started to go further into the study of style.

In studying style, I explored the historical development of research in style. Then, I completed a case study in early 1988 on Frank Lloyd Wright's style to examine how Mr. Wright created his style. Late in that year, the concepts of operational definition, measurement, and recognition of style studied from the product point of view were set up, tested, and justified through four psychological experiments. In 1989, the seventh experiment on studying the creation of style was conducted through observing a successful architect's eight design processes in real time. Again, protocol analyses were applied to get first hand data for proving the hypotheses set up in the studies. It was intended to develop a cognitive theory of style in design.

After my teaching career started at Iowa State University, I was invited to guest lecture at a number of universities. During these years of teaching and research, the same experimental techniques were applied to concentrate on why some designers are more creative and stylistic than others, what cognitive factors generate both style and creativity, and what differences are between the two in results. After the eighth experiment on exploring design expertise, completed in 2000, I started to use other techniques of case study to further explore design thinking. Of course, five case studies after 2001 on a number of master architects showed evidences of the connections between the phenomena of style and creativity, which are discussed here.

Design is a skill and thinking is also a skill. We have to learn how to design and how to think. Likewise, we also have to learn how to generate an individual style and how to be creative. That is why this book intends to provide the knowledge and is what this book is all about.

Ames, IA, USA   Chiu-Shui Chan

Acknowledgements

The fundamental part of the study that treats style and creativity as entities was done at Carnegie Mellon University (CMU) and further developed at Iowa State University (ISU). I was lucky to have a wonderful research environment at CMU, working with Omer Akin, John Hayes and Herbert Simon. I am grateful to Omer Akin for leading me through the exploration of applying psychology to architectural design. His critique on my concepts helped me shape the theory. His support on providing me with lab space and equipment was a great help.

I also want to thank Dick (John) Hayes for providing subjects from his psychology classes to make the experiments on the definition of style part possible. His numerous comments also gave me profound understanding on how to conduct good psychology experiments and how to analyze data. His wisdom and expertise on problem solving stimulated a lot of encouragement on my later study of creativity.

The late Herbert Simon was the one who influenced me the most—intellectually, philosophically, and spiritually. He introduced me to the area of models of thought through cognitive science. I learned from him that science is not necessarily independent of the fine arts. From him, I understand what scientific research is, while you expose yourself to the world of science. From him, I realized the meaning of the state of the arts. But, most of all, I learned from him what makes a good and respectable scientist—be able to grasp the insight into discovery.

I am greatly indebted to the subjects who participated in the seven series of psychological experiments done at CMU. Without their participation, it is impossible to collect the valuable mental data to justify the hypotheses developed that make the theory sound. Particularly, Richard Cleary, who participated in my study of the measurement of style and gave me valuable comments on my early study of Frank Lloyd Wright's Prairie House style. The other key subject provided me with a rich source of protocols. Leonard P. Perfido of L. P. Perfido Associates, an outstanding practicing architect from Pittsburgh, spent many hours in the college lab for a period of 3 months and deserves my special thanks.

The final theory of the study was developed and matured at ISU. I would thank my colleague, David Block, who attended my experiment on studying how designers' expert knowledge, which I termed seasonal knowledge, would affect design style. I must also thank students at the Architecture Group, Institute of Applied Arts, National ChiaoTung University (1997–1998), National Yunlin University of Science & Technology (2005), National ChengKung University (2005), and ISU for the many class discussions. Because of the talks and discussions, more thoughts on creativity and its connections to style were generated, particularly the "Design Thinking" course offered in the fall of 2011 at ISU, and at School of Architecture, Tianjin University, since 2011 to the present time. From offering the courses, I have the opportunities to refine, modify, and improve the theory and thinking in "thinking." I also want to acknowledge Jasmine Brown for her participation on the concept of repetition and offered her drawings to be included in Chap. 2, and Laurel Tweed for providing editing assistances for the manuscript. Supports from the ISU University Publication Endowment Fund 2011 and Faculty Professional Development Assignment Fund 2014 have made this book possible.

Certain images in this book have been made available for use under the Creative Commons license, available at http://creativecommons.org. Regarding other photo acknowledgments, many of the pictures for this book come from secondary or Internet sources. In every case, every effort has been made to contact the copyright holders of the figures and tables which have been reproduced from other sources. Anyone who has not been properly credited is requested to contact the publishers, so that due acknowledgment may be made in subsequent editions. I should also thank William Allin Storrer and Norman McGrath who offered their wonderful image(s) for free use in this book.

The final acknowledgements should go to my family for their patience while I spent so much time on the writing of this book, which is the extension of concepts written in dual languages of English and Chinese published in *Design Cognition: Cognitive Science in Design* by the China Architecture and Building Press.

Contents

1 Introduction 1

Part I Conceptual Framework

2 Introduction of Design Cognition 9
  2.1 Definition of Design 9
  2.2 Design Thinking and Cognition 10
  2.3 Development of Design Studies 11
  2.4 Cognitive Mechanisms and Functions 21
    2.4.1 Design Is Problem Solving 21
    2.4.2 Design Is Making Associations:
           The Nature of Knowledge 26
    2.4.3 Design Is Goal and Constraint Driven 33
    2.4.4 Design Is Reflective and Problem Structuring 37
    2.4.5 Design Is Looking for Representation 41
    2.4.6 Design Utilizes Cognitive Strategy 43
    2.4.7 Design Uses Certain Reasoning 51
    2.4.8 Design Applies Repetition for Design Generation 53
  2.5 Overview and Operational Definition of Design Cognition 59
  2.6 Other Affecting Design Thinking Factors 61
  2.7 Study Methods 62
    2.7.1 Controlled Experiment Type 62
    2.7.2 Interview Type 63
    2.7.3 Case Study Type 64
    2.7.4 Think Aloud Type 66
    2.7.5 Other Study Types 68
  References 72

Part II   Style

3   Development of Studies in Style.. 81
   3.1   Historical Development of Art Theories 82
         3.1.1   Theory of Art by Plato 84
         3.1.2   Theory of Art by Reynolds 85
   3.2   Modern Art Theories and Movements 85
   3.3   Historical Development of Studies in Style...................................... 92
   3.4   Other Approaches in Style 98
   3.5   Studies in Individual Style 103
   References 104

4   Style as Identities in Design Products 107
   4.1   Features Identification... 112
   4.2   Style Identification 114
   4.3   Operational Definition of Style: Experiment 1 115
         4.3.1   Experimental Materials... 116
         4.3.2   Summary of the Style Definition 128
   4.4   Degree of Style: Experiment 2.. 128
         4.4.1   Subjects and Materials 128
         4.4.2   Experiment Procedures 130
         4.4.3   Summary of the Strength of Style............................ 135
   4.5   Measurement of Style: Experiment 3 135
         4.5.1   Feature Frequency... 136
         4.5.2   Summary on Measurement of Style........................... 142
   4.6   Recognizability of Stylistic Features: Experiment 4 142
         4.6.1   Subject and Experimental Materials 143
         4.6.2   Experimental Procedures 144
   4.7   The Degree of Style Within Style and Across Styles...................... 148
   4.8   Conclusions... 150
   References 154

5   Style Approached from the Design Process 157
   5.1   Studies Approached from the Means Point of View 157
   5.2   Style Created in the Design Processes 160
   5.3   Case Study: Prairie Houses Style... 161
         5.3.1   Background of Frank Lloyd Wright............................ 162
         5.3.2   Method of Data Collection.. 162
         5.3.3   Replicated Features Appeared in Design 163
   5.4   Constant Concepts and Principles Used in Design 167
   5.5   Constant Constraints Used in Design 169
         5.5.1   Constraints Used in Making Floor Plans 170
         5.5.2   Constraints Used to Determine the Wall Location.... 170
   5.6   Constant Design Methods and Processes.. 172
         5.6.1   Design Methods 172
         5.6.2   Design Processes... 176
         5.6.3   Howe's Interpretation... 177

5.7   Reconstruction of Wright's Design Process.................................... 179
5.8   Matching Features with Design Methodologies 181
        5.8.1   Results from Design Methods.. 182
        5.8.2   Results from Design Principles.. 184
        5.8.3   Results from Design Constraints 185
        5.8.4   Results from Other Factors 186
5.9   Conclusions.. 190
References 193

6   Creation of Style in the Design Process... 197
6.1   Methods for Exploring Design Processes 198
6.2   Hypothetical Factors Generating Styles.. 199
        6.2.1   Personal Preferences on Certain Forms 200
        6.2.2   Design Goals.. 200
        6.2.3   Design Constraints 201
        6.2.4   Search Pattern and Order 202
        6.2.5   Presolution Models 203
        6.2.6   Mental Images... 203
6.3   An Empirical Study.. 204
        6.3.1   Subject... 204
        6.3.2   Experimental Tasks and Procedures 204
6.4   Experimental Results 209
6.5   Data Analysis 214
        6.5.1   Repetitious Features Shown in Products.............................. 214
        6.5.2   Observation on Repetitious Utilization
                    of Design Goals and Procedures.. 216
        6.5.3   Repetitious Constraints Used in Designs............................. 219
        6.5.4   Observation on Repetitious Presolutions
                    Used in Designs 226
        6.5.5   Observation on Repetitious Mental Images
                    Used in Designs 227
6.6   Discussion 231
6.7   Conclusions.. 237
References 238

Part III   Creativity

7   Development of Studies in Creativity 243
7.1   Historical Overview 245
7.2   Current Theoretical Development.. 252
        7.2.1   What Traits Make Creative People Creative? 252
        7.2.2   What Conditions of Society Affect
                    Creativity and the Creative Environment?........................... 254
        7.2.3   What Are the Cognitive Processes
                    Engaging Creativity? 254
        7.2.4   How Is Creativity Assessed?.. 258
        7.2.5   How Can Creativity Be Trained?... 259

7.3   Operational Definition of Creativity in Design 260
7.4   Methods of Creativity Data Collection 263
7.5   Conclusions 266
References 267

Part IV   Style and Creativity

8   Creative Processes and Style 275
8.1   Degree of Style and Creativity 275
8.2   Case Studies 276
       8.2.1   Personal Design Grammar and Rules—Wright 276
       8.2.2   Geographical Context of the Project and Building
                Typologies—Otto Wagner 276
       8.2.3   Design Intentions Developed for Each
                Project—Otto Wagner 277
       8.2.4   Other Examples of Renzo Piano and Tadao Ando 279
8.3   Factors of Creativity in Design 289
       8.3.1   Problem Representation 289
       8.3.2   Expert Knowledge and Unique
                Combination of Knowledge 294
       8.3.3   The Uniqueness of Design Constraints Utilized 295
       8.3.4   The Unique Design Methods or Grammar Applied 296
8.4   Group Big Creativity and Little Group Style 298
8.5   Creativity and Style 299
8.6   Conclusions 300
References 302

9   Cognitive Theory of Style and Creativity 305
9.1   Thinking and Cognition 306
9.2   Design Thinking and Processes 307
9.3   Design Cognition, Style, and Creativity 309
9.4   Cognitive Theory of Style and Creativity 311
9.5   Could Style and Creativity Go Together? 312
9.6   Improving Cognition, Creativity and Style 314
       9.6.1   Can Cognition Be Improved? 314
       9.6.2   Can Style Be Improved? 315
       9.6.3   Can Creativity Be Improved? 315
9.7   Validity of the Theory 316
9.8   Future Studies 317
References 321

Name Index 323

Subject Index 327

Chapter 1
Introduction

Design thinking processes have been studied since the 1970s, exploring the intellectual phenomena that occur while designers practice design. These intellectual phenomena include the formation of intelligence, operation of knowledge, and actions taken for implementing design, which are the cognitive activities explored in the field of psychology. Recently, these cognitive activities that occur in design were recognized as a special cognitive domain with special behaviors, named *design cognition*. Design cognition focuses on the processes and psychological phenomena of how humans understand, process, formulate, generate, store, retrieve, and recycle design related knowledge that leads to the creation of a design. These design processes and cognitive phenomena are not only different from other thinking domains,[1] but are also unique forces that turn design conceptual schemes into physical artifacts. Behind these psychological activities, forces exist that create certain "recognizable features" in design products. These features are recognized through perception to categorize the product as an *individual style*. Due to the existence of *design styles*, outstanding design products can easily be recognized by the public as cultural symbols.

The household product of a tea kettle, designed by **Michael Graves** for **Alessi** (1985) and Target (1999), demonstrated that features of products could cause popularity and change the way Americans look at design in their homes. Other product designers followed the same style, and the kettle revolutionized small household object design. Thus, a design style has eminent power that drives culture and design patterns. In March 2012, ***Time*** magazine published the spring issue of *Time Style & Design*, introducing advances in art and architecture, interior and automotive design, and food and drink. The editor's letter highlighted, "*Style is not something superficial but something intrinsic. It is not gilding the lily but the lily itself.*" This issue

---

[1] In design domains, designers usually utilize mental images when designing. This media is the mental representation of design information. Operational methods of handling such a representation differ from operating different media; for example, symbols, whether abstract or numeric in computational domains, would be treated differently.

© Springer International Publishing Switzerland 2015
C.-S. Chan, *Style and Creativity in Design*, Studies in Applied Philosophy,
Epistemology and Rational Ethics 17, DOI 10.1007/978-3-319-14017-9_1

displayed and illustrated the trend of modern style, and provided a lens to understand the cultural movements in style. Thus, it is the appearance of the lily that defines a style. Yet, the style discussed in this book is on how designers do design, what the cognitive activities are that create a style, and on how a style is recognized. As such, it is not just the lily itself, but on how a lily is created and identified.

Behind the psychological phenomena of mental activities and perceptions, there are other cognitive forces that trigger the creation of a unique design product. A unique design has the following characteristics: it is valuable, the act of creating it is original, and it is performed with special mental abilities that, in all, constitute creativity. Thus, creative products must be the work of initial and original creations. Outstanding creative works are generally treated as culturally influential, publicly recognizable, and socially beneficial products. Thus, creativity and style are two important cognitive phenomenon affecting design products and influencing civilization. If these cognitive aspects and causing factors can be understood, individual style and human creativity can be consciously improved accordingly, to improve design quality. For instance, the **Apple Company**'s digital products supervised by **Steve Jobs** are good examples. **Jobs** applied arts and aesthetics into engineering design, and insisted on simplicity for the user interface in Apple products. His way of design generated a style with an economical and unique appearance that dominated the market internationally. His style set a classic example for the fields of design, engineering, marketing, and management on how critical it is to think creatively. However, full studies on the cognitive factors that trigger style and creativity have rarely been completed. This book intends to explain the causal effects of individual style and creativity, approached from the perspective of design cognition and problem solving theory.

There are nine chapters in this book, concentrating on the study of the human thinking processes causing both style and creativity, approached from cognitive science. Methods used to discuss these two subjects have some differences. Studies on style focus on analyzing the stylistic features and factors that generate style. Studies on creativity focus on cognitive reasons generating creative actions in the design processes. Creativity research has moved towards the specific design methodologies used and specific factors triggering creativity, rather than to study the repetition of design patterns. Thus, the shaping of design representation, which is critical for design thinking, and its influence to creativity are covered more than style itself. This also is because representation has an indirect and intangible influence on style formation.

Chapter 2 introduces the historical development of design studies, concepts of design cognition, theories of problem solving, and all possible cognitive factors that relate to the generation of creativity and style. This chapter intends to set up the fundamental understanding of the studies in thinking approached from cognitive psychology, and outline all possible reasons causing creativity and style in the design process. Thus, cognitive patterns occurred, cognitive strategies applied, and cognitive mechanisms utilized in the design processes are itemized and concisely explained through examples. Particularly, the cognitive mechanism of repetition, which is the major factor causing style and rhythm, is explained by works done by

Chapter 8 introduces the cognitive factors causing creativity in problem solving processes, and then they are analyzed by case studies. Four master architects' design projects, including **Frank Lloyd Wright**, **Otto Wagner**, **Renzo Piano**, and **Tadao Ando**, are used as examples for illustrating the concepts. Data collected from publications has shown the correlations between degrees of style and degrees of creativity to be in positive relations. **Wright** has strongest style and is the most eminent creator with the highest creativity, because his Prairie House designs create and display many features. **Wagner** has fewer features created and repeated in his designs, thus, his style and creativity are ranked second. **Ando**'s design has shown very unique creativity and style with a certain number of features created. His designs have demonstrated the application of the major factor of representation that causes unique style and creativity. **Renzo Piano**'s designs should be seen as a part of group style and group creativity, for his projects are mainly team works. From these case studies, the correlations between style and creativity in design are explained clearly.

Chapter 9 is the concluding chapter that summarizes the studies in thinking and cognition, and the cognitive mechanisms causing style and creativity. Following the summaries on cognitive operations that generate style and creativity, a number of methods on how to improve cognitive skill to improve style and creativity are proposed as a part of the cognitive theory of style and creativity. The theory defined is also verified by ecological validity to see whether the findings from the studies are extendable and generalizable to real world problems. Future studies in this regard, particularly in interdisciplinary cooperation with neuroscience, are briefly outlined as a conclusion of this book.

Hopefully, concepts explained through case studies and theories developed from psychological experiments in this book should provide inspirations and stimulations to professional designers and academics in design related fields. Parts of the concepts explained in Chaps. 4, 5 and 6 have been briefly discussed in the journals *Design Studies*, *Environmental and Planning B: Planning and Design*, and the *Journal of Architectural and Planning Research*. But, this book provides comprehensive and thorough descriptions on the cognitive theory of style and creativity. Future studies will be conducted to further verify a number of details provided in this book, which will be published in both English and Chinese language versions.

Part I
Conceptual Framework

Chapter 2
Introduction of Design Cognition

2.1 Definition of Design

Design is the human conception and planning of virtually everything in the world. All man-made design has as its fundamental essence that everything is driven by certain intentions and is accomplished by a series of actions to generate results. Design is process, artifact, and discipline. As explained in the American Heritage dictionary, to design, seen from the action point of view, is: (1) to conceive or fashion in the mind, invent; (2) to have as a goal or purpose, intend; and (3) to create or contrive for a particular purpose or effect. Synthesizing these definitions into an integrated conceptual framework, design can be described as to conceive a purpose, contrive a goal, and formulate a plan for a purposeful intention in the mind. On the other hand, design seen from the perspective of an entity, is: (1) a drawing or sketch; (2) a graphic representation; (3) a particular plan or method; (4) a reasoned purpose; or (5) a deliberate intention (American Heritage Dictionary 2013). Here, a purpose and an intention are both treated as entities, for they are the products of creative actions. Thus, design is a created object, a generated method, a developed purpose, or a conceived intention for everyday routines conducted through mental efforts. Design should be recognized as an essential part of human life—in fact, a critical component of human intelligence—and, as such, deserves critical discourse.

If looking at design from the fields of making artifacts (e.g., architectural, graphic, fashion, industrial, interior, and engineering design, to name just a few), design can be defined specifically as *all human creative endeavors of shaping objects to meet purposes or constructing a structure adapted to objectives, which require professional consideration on aesthetic beauty, functional uses, social symbols, and market demands and supplies.* All these endeavors can also be seen as solving certain problems with satisfactory solutions. Thus, to make the long definition short, design can be seen as *human endeavors of creating satisfactory solutions or beautiful artifacts to fulfill certain functions*, which explains the essence of design

© Springer International Publishing Switzerland 2015
C.-S. Chan, *Style and Creativity in Design*, Studies in Applied Philosophy,
Epistemology and Rational Ethics 17, DOI 10.1007/978-3-319-14017-9_2

activities—the intellectual creation of feasible solutions or aesthetic artifacts—and these creation efforts must be devoted to meeting functional requirements. This explains a part of design thinking.

In fact, an astounding array of design products are created by designers and shared and used every day by the public. Constant use of any product has dramatic impacts on many aspects of human life. These product impacts are causal effects from design and by design. For instance, usability provided in household products will impact users' productivity (Norman 2002). Phenomenon of patterns created in building forms will yield different perceptions in the built environment to the beholders (Holl 2006). Color and materials used in working environments will affect human cognitive performance (Chan 2007). Since the impact of design products influences our well being and everyday performance, a comprehensive understanding of design, especially design thinking, would help to explain how design is generated to improve design ability and enhance design products. Indeed, design thinking is a part of problem solving activities and is an ability that can be utilized in the creation of artifacts and for other everyday tasks. Techniques used in formulating and solving design problems can also be used to solve relevant problems in our own lives.

2.2   Design Thinking and Cognition

If we see design as a series of mental activities generating entities, then we should see these activities as intellectual processes, the parts of human thinking operated by consciousness. In cognitive psychology, *thinking* is defined as the phenomena of human cognitive operations. Thus, *design activities* can be defined as thinking activities executed by cognitive operations. On the other hand, design products could also be seen as the physical results coming from the design thinking process operated by cognition. Thus, whether interpreting design from the process or from the product point of view, the bottom line is that design is created by human cognition.

For decades, scholars have studied the nature of and attempted to define *design thinking*, which has been considered a "way of thinking." Peter Rowe was the first one who used the term in the title of his book, *Design Thinking,* to explain problem solving procedures used by architects and urban planners (Rowe 1987). From combining numerous scientific studies conducted in various fields, a clearer picture has emerged and the term **design cognition** used to categorize the activities that occur in the design process. Design cognition has gradually been developed into a discipline. For instance, the nature of design cognition has been studied through looking at design from the computational (Cross 1999) or problem solving perspective (Cross 2001). Chuck Eastman used the term to refer to the study of human information processing by using different theoretical and empirical paradigms to describe the process of design information (Eastman 2001).

Given the rapid development of new technologies and new study tools available, it is prudent to re-examine and rethink this topic to incorporate new scientific developments occurring in other related fields. This chapter takes this new approach by looking at the conventional studies, rethinking the methodologies applied, reviewing the directions, and suggesting a new approach, particularly from the architectural design perspective. Once design cognition is well defined and fully understood, it is appropriate to analyze how designers stylistically and creatively do design, and identify the causes of style and creativity that occur in the design process.

2.3 Development of Design Studies

If *design research* is defined as a scientific approach to design or regarded as a study approached from the scientific perspective, as **Herbert Simon** explained in his theory of the "Science of Artifact" in 1969, then all the works done through methodological and rigorous sequences by artists in their creative processes or by scholars in their studies should be recognized as design research. In his book, Simon proposed that a science of design is a body of intellectually tough, analytic, formalizable, empirical, observable, and teachable doctrine about the design process (Simon 1969). Thus, all design works performed by a system of inquiry procedures can be taught, recorded, and studied. Artists, in their creation processes, are researching the ways of creating artifacts through implementation and execution, which can be duplicated. Scholars, in studying artists' creative processes, explore the various methods used for creation and expect to summarize the knowledge of creation that can be learned and replicated. Therefore, both artists working systematically on creating a work of art or a craftwork, and scholars on studying how a designer thinks or creates a design, should be recognized as conducting design research.

Following this premise, the geometrical perspective drawing method, developed in the Fifteenth Century in Florence as Quattrocento Art (Hartt 1994), is a result of design research. The use of a physical model to represent design, such as the scale model of the Basilica of St. Peter's, founded by the Emperor Constantine, should also be recognized as design research. However, the modern development of design research with a systematic approach only traces back to the works done by the **De Stijl** group in the beginning of the twentieth century (Cross 2000). **De Stijl** (Dutch term for "The Style") was a movement led by architect **Theo van Doesburg** in the early 1920s. Their design principles were based on functionalism with the following characteristics: (1) using rectilinear planes in the way that is similar to slide the planes across one another; (2) eliminating decoration on surfaces; and (3) limiting color schemes to pure primary hues together with black and white. Their works had clear principles and procedural routines to follow, which were the tendency of methodological approaches of the time on artistic creation. Because these rules and procedures were followed by a group of artists and designers in various design fields, their creation endeavors are recognized as the beginnings of design research that applies scientific principles to systematically arrange design composition.

Bauhaus, the school of Art and Craft, also incorporated concepts around 1920 from the **Arts and Crafts Movement** in England to develop structural concepts in their design curriculum. By integrating arts and science and combining crafts and practice, Bauhaus, led by **Walter Gropius**, created a new direction for architectural design education. They emphasized the honest and direct use of materials as the most functional path to design. At Bauhaus, students were required to take craft courses as well as painting, drawing, and theoretical studies in design and color. Faculty at Bauhaus also conducted research on mathematical and representational analysis of the concept of how sensations were organized into a unified perception, which was the "field" theory developed by **Gestalt psychology**[1] (Wertheimer 1923). **Paul Klee**, the Swiss artist who joined Bauhaus from 1921 to 1931, utilized the rules of organizing patterns to explore sensation and perception in his course materials (Teuber 1973). In the period from 1920 to 1930, scientific methods for exploring design emerged, which became a discipline, and was termed *design studies*.

During the two world wars (1914–1918, 1939–1945), the war activities drove the need to automate production. They resulted in mass production of bombs, ships, and weaponry. Equipment in factories used by the war industries were thus of the highest possible level of sophistication at that time. All the rest, for example, housing, food, or transportation was in need of mass automation after two world wars. So mass production during the world wars was set as an example to drive automation in civil industries. Thus, studies in design started to focus on mechanical efficiency to improve the performance of products. However, an individual designer had limited ability to handle the increasing complexity of the industrial manufacturing of products. Thus, methods of production through systematic design procedures became the research focus. In the post war era, the shortage of labor required technological development on mass production and product automation. Labor issues, together with the swift transformation from military wartime equipment to civilian demand for consumer products required the development of new production methods in manufacturing. For instance, in industrial design, studies began to utilize systematic methodologies in design around 1950 to make processes more efficient and effective.

From 1950 to 1960, design studies were mainly influenced by system theory and systems analysis on design, which set up the basis for a "**design method movement**" (Cross 1984). The system theory and analysis were coming from well-developed information processing theory and operations research, which inspired scholars to study design methodology. At that time, design methodology was regarded as a prescribed and rigid approach to design. This movement called designers' attention

---

[1] Gestalt Psychology involves recognizing how sensations are organized into a united perception by human beings. Its major concentration is, in vision, on understanding how the whole is different than the sum of its parts while we are perceiving things. The major concepts of this theory are that the perceiving behavior of an object is determined by the spatial-temporal configuration of the objects in our visual sense. Because of the good understanding of the phenomenon of perception, it has big influence to the fields of industrial design, painting, graphic design, and typography after it is developed from 1920 to 1930. Details of the notions applied in design are explained in Chan (2008).

towards exploring systematic design procedures, and proposing systematic methods for designers to apply. Two classical examples presented in the first Conference on Design Methods, held in London in 1962, attracted designers' interests.

The first classic example is the works of **J. C. Jones**. Jones, as an industrial designer for a manufacturer of large electrical products in the 1950s, applied **ergonomics** as a means to understand the design process of engineers on designing electrical equipment that better responded to user requirements. When his ergonomic studies found that the engineer had not considered user behavior, he redesigned the engineer's design processes to respond to the human requirement first, and the machine requirement second. A procedural sequence applied in design to allow intuition and rationality to co-exist was called *design methods*. Other than generating and utilizing design methods in the design process, Jones proposed that the design process has cycles, or stages of analysis, synthesis, and evaluation. The analysis component is the problem analysis stage that requires making all design requirements on a list, and carefully studying their interactions to make a set of logically related performance specifications. The synthesis stage is to find solutions for each specification and build up an integrated design solution with the least conflict among the specifications. The evaluation stage is to test the accuracy of each alternative solution in order to select the final one. The processes could be recycled and repeated whenever necessary (Jones 1963). He described these three essential stages of design in simple words as "breaking the problems into pieces (analysis) … putting the pieces together in a new way (synthesis) … and testing to discover the consequences of putting the new arrangement into practice (evaluation)." The purposes of his study were: (1) to reduce the amount of design error, re-design and delay; and (2) to make possible more imaginative and advanced designs (Jones 1963, p. 63).

In architecture and urban design, **Christopher Alexander** (1964) developed a "**pattern language**" methodology, which serves as a second classic example from the 1960s. The fundamental notion of pattern language was to solve design problems through combinations of selected design patterns. To Alexander, a language had vocabulary, a collection of named solutions to problems, called *design patterns*. The conceptual framework was that our built environment included the geometry of physical objects, which was a 3D embodiment of a culture, and an organization of social institutions defined by human activities. The human activities were used to categorize the social institutions and be anchored in spaces. Thus, spaces categorized social institutions that attributed to human activities. For instance, categories of space inside a house embodied the culture of its family; whereas the categories of space in a city embodied the culture of the people in the city. Different environments had their own morphological laws given by the acts made by builders, and the formations of acts were guided by the combination of the builder's mental images. Such combinational systems of images were exactly, to Alexander, like human languages. These systems allowed a person to produce various combinations, and every environment was formed from such combinatorial systems of images representing patterns, much as in combining words in languages. That is why Alexander called such systems "Pattern Languages" (Alexander et al. 1977).

The patterns in pattern language correspond to the rules of grammar in natural language. Each pattern is a fluid image that can be combined with other patterns. In all, patterns can be described as the overall layout of a building, ecology, large-scale social aspects of urban planning, regional economics, building components, structural engineering, or building construction. The rules of grammar and their meaning in pattern language systems are not given predigested. Every designer can generate and share his or her own language. Then, shared languages gradually evolve towards greater and greater wholeness. To **Alexander**, the good patterns would spread widely; the bad patterns would eventually drop out. Therefore, the environment so created by designers, though changing constantly, were as coherent, whole, and real, as a totality. In 1996 at an Association for Computing Machinery (ACM) Conference, he reinforced that the purpose of a pattern language is to create morphological coherence in the things that are made with it.

The definition of each pattern in the language, explained by **Alexander**, includes a problem that occurs over and over again in the environment. A solution to this problem is described so that it can be used many times without solving it the same way twice (Alexander et al. 1977, p. X). Following this definition, each pattern has the following standardized format:

1. A picture shows an architectural example of that pattern.
2. An introductory paragraph sets the macro context of the pattern by explaining how it helps to complete certain larger patterns.
3. A short format of a headline with less than two sentences explains the essence of the problem.
   *** (Three diamonds mark the beginning of the problem)
4. The body of the problem describes the empirical background of the pattern, the evidence for its validity, and the range of different ways the pattern can be manifested in a building.
5. The solution part describes the field of physical and social relations that are required to solve the stated problem, in the stated context. The solution is stated instructionally on what is needed to do to build the pattern.
6. A diagram shows the solution with labels to indicate its main components.
   *** (Three diamonds show that the main body of the pattern is finished.)
7. A paragraph ties the pattern to all smaller patterns needed to complete this pattern, to embellish it and to fill it out.

Therefore, pattern language consists of problem statements explaining the problem context, solutions of the problem with rules embedded, a sketch diagram showing the solution schematically, and contexts to combine the solution with other patterns globally and locally. Table 2.1 is a general outline of patterns. An example given in *A Pattern Language* (Alexander et al., p. 854, pattern 184) on a cooking layout that could be located in a kitchen (pattern 139, p. 663) is simplified in Table 2.2. Figure 2.1 is a diagram showing the cooking layout solution (p. 856).

As shown in the format, we can present the problem statement component as X describing the context of the pattern, Y is the solution, and Z is the problem to be

Table 2.1 Format of patterns

Format of a pattern (problem):
1. Example picture
2. Introductory paragraph, global context
***
3. Problem headline
4. Problem body
5. Solution grammar
6. Sketch diagram
***
7. Links with other patterns, local context

Table 2.2 A cooking layout example

Name:	Cooking layout pattern #184:
Pattern context:	Within the Farmhouse Kitchen (#139), or any other kind of kitchen. The character of a good kitchen comes from the arrangement of the stove and food and counter
Problem:	Cooking is uncomfortable if the kitchen counter is short and also if it is too long
Problem statements:	The best arrangement for a kitchen design is the one that saves the most steps; and this led to compact kitchens. These compact layouts save steps but don't have enough counter space. Studies have shown that there is insufficient counter space in many kitchens
Solutions:	To strike the balance between the kitchen which is too small, and the kitchen which is too spread out, place the stove, sink, and food storage and counter in such a way that:
(Rules:)	1. No two of the four are more than 10′ apart
	2. The total length of the counter – excluding sink, stove, and refrigerator – is at least 12′
	3. No one section of the counter is less than 4′ long
Solution context:	Put the pattern with Thick Walls (#197), Sunny Counter (#199), Open Shelves (#200)

Fig. 2.1 A diagram of cooking layout (Source: Image was adapted from *A Pattern Language*, 1977, page 856 by Alexander et al., Oxford University Press)

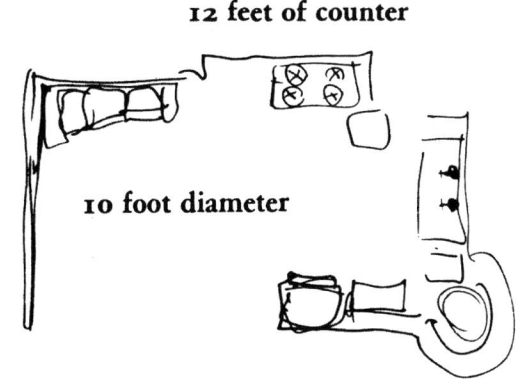

12 feet of counter

10 foot diameter

resolved. Yet, the most essential nature of the language is the various contexts associated with the problem Z and solutions Y. Contexts X are not only paired with problems, but also are required to be considered in the solution syntax. For instance, the larger context of the cooking layout pattern is in the farmhouse kitchen (pattern #139) or any kitchen area, whereas the solution contexts are that it will be in the working surface of sunny counter pattern (#199), thick walls pattern (#197), and/or open shelves pattern (#200). While designers apply the patterns in design, various combinations can be generated due to the flexible rules provided. In execution, the format of the language can be seen as an hierarchy of parts, linked together by patterns that have a carefully defined set of design rules for solving generic recurring problems associated with the parts.

A famous example given by **Alexander** to explain the method of running the language is the Indian village design, published in 1963. In that example, he observed the problem context, made a list of 141 design requirements, then studied these requirements by pairs and determined by graph theory through computer programs whether they were dependent or not. The interacted pairs of requirements were further connected by links and grouped into twelve subsystems, which were combined further again into four major subsystems. The set of requirements and the set of links together defined a linear graph that served as a complete structural description of the village. Of course, each subsystem had its diagrammatic concept to show their images (Alexander 1963).

Pattern language has had a big influence on design education. Many scholars have discussed its applications in studio teaching, but there also are concerns that: (1) the problem situation encountered by the designer might be different from the given context (X), but the designer might pay attention to the factors of the problem situation that are similar to X and assume the same solution Y, rather than to dig into the problem further; and (2) the context (X) may not be an accurate representation of the situation reflecting the relationship between the solution (Y) and problem (Z), thus solutions might be misguided (Lang et al. 1974). Despite concerns expressed in the field of design, pattern language concepts have influenced other fields, because the notion of pattern languages helps problem solvers to tackle the complexity of systems through patterns; for example, it shares the same notion in object-oriented work on patterns in computer science. Each object in the **object-oriented programming** (OOP) paradigm has context, problem, and solution, many of which can be shared and evolved as entities. Thus, after the original publications in design fields, notions of pattern language have since been applied to the fields of software engineering (Gamma et al. 1994), computer science (Buschmann et al. 1996), software design (Fowler 2002), and user interaction design (Tidwell 2005).

Similar studies done in the 1960s and 1970s on developing procedural techniques to be used in design to systematically manage the progress of the design processes include work by **Bruce Archer**. Archer developed a model to illustrate a systematic method for solving industrial design problems (Archer 1965). He indicated that a design had six sequential stages of: (1) receive brief, analyze problem, prepare detailed program and estimate; (2) collect data, prepare performance specification, reappraise proposed program and estimate; (3) prepare outline design proposal(s); (4) develop prototype design(s); (5) prepare and execute validation studies; and (6) prepare manufacturing documentation. These stages logically covered the design from the beginning to the ending stage. Within the process, there were 227 activity items formulated in a completed checklist for product designers to consider. These sequential stages, to him, were sometimes overlapping, sometimes confusing, and sometimes required returning to early stages when difficulties were found. He explained that the art of industrial designing was essentially the art of reconciling a wide range of factors drawn from function, manufacturing, and marketing. In design practice, some projects were quite complicated, and involved contrasting skills and a wide variety of disciplines. Therefore, designers must make some assumptions or judgments, which might not be supported by collected data. Under such circumstances, any proposed design solution that constituted a hypothesis based upon imperfect evidence had to go through tests of the marketplace, or an indirect feasibility analysis, before it was completed (Archer 1965). Archer also used management science and operations research with logic operations to frame the overall structure of the design process that fit textile, clothing, jewelry, ceramics and interior design. The essence of the overall structure of the design process was in applying a reiterative problem-solving operational model to a goal driven system that was appropriate to the design problem (Archer 1970). Archer's studies have shown the influences of problem solving theory and operational research, in addition to the characteristics of systematic methodology.

This 1960s–1970s period, categorized as the **methodological approach to design** (Cross 1984), generated a number of studies on design methodology, yielding significant results on exploring the steps to complete a design. A number of special groups were also formed. These groups, including the **Design Research Society** (founded in London, 1966), **Design Methods Group** (in Berkeley, 1967), and **Environmental Design Research Association** (in North Carolina, 1968), have hosted conferences and published conference proceedings or newsletters since 1962 (Cross 1984; Bayazit 2004). Researchers in this period mainly concentrated on observing design experience and phenomena occurring in the process. Through observing these phenomena and from applying simple mathematical diagrams and flow-chart type models, certain patterns and structures in the design processes were analyzed and reported descriptively. However, these approaches were criticized as merely studies on the general interpretation of design logic or on general explanations of processes. These methodological studies disenchanted professional practitioners. Scholars were also aware of the weaknesses of these approaches (Archer 1979). Interestingly, **Alexander** indicated that design methodology would not help design because it lacked the motivation for making better buildings (Alexander 1971). Similarly, **Jones** also rejected design methods and changed his research direction (Jones 1977). Thus, first generation design methods were not capable of handling the complexity of real-world problems.

At the same time, problem solving theory guided studies to understand the nature of the complexity of design problems, particularly the structure of "ill-defined" problems (Reitman 1964; Rittel and Webber 1973; Simon 1973). Moving forward to the early 1980s, architectural research on "design methodology" had been changed by the inclusion of three influential factors to the study of the nature of design activity and "design thinking." The first factor, scientific research method, was inspired by **empiricism** (or empirical studies) and **positivism**. Empiricism posits that the origin of all knowledge comes from sense experience. This theory requires running observations or experiments to test all hypotheses and theories for obtaining knowledge. Parallel to empiricism is positivism. Positivism, similar to empirical study and inspired by natural science, says that knowledge is basically constructed by observable and measurable facts. These facts exist around us and are waiting for us to systematically and objectively discover their nature. Scholars must define the cause-effect relationships among the variables to synthesize the relationships into general principles that can be applied by others for further prediction and developments.

The second factor was based on the notion that designers' thinking had close connections to psychological sequences, and design was the psychological process of solving problems. Scholars had mutually recognized the basic premise that in everyday life, the ways of design—treated as solving design problems—came from human nature. For the purpose of solving complex design problems and meeting user requirements, design was considered as a problem-solving and decision-making activity. Thus, studies on design started to center on fundamental human intellectual capacities and cognitive faculties.

The third factor was on the adequacy of applying systematic theory in design. Systematic design as studied from 1950 to 1980 had the tendency of applying systematic analysis-synthesis procedures, which do not fit the character of design as ill-defined problems. Systematic design and its associated systematic procedures could not adequately cover the entire complexity of modern design problems. Particularly, systematic procedures are mostly linear sequences. Thus, studies on systematic procedures and methodologies changed to understand the nature of design activities through studying the design processes. On the other hand, if designers needed to improve on their design procedures, their design thinking also needed to be explored more.

Following this trend and based on positivism and empirical research methods prevailing after World War II, studies in architectural design thinking in the 1970s started to move from using flow-chart diagrams to analyze designing patterns to concentrate on designers' ways of designing and decision making. They identified their research, conducted psychological experiments, collected data for analyses, built abstract and simple models, generalized complex psychological activities, and investigated cognitive phenomenon in the design processes. That led to the new scientific exploration of "thinking processes" in design, and studies in this regard have investigated what designers do when they design, from open-ended interviews to controlled laboratory experiments. Methods applied were very scientific in nature with several techniques for data collection, including interviewing designers with specific questions asked, performing case studies of designers' individual projects, and asking designers to think aloud and record their verbal data. Details of each technique are explained in the section on study methods in this chapter.

After the late 1980s, studies in design thinking focused on understanding the cognitive mechanisms applied for and involved with implementing design activities. A number of conferences have been organized since then. The most significant one is Design Thinking Research Symposium (DTRS), which was initially founded in 1991 at Delft University of Technology by Cross et al. (1992). Because of the success of its first meeting, it was hosted sequentially in following years to study the nature of design ability, how designers think, and ways of knowing design in general. Each symposium has its own theme: the application of protocol analysis in 1994 at Delft (Cross et al. 1996); descriptive models of design in 1996 at Istanbul Technology University (Akin 1997); representation in 1999 at MIT (Goldschmidt and Porter 2004); interdisciplinary approach in 2001 at Delft (Lloyd and Christiaans 2001); expert designers' design nature in 2003 at University of Technology, Sydney (Cross and Edmonds 2003); analyzing design in 2007 at University of the Arts, London (McDonnell and Lloyd 2009); interpret design thinking in different disciplines in 2010 at University of Technology, Sydney (Stewart 2011); and analyses of different responses to a given design task in 2012 at the University of Northumbria, England (Rodgers 2012). These studies have brought the research in design thinking to a new age and given credibility to design thinking research as a distinct field of inquiry across different disciplinary backgrounds.[2]

Looking back through history, the first generation of design research was influenced by mathematics, operational research, and system theory around the 1960s to posit a condition with a set of actions to follow. When the condition that matched the problem situation were met, consequences followed to accomplish the design. This first modern approach after World War II, however, was recognized by scholars as being unable to solve complicated design problems. Then, in the 1980s, research directions changed to a second generation on exploring the nature of design. This change adopted notions from cognitive science to set up hypotheses about some design phenomenon and behavior, and then conducted psychological experiments to test, prove, and revise the hypotheses. This second movement of design study has been a major research trend that prevails to the present time and generates a lot of insight on how designers think.

When information technologies bloomed in the 1990s, design practice changed from using traditional pencil-and-paper to the applications of computer systems. Scholars realized that conventional graphic thinking mode and drawing representation would eventually be replaced by digital representation; studies must be performed on the impact of the change. They also realized that psychological experiments applied in the second generation of design research could provide data to develop computational models. Thus, a new conference series, Design Computing and Cognition (DCC), was initiated in 2004 at MIT (Gero 2004) and hosted in the following years (Gero 2006, 2010; Gero and Goel 2008). DCC has focused on

---

[2] Details can be seen on the Design Group, Open University Web pages: http://design.open.ac.uk/ cross/DesignThinkingResearchSymposia.htm

computational theories and systems that enact design. Research efforts have focused on predicting design processes, developing computational models of these processes, and examining processes from their computational results. Expectations of combining computation and cognition in DCC conferences were not on exploring the essential relationship between aspects of human cognition as computation, but on how computation could inspire cognition, which was influenced by theories of artificial intelligence. The Deep Blue II chess-playing machine is a good example to describe the connections between artificial intelligence on computing and natural intelligence on thinking (Cross 1999). As concluded by Casti (1998) on the chess-playing program, nothing has been learned about human cognitive capabilities from it or about methods from the construction of the machine point of view. But the fact that the program defeated world chess champion Garry Kasparov in 1997, has taught us something of the cognitive strategies of human chess players. Similarly, it is possible to also learn the nature of design cognition through looking at design from the computational perspective.

After studies gained some understanding on mental activities in design, scholars started to explore particular areas to improve human cognition. For instance, the ACM Creativity & Cognition (C&C) conference brings artists, scientists, designers, and researchers from various fields together to understand human creativity in various aspects. Their program[3] explains that they "seek to understand human creativity in its many manifestations, to design new interactive techniques and tools to augment and amplify human creativity, and to use computational media technologies to explore new creative processes and artifacts in all human endeavors ranging from the arts to science, from design to education."

Similarly, spatial cognition is another interdisciplinary research field with organized symposia. Spatial cognition concentrates on the acquisition, organization, modification, and revision of our knowledge learned from perception about spatial environments. It has been considered an imperative building block of general cognition for the reason that it is the procedure by which we obtain, modify, memorize, and retrieve information from memory in relation to spatial images. In 2003, the "Spatial Cognition: Reasoning, Action, Interaction program" was established by German Research Foundation (DFG) at the Universities Bremen and Freiburg. The concept of spatial cognition defined by the group is that "Spatial Cognition is concerned with the acquisition, organization, utilization and revision of knowledge about spatial environments, be it real or abstract, human or machine. Research issues range from the investigation of human spatial cognition to mobile robot navigation. The goal of the SFB/TR 8 is to investigate the cognitive foundations for human-centered spatial assistance systems."[4] Their research issues range from the investigation of human spatial cognition to mobile robot navigation, which provide understanding of human perception in space.

Through these years of design studies, starting from the systematic design method movement initiated in the 1960s, to the utilization of theories applied from

---

[3] ACM Creativity & Cognition Conference: http://dl.acm.org/event.cfm?id=RE326
[4] URL of SFB/TR 8 Spatial Cognition: http://www.sfbtr8.spatial-cognition.de/

cognitive science to explore the design nature in the 1980s, and further to the computational approach developed in the dawn of twenty-first century, theories of human cognition in design have been developed to describe essential patterns of design activities. Some theories, demonstrated by experiments, did explain aspects of cognitive operations applied, factors involved, and mechanisms utilized in the design processes. Thus, summarized from the cognitive science perspective, all cognitive mechanisms and cognitive functions as reported in published literatures can be categorized into patterns: (1) design is problem-solving; (2) design is making associations; (3) design is goal and constraint driven; (4) design is reflective and problem structuring; (5) design is looking for representation; (6) design utilizes cognitive strategies; (7) design uses some reasoning; and (8) design applies repetition for design generation. These patterns of activities rely on the use of cognitive mechanisms to fulfill cognitive functions.

2.4   Cognitive Mechanisms and Functions

By organizing patterns of theories into categories, the characteristics of design thinking can be better described and well presented. By doing so, an overall picture of design cognition would be further developed, much like the "State, Operator And Result (SOAR)" system proposed. The SOAR was a candidate unified theory of general cognition developed by Newell (1990) in the field of artificial intelligence to handle the full range of capabilities of an intelligent agent. It has been used by computer science and AI researchers to model aspects of human behavior. Similarly, if a unified theory of design cognition could be developed, then it could help designers to understand more on design thinking, style, creativity, and how to do digital simulation better. In fact, these categories of concepts, as discussed in the following sections, describe cognitive aspects of how design processes are moved forward, knowledge acquired, solutions searched, ideas obtained, and forms created across many design fields. These notions explain the involved cognitive factors that are the fundamental elements of style and creativity in design.

2.4.1   Design Is Problem Solving

The new movement of scientifically studying design thinking in the 1970s was influenced by Herbert Simon, who proposed to discover design processes that would provide the means by which mental phenomena, including human design behavior and cognitive patterns, might systematically be predicted and explained (Simon 1969).[5] Since then, research shifted from developing systematic design

---

[5] In his book of the "*Sciences of the artificial*" published in 1969, Herbert Simon gave a very good definition of the knowledge of design.

methodologies for managing overall design processes (Alexander 1964; Broadbent 1969; Jones 1970) to understanding how designers tackle problems with conventional procedures (Wade 1977; Cross 1984; Akin 1986). With this change, the research trend started to treat design activities as ways of solving problems (Simon 1969), which was led by **problem-solving theory**.

Problem solving theory was developed by Newell and Simon around 1956. In their view, a human is a processor of information, and a computer is also an information processor. Thus, a man could be modeled as a digital computer that processes information. Similarly, human problem solving activities could be modeled by an information processing system executed in computers to explain how humans process task information (Newell and Simon 1972). Later on, such an information processing approach to study the human factor combined the areas of computer science and psychology into the field of cognitive psychology. The theory studied the ability to solve problems, which is one of the important psychological behaviors associated with intelligence. A problem exists when a person is confronted with a situation where he or she wants to accomplish something but does not know immediately what series of actions to take or how to find the means to achieve the ends. Thus, a problem occurs and the process of solving the problem involves high-level cognitive processes.

Historically, the study of problem solving was first approached in the 1930s by **Gestalt psychologists** conducting experiments to study visual behavior. Their studies created interesting results on perception (Kohler 1930; Koffka 1935). In 1960s, Newell et al. (1958) generated systematic outlines on how humans responded when they were challenged by unfamiliar tasks. Their studies on problem solving applied experiments conducted in a laboratory setting using problems that could be solved in short periods of time and sought maximum data on the solution processes. Experimental tasks involved asking subjects to solve structured, puzzle-like problems that were also simulated by computer programs written at that time to clearly explore the subjects' problem-solving strategies. These problems included the Missionaries and Cannibals problem,[6] Tower of Hanoi puzzle[7] chess moves,[8] proofs for theories in Euclidean geometry, and crypto-arithmetic problems.[9]

---

[6] Three missionaries and three cannibals come to a river and find a boat that holds two. If the cannibals ever outnumber the missionaries on either bank, the missionaries will be eaten. How can all six get across the river? See URL: http://www.learn4good.com/games/puzzle/boat.htm or URL: http://en.wikipedia.org/wiki/Missionaries_and_cannibals_problem

[7] The Tower of Hanoi was invented by Edouard Lucas, a French mathematician in 1883. The original tower has eight disks stacked in increasing size on three pegs. The puzzle is to move the entire stack to another peg. The rules (or constraints) are to move only one disk at a time and never a larger one onto a smaller. The number of disks could range from three to eight. See URL: http://www.cut-the-knot.org/recurrence/hanoi.shtml or URL: http://en.wikipedia.org/wiki/Tower_of_Hanoi

[8] In 1957, Irving Chernev wrote the famous book "Logical Chess: Move by Move" which uses unique style of presentation to explain every move of 33 games from the masters.

[9] A crypto-arithmetic puzzle is made of adding two integer expressions to an equation, and numbers in which are replaced by letters. Famous examples are "SEND + MORE = MONEY" and "GERALD + DONALD = ROBERT." The puzzle is to find the integers. See URL: http://en.wikipedia.org/wiki/User:Paolo_Liberatore/Crypto-arithmetic_puzzle

In the 1970s, research on problem solving further classified the types of problems that human beings encountered, and explored the natures of those problems. Defined by task domains and their complexity, problems could be categorized to five types. One was the everyday routine problem that could be solved by common sense or educated guess. The second type was procedural problems with certain rules or that had fixed steps to resolve (accounting, math, or engineering problems). The third type was an open-ended problem that had many satisfactory solutions available (design problems). The fourth type was insight problems, where a problem solver must discover a critical element, and once this element was found, all the other elements would fall into place and the problem would be resolved (scientific discoveries). The fifth type was the problem that has a changing nature over time, and the solution given not only has its own consequences, but might also need to be revisited again and again (social policy problems).

After studies on the nature of problems generated interesting reports, research started to evolve in two directions. One was finding methods for working with large bodies of **semantic information**[10]—for instance, medical diagnoses and interpreting mass spectrogram data. This research direction focused on tasks that were well structured, with clear-cut goals and **constraints**. With this level of understanding of problem solving related to real-world problems, an *expert system* emerged that combined computer science and artificial intelligence. The other research direction focused on understanding problem-solving tasks when goals were complex and sometimes ill defined, and when the very nature of the problem was successively transformed in the course of exploration. The processes of architectural design were one area of interest. Especially, the nature of architecture design problems may change from time to time in the designers' design processes. During that time, the term *ill-structured problem* (also called ill-defined) was used to differentiate it from a *well-structured* (well-defined) *problem*.

Problems in the arts, humanities, social sciences, and engineering areas were categorized as ill-structured problems (Reitman 1964; Newell 1969). An ill-structured problem had a large problem space and no fixed problem-solving sequences or definite goal sequences; any procedure could lead to a possible satisfactory solution, but not an optimal one. On the other hand, well-defined problems were those existing in crossword puzzles, chess-playing, standard calculations, or the fields of natural sciences or mathematics. Well-defined problems had clear goals specified, which could be resolved sequentially by a series of fixed rationales. These rationales responded to certain formulas or rules. The steps to solve well-defined problems were limited, with limited solutions. When the correct formulas and rules were found and fixed routines applied, problems were resolved (Simon 1973). One example of a well-defined problem solving technique is troubleshooting. Troubleshooting involves a systematic approach, usually a checklist application, to

---

[10] *Data* is original raw information that has not been organized nor analyzed. *Semantic information* is organized information that is processed with certain methods or procedures, which relate to the semantic considerations, to turn the data into a certain format that can be used for decision making.

locate and correct problems in existing engineering systems. By the same token, an ill-structured problem can be resolved similarly by dividing the problem into sub-problems, which can again be divided further to other levels of manageable sub-problems until the problem is solved.

Regarding the nature of a problem, Newell and Simon clearly defined that a problem has a starting stage and a final goal stage with a number of in-between stages. In order to achieve the final solution stage, actions on selecting operators must be taken step by step. The available information and actions to be executed for each problem constituted its unique universe, which was called the *problem space* (Newell and Simon 1972). Regarding the nature of solving a problem, a number of studies were conducted through experiments to explore the characteristics (Newell et al. 1958; Newell and Simon 1972; Eastman 1970; Akin 1979; Chan 1990). In general, problem solving activities could be outlined as: identify the problem, postulate possible solutions, test the best solution, and determine if the problem is solved. More specifically, the problem solving activities included the following eight cognitive sequences: (1) identify and select problem; (2) analyze the selected problem; (3) generate possible solutions; (4) select and plan the solution; (5) implement the solution; (6) evaluate the solution; (7) determine if the problem is solved; and (8) memorize the final solution as a knowledge schema for future use. These eight cognitive stages are common thinking sequences for addressing a problem. In design, each of these stages has unique procedures that characterize design cognition, which are elaborated in the following on knowledge and human intelligence.

In the 1970s, studies in social policy by Horst Rittel and Melvin Webber posted a good question on whether social scientists could solve social policy problems in the ways scientists and engineers solve their sorts of problems (Rittel and Webber 1973). The question is based on the reasons that: (1) policy problems cannot be definitively described; (2) the classical paradigm of science and engineering problem solving is not applicable to resolve problems of open societal systems; and (3) the cognitive and occupational style of the social policy professions could not work on a much wider array of social problems. They used the term *wicked problems* to differentiate *tame problems* occurring in the areas of natural sciences and engineering.

To Rittel and Webber, natural science problems are definable and solutions are also findable, thus these problems mostly are tame ones. Governmental planning problems are ill-defined, and the resolution relies on elusive political judgment, which might lead to the situation of solving the problem over and over again. For instance, what should we do to reduce street crime? Street crime is a problem that has infinite associated variables. Any solution proposed might trigger other problems. Thus, social policy problems are inherently wicked. Unlike well-defined problems, which have true or correct solutions, solutions to wicked problems are not true-or-false, but good-or-bad or better-or-worse. In comparison to well-defined problems that have many applicable and potentially satisfactory solutions, wicked problems have the characteristic that the problem might be a symptom to another problem and it may reoccur. The solution to be implemented is obligated according to law and consequential in action. It is also true that the implemented solution might have bigger impacts to a larger population. As such, one cannot build a subway to see how it works and correct it after its unsatisfactory performance is found. As Rittel and Webber argued, every attempt of the solution counts significantly even though the problem may re-occur and the solution may be revisited. This explains the special type of social policy problems that differ from scientific discovery, natural science, and design problems. They concluded that the kinds of problems that planners deal with—societal problems—are wicked ones, for they are liable for the consequences of the actions they generate; they have no right to be wrong. For governmental public policy problems, it has been argued that policy problems are defined, and not identified or discovered; they are rather the products of imposing certain frames of reference on reality (Dery 1984).

Differences between problem solving activities of well-defined and ill-defined problems are that in the well-defined domain, the problem most of the time is known, solutions are limited, steps used to evaluate the solutions are limited and goal state is clear. Legal impacts to the public on solving well-defined problems are not big in scope. Wicked problems are very much the same as ill-defined ones, but actions executed in wicked problems are liable and solutions executed should not be wrong. Table 2.3 briefly summarizes the sequential steps of the thinking activities that occur in the three different kinds of problems. In the table, a checkmark indicates the type of activities that would occur in the category of problems. For instance, analyzing the selected problem in the second row of the table would definitely occur, as checked, across all problems. Design problems, classified as ill-defined problems, are similar to wicked problems to some extent that design solutions should also be right or correct. Otherwise, it shall go through a number of revisions. For instance, the bridge design on the Canal Grande in Venice by Santiago Calatrava in 1996 had gone through numerous structural changes, because of the mechanical instability of the structure and the excessive weight of the bridge, which could cause the banks of the canal to fail.[11]

In studying design thinking, problem-solving theory has inspired scholars to work on psychological modeling of the way designers think and do design. In fact,

Table 2.3 Cognitive sequences and characters of three major problem solving activities

Problem solving stages:	Well-defined	Ill-defined	Wicked
Identify and select problem	Known	Unclear	Unknown
Analyze the selected problem	√	√	√
Generate possible solutions	Limited	Unlimited	Unlimited
Select and plan the solution	√	√	√
Implement the solution	√	√	√
Evaluate the solution	√	√	√
Determine if the problem is solved	Clear	Unclear	Unclear
Develop a schema for future use	√	√	√
Consequence of actions	Limited	Limited	Liable
Problem reoccurrence	No	No	Yes

---

[11] See the criticism of Santiago Calatrava's bridge design on Wikipedia Web page, URL: http://en.wikipedia.org/wiki/Santiago_Calatrava

the problem solving theory, with its connection to cognitive psychology, provides a good structure for scientifically explaining how humans learn, process, store, and search for information for design. Because of the research methods developed in the field since the 1950s on modeling human thinking patterns, problem solving theory provides an essential research platform for exploring style and creativity. For example, one early working hypothesis on creative thinking is that creative thinking is simply a special kind of problem-solving behavior (Simon et al. 1962), and it applies that creative design thinking is also a special type of problem solving behavior.

Research methods used to explore design thinking are to identify the characteristics of human information processing. Particularly, it is critical to recognize mechanisms and operators that designers use to process information mentally while they are working on design, and to have appropriate **conceptual representations** used to model the process. **Conceptual representation** is to use some means to appropriately represent the ideas created in the design process. A number of studies have been generated with this approach. For example, Chuck Eastman (1970) observed the cognitive processes in a space-planning task and explained the characteristics of "generate-and-test" processes. Foz (1972) observed design behavior in developing a parti design. Omer Akin (1979) developed a number of processing models to describe the cognitive aspects that occurred in design processes. Jane Darke (1979) intended to find whether architects had in mind an image or expectations about users during their design period. Stiny and March (1981) developed a design language starting from the information-processing theory to the creation of shape grammar. Other research studied Frank Lloyd Wright's design process by observing Wright's works from 1901 to 1910 and theorizing the resulting Prairie House style (Chan 1992).

Most studies in this trend focused on exploring design actions with the purpose of finding an appropriate model or procedure that could explain the processes and the phenomena generated by the actions. The efforts of these researchers are based on treating design activities as mental processes of problem-solving, which could be further applied to explore: (1) what kinds of external information are attended to in the design problem solving processes? (2) how is internal information (knowledge) applied efficiently on solving problems? (3) how is design knowledge structured in memory? (4) what are the cognitive sequences on the use of design representations in drawing, digital models, or physical models? and (5) what design tactics have been used in the processes? These inquiries, as explained in the following sections, relate to the cognitive processes and the internal knowledge structures stored in memory, which are specifically addressing the essential nature of how knowledge is created, maintained, utilized, processed, and presented in handling design tasks.

2.4.2 Design Is Making Associations: The Nature of Knowledge

In design education, when a project is assigned as a studio exercise, instructors encourage students to search for related and similar precedents and learn from examples to solve the design problems at hand. This "learn by example" is a part of the theory of making associations. But how could the learning processes be scientifically explored to verify the theory? In the 1950s, **Alan Turing** proposed the famous brain-as-computer metaphor, that the computer is an information-processing unit, and the human brain also is an information-processing unit for they both share similar mechanisms of input, output and central processing unit (Turing 1950). The acquisition, storage, retrieval, and utilization of knowledge in the human mind could be treated as a number of separated processing stages. If knowledge could be represented numerically and programmed in computers by sequences, the processes could be simulated to let machines execute intelligent human tasks by implementing step-wise algorithms. Thus, digital computers could perform the same tasks humans do. This notion relates to earlier studies on understanding how knowledge is processed in the brain and controlled by consciousness.

How human beings process knowledge is the study of human cognition. *Cognition* is recognized as the "mental process of transforming, reducing, elaborating, storing, retrieving, and utilizing **information**" (Chan 2008). **Information**, in this regard, is knowledge. The ways of organizing knowledge and combining the knowledge with reality is cognition, which defines intelligence. In psychology, knowledge has long been classified into two major types—*declarative knowledge* and *procedural knowledge*. Declarative knowledge is concerned with knowledge as static information that comprises the facts and concepts we know; whereas procedural knowledge is knowing how to perform a task that comprises the knowledge of procedures and methods of performance (Posner 1973; Winograd 1975; Anderson 1980). After procedural knowledge is practiced or performed for a while, it becomes an automatic skill. On the other hand, when performing certain tasks, declarative information will be transformed into a procedural form. Drawing is one example. After basic drawing skills are learned and practiced, they become self-acting or self-regulating under fixed conditions. When thoughts (declarative knowledge) are put into drawing routines (procedural knowledge), higher levels of expression are created. For instance, Picasso's cubism paintings, Dali's surrealistic paintings, or perspective drawing used to highlight certain aspects in architectural design are results of adding declarative knowledge into procedural.

How is knowledge established? Jean Piaget, a major figure in developmental psychology, explained the learning process of establishing knowledge from exploring the cognitive development of children. Piaget theorized that children are constantly creating and re-creating their own model of reality, achieving mental growth by integrating simpler concepts into higher-level concepts at each stage. Intellectual development has two processes (Piaget 1953). One is assimilation. Piaget assumed that whenever one transforms the world to meet individual needs or conceptions, one is assimilating it. It occurs when a child responds to a new event in a way that is consistent with an organized pattern of thoughts already stored in memory. The other process is accommodation, where one would modify some of the saved mental structure of patterns to meet the demands of the environment. A child would either modify an existing pattern, or form an entirely new pattern to deal with a new object or event. Piaget argued that intelligence develops in a series of stages because one stage must be accomplished before the next can occur. For the last stage, the child would form a view of reality for that period of age. At the next stage, the child must keep up with the earlier view by reconstructing or revising it to form a new version of the concept from an older stage (Piaget 1967).

Such knowledge generation is a part of the evolution process of human intelligence. Putting this theory into a larger picture, knowledge is accumulated from experience and learned knowledge is connected to a-priori ones. Such learning that involves relations between events, for instance, between a stimulus and a response, or between a response and its consequence, is called *associative learning* (Gluck et al. 2007). Associative learning, as explained in neuropsychology (Frith 2007), is an important mechanism for acquiring knowledge about the world. The concept is that "what is learned is an association between an arbitrary stimulus and a rewarding stimulus or a punishing stimulus (Frith 2007, p. 88). An example given by Frith to explain such a phenomenon of learning is the color of a fruit signifying whether it is ripe. As a fruit ripens, it tends to be less green or redder, and people prefer nice ripe (redder) fruit over the nasty unripe (greener) fruit. Therefore we learn about which are the nice fruit from their color. The association between color and ripe (arbitrary stimulus) and nice fruit (rewarding stimulus) provide us with enough knowledge for decision making on selecting nice fruit.

In the professional practice of design fields, procedural knowledge also relates to the professional knowing that is engaged by practitioners in everyday practice over various projects, different from learning the knowledge presented in textbook or journal papers. Designers could learn more through reflection in the midst of action and get intuitive knowing from this reflection. This reflection-in-action notion (Schon 1983) also explains the cognitive interaction between procedural knowledge and declarative knowledge, which is what designers could learn from executing procedural knowledge to earn new knowledge. And, the new knowledge learned from professional practice would be the new facts or concepts generated from actions of doing, which would become a part of the designers' declarative knowledge. Often times, the newly generated knowledge that is inspired by the old knowledge would have some connections between the two, which also relates to how knowledge is structured in mind and stored in memory.

How is knowledge structured in memory? In psychology, the structure of knowledge in memory was theorized as the mind composed of a network of elements—usually referred to as sensations and idea—that are organized by various associations. The concept of associations between events or elements are formulated by chronological contiguity, frequency of connection, similarity and contrast, cause and effect, or by associated meanings or experiences formed through learning. Philosophers have discussed the theory in detail. But, **Hermann Ebbinghaus** (1850–1909) was the first one who scientifically studied the association of memory through applying a set of nonsense syllables to explore how memory is formed and recalled. He theorized that participants would not have previously associated meanings and experiences with the provided stimuli, and when asked to remember them, they would be forming new memories. The goal of his study was to explore how associations between these nonsense syllables could be created without the use of previous knowledge, learning, and experiences that are normally available to humans (Davis and Palladino 2002). Results confirmed in his experiments that the most commonly accepted law of association is by contiguity (the items next to one another are associated and memorized). Ebbinghaus argued that memory was a process of receiving experiences and storing them away to be recalled later. The processes of memorizing something involved the formation of new associations, and these associations could be strengthened through repetition. More repetitions of exposure to the nonsense syllables generated a stronger memory of them. Thus, Ebbinghaus proposed that after an association is established between two elements, these elements are linked, stored, and could be recalled through the built associations (Ebbinghaus 1913). Together with other associations on meaning and experience between memorized events, the concept of human memory structure was further developed into **semantic network theory** (Collins and Quillian 1969; Collins and Loftus 1975), **schema theory** (Rumelhart and Ortony 1977) and **network theory** (Anderson 1980). All these theories propose that memory is treated as a hierarchical network of nodes or chunks. Chunks are connected by various associations. Personal knowledge is stored in information chunks or schemata that compose the mental constructs for ideas.

How is knowledge stored in the brain? Metaphorically speaking, the brain could be seen as the hardware and the mind as the software representing the **consciousness** that manages information stored in the brain. When information comes into our memory through sensory input, it needs to be changed into a form that the memory system can store, which is the encoding process. Details of the encoding process are unknown, but formats of information code stored in the brain have been explored in psychology. For instance, music input will have acoustic coding, whereas pictures have visual coding and concepts semantic verbal coding. Thus, after information is encoded and passed through short-term memory, it will be stored in long-term memory by chunks. People could remember the pictures, sounds, or words and present them immediately after they are recalled. In cognitive psychology, **Paivio** (1971, 1986) proposed a dual-coding theory, which posits that an encounter with either a picture or a word may develop both a "visual" and a "verbal" code. But the visual codes are formed for pictures and verbal codes are for words. This suggests that knowledge is coded by certain format and humans could handle various codes developed during the process of perception.

The dual-coding theory has strong implications for design visual thinking. This is due to the fact that the use of mental images to help thinking is not new to most designers. Designers often mentally manipulate the spatial relationships among fundamental geometric shapes and then represent results of this manipulation through graphic drawings or physical models. Sketches are tried out in the mind's eye before they emerge onto the drawing board. The concepts of visual coding were explored in a study of mental images of fireplaces, which applied dual-coding theory to detect the internal knowledge representation and further investigated the differences of knowledge structure between expert and novice designers (Chan 1997, 2008). That study discovered that architects seem to organize their architectural knowledge by architectural functions. In the experiments conducted on drawing fireplaces located in different environmental contexts, subjects were asked to mentally recall a fireplace image they had seen before and draw on paper. The theory was that the sequences of the fireplace drawing suggest the sequences of recalling the fireplace coding from memory, which will reveal how the information chunks are stored in memory. If the sequences between two consecutively retrieved image components have architectural functional connections, then the chunks stored in memory would also be associated with architectural functions. The collected experimental data showed that expert architects had 75 % of the links that were functional, whereas 52 % of the novices' links were functional. The study also found that expert architects have larger chunks of architectural symbols (M = 73) than do novices (M = 40.5) in the collected memory data. Results of the study suggest that architectural knowledge in memory is hierarchically built up through functional associations that link architectural components together.

Expert designers seem to organize their design information in a functional way such that they are able to store more information, retrieve it more rapidly, and more capably fit known objects into new contexts. Studies in chess (Chase and Simon 1973), painting (Gombrich 1960), and architectural design (Chan 1997) demonstrated that expert chess players, painters, and architects tend to develop large sets of domain-specific knowledge chunks during years of practice, which are stored in long-term memory. Among these knowledge chunks, some are the patterns of the solutions generated before, termed pre-solution models or cases, which could be used to solve new problems. All these knowledge chunks shaped their chess-playing strategies, painting skills, and design expertise into a set of **repertoire** in memory, which forms master chess player, painter, and architect resources of creativity. This body of knowledge repertoire or pre-solution is the first notion of design creativity (DC1) explained from the cognitive psychology perspective.

Similar notions that knowledge is stored by chunks in memory have also been demonstrated in the field of neuroscience.[12] In neuroscience, scholars have categorized knowledge storage through memory types. For instance, *semantic memory* refers to the memory of meanings, understanding, and acquired general knowledge. *Episodic memory* refers to the memory of events, times, places, emotions, and knowledge that relates to an experience. *Procedural memory* refers to the memory of skills and procedures. Combining semantic and episodic memory, *declarative memory* is formed in the brain. Neuroscientific research suggests that procedural and episodic memory use different parts of the brain working independently. Some believe semantic memory resides in the **temporal neocortex**, which is involved in auditory processing and is home to the **primary auditory cortex**. The temporal lobe (see Fig. 2.2) in the brain is also heavily involved in semantics, both in speech and vision. It is suggested by evidence that the temporal lobe contains the **hippocampus**,

---

[12] Design thinking is the information processing (IP) sequences in our brain and mind. Metaphorically speaking, the brain is the hardware and the mind (consciousness) is the software of the human IP system. At this point of time, neuroscience studies the hardware (brain) and cognitive science focuses on the software (mind). In the future, we have to explore how the two components align together. The brief introductions provided in this chapter intend to explain the connections between the brain and the mind during thinking. More details could be found in the well written book entitled "Making up the Mind, How the Brain Creates our Mental World" by Frith (2007).

Fig. 2.2 The temporal lobe in the brain (Source: Image was adapted from Wikimedia public domain, http://en. wikipedia.org/wiki/Human_ brain#mediaviewer/ File:Gray728.svg. Accessed 8 July 2014)

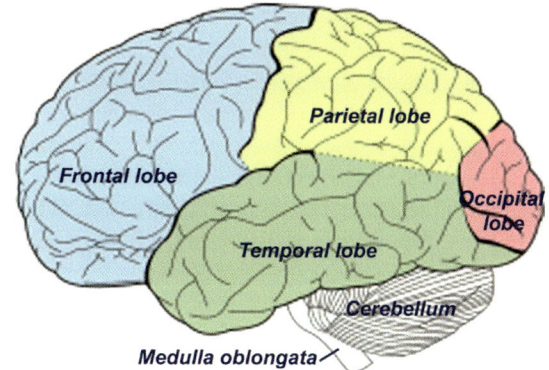

which is essential in the formulation of new memories about episodic and general declarative memory, and it is used for storing and processing spatial information. Therefore, the temporal lobe that contains the hippocampus should be involved in formulating memory as well. This brief explanation describes the fundamental structure of how knowledge is stored by types in different brain areas, and different areas of the brain process different types of intelligence.

How is knowledge retrieved from memory? In semantic network theory (the first modern theory developed and the most influential one), each node is a symbol that can be interpreted as representing a specific concept, a word, or a feature. The link between nodes is a relation or "**association**" that can hold between any two nodes. Processing in a semantic network is the form of *spreading activation* or search via associations. Spreading activation suggests that the activation of a concept can lead to the activation of other related concepts as the activation spreads along the paths of the network. This explains the notion that knowledge can be learned by association, stored in memory by groups, and better retrieved by association as well. For instance, in the psychological study of language, some psychologists have explored that humans can learn language by learning the associations between adjacent words in a sentence. Blank and Foss (1978) studied how an appropriate semantic context facilitated the comprehension of sentences by providing relations. In other psychological experiments, studies on organized and random words recall have shown that words linked together with associations are recalled better than randomly linked ones (Bower 1970). Recent experiments also show that people are biased by past experience to make new decisions between options that were never previously rewarded (Wimmer and Shohamy 2012). In neuropsychology, it has been proposed that the hippocampus encodes relationships between items and events that appear together, forming an associative link between them. It is because of the built associative link, when a person encounters one item; the hippocampus completes the pattern and reactivates the neural representation of the other item to integrate the old memories with new ones. This explains the association aspects of memory construction, storage, and recall in both the hardware (brain) and software (mind) sides of human cognition.

Lastly, how is knowledge presented in the mind when we recall and process it? This relates to the ways of describing the nature of ideas and concepts appearing in thinking, and the format of internal knowledge representations. Here, the format relates to the structure of a code, as defined by the nature of the elements and the nature of the relations between them (Kosslyn and Pomerantz 1977). It has been argued that mental objects exist that are abstract structures of propositions expressing precise relations between entities, and they have evaluable semantic properties of content, reference, and true-value (Minsky and Papert 1972; Anderson and Bower 1973). This propositional format of knowledge explains that material in memory is represented by networks of propositions (or called nodes), which are entirely abstract objects (Pylyshyn 1973). On the other hand, visual modality is another significant set of information in our mind, and visual image has been described as a picture in the head, stored in memory and organized into meaningful parts, which in turn are remembered in terms of spatial relations among them (Reed 1974). Mental images, in fact, are easy to recall and utilize to help solve certain types of problems requiring the use of images.

Beyond the network theories, studies on knowledge representation in the 1980s turned to the notion of "**connectionism**," which proposed another network modeled after neural networks in the human nervous system. The concept is analogous to neural processing in the human brain. Every node in the network is a **neuron**-like (see Fig. 2.3) processing unit with weights that measure the strength of connections between the neuron units.[13] Connectionism also is called the Parallel Distributed Processing (PDP) approach, as the processing takes place in a parallel format and the output is distributed across many units (McClelland et al. 1986). This approach generated another conceptual framework for understanding the nature of the mind and its relation to the brain, which also applies to the study of design thinking.

In sum, we organize our knowledge in memory by association and retrieve information by association as well. Association is inferred from the tendency of one item to evoke another, which is triggered by resemblance, by contiguity in place or time, by frequency of connection, and by cause and effect between two items. The theory of association has served as the central core in the history of publications on memory, and has become a theoretical view embracing the whole of psychology.[14] But the associative properties among chunks (or nodes) in regard to the linguistic information stored in both semantic network and network theory have been criticized, as they might not adequately encompass all aspects of meaning of the information (word or feature) in the chunk (Harley 1995), and the search for information through the network hierarchy is not economical (Rosch 1978). Similar considerations were

---

[13] A neuron is a functional unit of the nervous system. Each neuron could be seen as a powerful microprocessor, for it has three components analogous to a computer. One component of cell body (*soma*) contains information for managing cellular functions. One of the cellular functions is receiving and sending messages. The second component is *dentrites* that serves as the reception sites for incoming messages. The third component is the *axon* that is the mechanism for sending messages on to other neurons.

[14] Concepts of association could be found in the Encyclopedia Britannica online edition. See URL: http://www.britannica.com/EBchecked/topic/39421/association

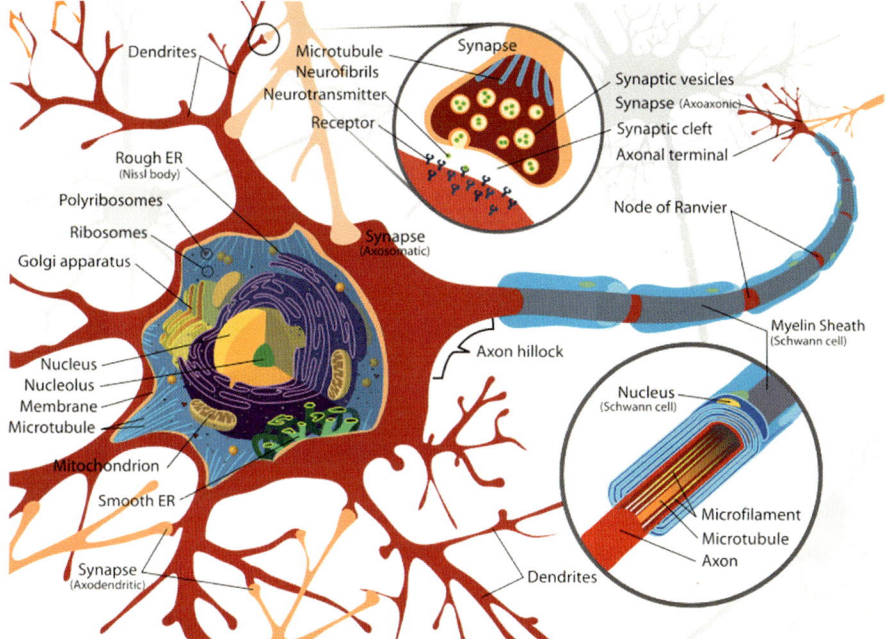

Fig. 2.3 A diagram of neuron (Source: Wikimedia public domain, http://commons.wikimedia.org/wiki/File:Complete_neuron_cell_diagram_en.svg. Accessed 10 Oct 2013)

posted to the schema theory as well. However, association has long been recognized as the major theory that the mind is composed of elements of knowledge chunks organized by means of various associations. Thus, association is a major cognitive factor in human intelligence operation. Therefore, if a designer makes an unusual link to a chunk of knowledge in memory, retrieves it, and applies in design, then the result might be unconventional, and suggests one of the causes of creativity. This association of diversified knowledge information in design is the second notion of design creativity (DC2).

2.4.3 Design Is Goal and Constraint Driven

Problem solving is a sequence of moving the problem forward stepwise to reach the solution. If the problem solver does not have goals in mind to accomplish, he/she would not be able to strategically move the problem ahead to complete the mission. Real-world problems tend to be complex, which cannot be solved with a simple one-step solution. Problem solvers must set up sub-goals that must be accomplished on the way to achieve the overall goal (Thevenot and Oakhill 2006). Therefore, problem solving is goal driven.

Similarly, designing is proactive problem solving (Baker and Dugger 1986). It includes defining a concept, refining the concept through research, experimentation, and necessary development to prepare the product for production. When designers develop the concept, they have to understand the nature of the problem, explore the issues to address, formulate the problem, and utilize strategies to come up with a workable scenario. These activities usually consist of various formats of thinking patterns of setting up design hypothesis, deciding issues, providing intentions, and putting up proposals. If occurring at the beginning of the problem solving processes, the stage of having these cognitive activities activated could be called the *design intentions development stage*, which is the fundamental stage determining the outline of the solution. At the end of this stage, a concept is created and a series of following stages would be further developed to complete the concept. Thus, the consequential following stages represent goal stages. As such, design activities are driven by sequences of achieving certain goals (Akin 1986, p. 20), and each goal has constraints attached to narrow down the scope of search for solutions (Simon 1969).

The goal sequences appearing in the design processes can be seen as a game plan for solving a design problem. A study on exploring the cognitive processes in architectural design problem solving has proven that a goal plan does appear in a design course (Chan 1990). In that study, a designer was asked to design a three-bedroom dwelling for a single family on a large site. The total floor area was limited to 2,200 square feet. The owner was a professional architectural perspective draftsman. Design requirements were set to minimal to explore how the designer would generate the design and what personal design knowledge had been applied within the intensive period of design creation. Protocol data collected in that nearly four hour (232 min) lab experiment shows a total of 22 goals that were achieved. Among them, 14 were major goals, and eight subordinate goals were developed to help solve their primary major goals. Data also showed that when a goal was achieved, the subject had a clear idea about what the next goal state was and gave keywords to describe the next move. The new state was usually discrete from its previous state. That study explained that at the first stage of problem solving, the problem understanding stage, a general goal sequence would have been prepared to guide the sequence of actions, and then executed one after another. If a goal cannot be achieved and the problem is not solved, a sub-goal is strategically developed immediately (Chan 1990). This phenomenon of setting up goals in the beginning is the process of understanding, formulating, and structuring the problem. During design, the structure needs sub-structures through developing sub-goals to find a satisfactory solution.

Another design thinking study found that a designer knows the general method of approaching a design, for that designer applies his own algorithms skillfully on solving the problem (Chan 2001a). In that experiment studying an architect's individual style, protocol data showed that the architect used the average constructional cost per square feet as a hypothetical assumption for determining the overall building size. He divided the given budget by the unit cost; a gross total floor area was created and used to generate a list of suitable rooms and their dimensions. Again, from modifying the gross area through adjusting room sizes, the final building size and footprint were determined. After that, the architect developed a scenario that summarized the design assumptions and major considerations into a conceptual framework that served as a potential solution (Chan 1993). Such steps of sequences were used in his other designs in the experiment. Thus, a general method or an algorithm was developed, and a design routine employed, formulated from years of practice working on a particular building type (Chan 1990, 1993). When designers encounter a new problem that has not been solved before, a new plan will be tried, new approach will be explored and memorized. As indicated by Heath (1984), such a new method developed might be individual and personal. It is also possible that every architectural problem requires a method specific to its building type. For instance, techniques used for generating a hotel design are different for designing a residential building or an airport terminal.

In accomplishing a goal in the design process, there usually is some type of information applied as a guide to limit the search efforts for solutions, or as operators used for creating solutions. These pieces of information, as add-on information called *constraints*, imposed in the design processes externally and internally, become limiting parameters. For instance, external constraints are the design requirements given to the designers by the clients, users, regulations, and the environmental context as parts of the problem components. In practice, clients present their expected ideas to designers when they commission the designers to work on the design project. Users, sometimes neither the clients nor the designers, have different perspectives or functional requirements on the use of the products than the designers, and designers must receive information relating to user-friendly and accessibility issues. In industrial design, legislators pose considerable restrictions to the health issues on the products that industry might create. In architecture design, the context of the site environment of the building provides forces to influence the orientation, circulation, and façade designs. Material specification, building code, and urban ordinances have certain regulations applied to the building design as well. All these requirements and expectations are part of the information given externally to the designers for design considerations, and, thus, are termed *external constraints*.

When the given external information makes the problem more complicated, problem solvers have to sort out the complexity and search for the appropriate solution among a huge number of alternatives. In order to effectively solve the problem, designers formulate the problem, identify key issues to consider, address only related information to guide the search effectively, and propose possible solutions. In this framework, some information from knowledge stored in memory or information accumulated from research is retrieved and collected to analyze the problem, shape the structure of the problem, develop a solution scenario, and generate alternative solutions. Such information composed from knowledge in memory is the *internal constraint*. Internal constraints are imposed by problem solvers resulting from cognitive mechanisms of reasoning operating in the processes. In architecture design, the internal constraints frequently comprise the majority of the design program (Lawson 2006). Architects generate internal constraints in the early stages to determine the number and sizes of spaces of various kinds and qualifications from the clients' given external constraints. In the same study of cognitive processes in architectural design (Chan 1990), the collected protocol data had clearly shown that the designer called from his experience to develop a set of internal constraints to respond to the given external constraints in the beginning of the design stage. These established constraints became *global constraints* that were constantly addressed in the entire processes. Based on the set of constraints, a conceptual framework was formulated; solutions were generated as design conventions do, through bubble diagrams and flow charts to represent the relationships of the spaces. Thus, internal constraints are the pieces of information retrieved or gathered from knowledge internally by designers for solution generations.

Regarding the characteristics of constraints applied in design, Lawson has categorized four types of radical, practical, formal, and symbolic constraints (Lawson 2006). *Radical constraints* are the primary and fundamental purpose of the object or system being designed. *Practical constraints* are the technical problems of making the design in reality. *Formal constraints* are the rules of proportion, form, color and texture applied to organize the form and appearance of the design objects. *Symbolic constraints* are symbolic meanings used to generate the object. These categorized constraints could be external or internal ones utilized in the design process, depending upon who provides the information. If the constraints are initiated and imposed by designers personally, then they are internal ones and external otherwise. In sum, it is clear that design problems are constructed by external constraints given by external forces, formulated by internal constraints imposed by the designers, and proceed from executing the goal sequences of actions to generate a solution that satisfies all these constraints. If designers can either utilize the external constraints uniquely, impose internal constraints innovatively, or take extraordinary actions in goals to sequentially create a solution that meets the constraints unconventionally, then a creative design emerges. This unique development of goal sequences, or utilization of external and internal constraints explains the third notion of design creativity (DC3).

As explained, designers apply internal constraints either to minimize their decision making, or to narrow down the space for searching for solutions. From the design strategy point of view, the utilization of internal constraints would make the design processes more efficient and effective. However, external constraints generate different effects. More given external constraints set up more limitations for decision making, for solution search, and for knowledge application, and are less flexible for considering creative components for solution generation and on making special links for innovative associations. For example, museums and public buildings might have fewer constraints than commercial buildings. Most of the classical designs are public buildings of theaters, museums, cultural, or convention centers done by master architects. In designing public buildings, architects may have more internal constraints to utilize and less external constraints to restrict their imagination. The Pritzker Prize laureates, for instance, have designed more public than commercial buildings, which make their designs outstanding. It is due to the fact that most commercial buildings, particularly shopping mall designs, have many constraints on the number of stores, the operational format, the business types, and the type of the image to be provided to the developer. These constraints are mostly imposed by clients and are more critical than the constraints given in public building designs. Thus, public building types provide better opportunities for creative innovation. Another example of how fewer external constraints may promote creativity is the Beijing MOMA project designed by Steven Holl. Holl explained that his client (Wanker) would accept any ideas he provided and increase the building budget to cope with all the concepts proposed.[15] The real estate developer had indicated to him, as explained in an interview, "We are going to build all of these. We know we can sell all these apartments. What is important is the spiritual dimension." Thus, creativity appears more frequently with less extrinsic constraint circumstances. This quantity of design environment constraint is the fourth notion of design creativity (DC4).

2.4.4 Design Is Reflective and Problem Structuring

In life, people have to think to resolve problems that might happen every day. In some cases, there might not be problems waiting for them except in their daily life routines. Under this situation, some people would ask questions to either find out what they should be doing to continue the life routine, or to identify potential ,solutions that might exist to make their life better. In other cases, people might have problems waiting, and they would also self-ask questions about the importance, critical state, and deadline of each problem to decide a priority list of the problems to be resolved. Considerations of the criticality and importance of problems are, in fact, setting up constraints to determine the goal sequences before handling the problems. Such a self-asking question phenomenon is a part of cognitive procedures utilized to generate actions for life routines, to find new problems, or to determine a problem priority list.

After a problem is taken into consideration, the problem solver needs to understand the problem, find related information, decide the requirements, recognize expected outcomes of the problem, formulate the problem in the mind, and find actions to implement the outcome until a solution is generated. This series of procedures could be seen as standardized problem solving routines for handling general problems. Yet, during the entire process, people would keep asking hypothetical questions as a technique to narrow the search efforts for best results. This phenomenon of asking questions to understand the problem, to set up sequences of moves, to impose constraints strategically for solution generation are the same as what happens in design thinking.

In design, a design project commissioned by a client has many sub-problems involved, which are more complicated in nature than well-defined problems. Designers handle sub-problems by managing the relationships among sub-problems.

---

[15] In the interview by Charlie Rose on July 23, 2007, Steven Holl indicated the real estate developer of the MOMA complex had handled the design ideas proposed by Holl with fewer constraints. The developer would take everything proposed by Holl and build them.

The relationships among sub-problems and the overall framework that holds the design project together define the structure of the problem. Explained from the cognition point of view, a problem structure can be seen as the format that consists of goal sequences, knowledge representation used to make design components, and design constraints developed to achieve goals. Such a format, formulated in the design processes, is the result of overall problem structuring (Chan 1990, p. 69). The problem structure signifies the problem context that characterizes the situation of a problem. When the problem structure is established, a schematic solution shall be reached. Usually, the primitive problem structure is formulated in the very beginning of the problem solving (Dunn 2004; Chan 2008), which is also called *problem definition* (Bardach 1981, 2000) or *problem delimitation* (Vesely 2007). At the stage of initial problem structuring, the problem solver would go through a series of questions to better understand the problem and to establish scenarios. After the problems are clearly understood and acted on, the basic structure would be refined or modified as needed. Analyzed from the design point of view, a problem structure is developed through the uses of drawings, sketches, scribbles, bobble diagrams, or any other representations; designers would mostly either keep working on the same structure or some would redefine the structure until a product is generated. Logan and Smithers (1993) have described that the formulation of the problem at any stage is not final. Whenever new aspects or inconsistencies are revealed in the formulation of the problem, the problem and the solution are redefined. To them, the process of designing is the process of exploring the formulation of the problem, or the exploration of structuring the problem. Nonetheless, while working on the problem structuring, designers would post hypothetical questions for implementing, maintaining, and modifying the problem structure from time to time.

Since the problem structure consists of a number of sub-problems, the corresponding solutions for sub-problems also generate a relationship structure, which can be seen as a *solution structure*. A solution structure is the sequential order of creating solutions. Design is an accumulation of a series of solutions to yield a big solution product. The timing for solution generation might affect the problem structure or affect the solution for the next sub-problem in line. This notion of problem and solution contexts can be identified through a **problem behavior graph** (Newell and Simon 1972; Chan 1990, 2008). Both the problem structure and solution structure compose the complete problem situation. Cognitively, designers would constantly check the problem situation on both problem and solution structures for accomplishing a final solution. Thus, the thinking activities in a design have the following patterns:

1. Defining problem stage: The initially presented set of external constraints by the clients for a design is translated into a set of design problems, issues and questions sufficiently enough to well define the project situation to allow specific actions on research (Woolley and Pidd 1981). At this initial stage of problem solving, designers would review the context of the problem and ask questions to clarify the problem context. The original problem statement would have been considerably widened and recast. In some cases, it might have the whole set of problems defined.

2. Creating solution stage: When the problem solving is moving ahead, the designer would ask what-if questions to simulate the problem situation to generate a number of alternative solutions. This is similar to the play in chess. Master chess players would consider a number of moves in advance and select the most winnable move. In design, designers would also ask what-if questions, responding to the problem context for simulating a series of hypothetical moves, and generating a number of alternative solutions.

3. Deciding and evaluating solution stage: When decisions on solutions must be made, designers would think of all possible questions to evaluate the situation of the problem context. They will simulate the future aspect and perceive any conflicts or potential problems that might arise for selecting the best one to fit the solution structure. Designers must be aware of the problem situation from time to time by asking questions.

These cognitive phenomena of asking questions for communicating the problem situation in the processes of designing has been explained by Donald Schon (1983) in his theory of reflection-in-action. Schon explained that design processes are actions of reflection, and designers learn from asking "what-if" questions in design. Schon indicated that there are more variables than can be represented in a finite model to simulate the design process. Any of the designer's moves tend to produce consequences other than expected and intended. When it happens, the designer may think on the unexpected changes by forming new understandings, new appreciations, and by making new moves. In this case, when a designer makes a move, the original problem situation changes and talks back to the designer, and he must respond to the situation's back talk. In design, the conversation with the situation is reflective. Whatever moves the designer made, he must reflect on the unexpected consequences and implications, listen to the situation's back talk, form new appreciations and reframe the problem to guide his further moves. While answering to the situation's back talk, the designer makes a reflection-in-action on the construction of the problem, shifts stance as they do so from "what if?" to recognition of implications, from involvement in the unit to consideration of the total, and from exploration to commitment, which are the characteristics of reflective conversation with the problem situation (Schon 1983, p. 103). Therefore, Schon concluded that our thinking serves to reshape what we are doing while we are doing it (Schön 1987). This also explains the nature of problem structuring and restructuring in the processes.

Lawson further explained the concept of reflection in the design processes as *reflection in action* and *reflection on action*. Reflection in action relates to the activities to formulate the problem, move the problem ahead, and evaluate the solutions. The designer continually reflects on the current understanding of the problem and the validity of the solutions (Lawson 2006). Reflection on action relates to the asking of whether the process is moving correctly, which problems have been examined, and the processes involved in representing, formulating and moving. These activities are acting upon the checking of the problem situation. In a design process, there might be the chance that a new solution created would totally change the problem situation, which is then called a *critical problem situation*. In a study on design problem solving, it was found that when the designer generated a solution (in that study, the problem was to solve a bay window for a façade design) and put the solution on a drawing, a critical problem situation was recognized and forced him to make design decisions. The critical problem situation was defined as having the possibility of changing the problem structure or solution structure, and is the state of affairs that will lead to a possible restructuring of the problem (Chan 2008). In that study, the designer perceived that the solution would cause the change of the interior form, material, structure and the character of space; thus, he abandoned the solution and changed to a different one (Chan 1990). This explains the phenomenon of reflection on action that relates to the activities in the process of problem structuring.

In summary, the processes of actions in reflection would have the potential of finding new problems in the context of designing, or redefining the problem structure while evaluating the solution structure. In industrial design, Dorst and Cross have explained in their studies that the restructuring of the problem in the design processes do lead to the generation of creative solutions. From the protocol data of asking nine industrial designers with 5 or more years of professional experience to create a concept for a "litter disposal system" in a Netherlands train, they identified the aspects of creative solution in design. In their studies, they concluded that creative design does relate to the formulation of the design problems and to the concept of originality. Creative design seems more focused on developing and refining together both the formulation of a problem and ideas for a solution (Dorst and Cross 2001). Eberhard (1970) gave an example on designing his client's office door. During the design process, the designer asked whether a doorknob was the best way of opening and closing the door, and further questioned whether the office needed a door. Eberhard indicated the phenomenon as regression, and it could be explained that after the problem was analyzed further and more information obtained about it, the designer became aware of the necessity to change the problem structure for a creative solution (Eberhard 1970). These aspects of the cognitive activities of problem structuring (in the beginning of the problem solving) and reconstructing, (from what-if reflection) attributing to creative leap, do explain the aspect of possible resources of creativity in design thinking. Thus, the problem structuring and restructuring is the fifth notion of design creativity (DC5).

In the fields other than design, problem structuring, or what Schon called problem framing, has been recognized as an important problem solving process. Many studies have explored the techniques of problem structuring for systematically developing the relationship among sub-problems to make the problem solving more effective (Vesely 2007). In operational research (OR), problem structuring methods (PSM) have techniques developed to provide decision makers with systematic help in identifying an agreeable framework for their problem (Rosenhead 1996). In civil engineering, the processes of problem re-structuring to match with the development of solution structure have been applied as the basis of computer algorithms for the co-evolution of a design problem with the potential solutions (Maher and Poon 1996).

2.4.5 Design Is Looking for Representation

Knowledge, as explained in the previous section, is saved in memory by chunks. When knowledge is accumulated to a shelfful, a repertoire of skills is developed as well. The notion of repertoire symbolizes the chunks of specialized knowledge in memory. In order to effectively utilize these chunks of knowledge for solving a problem, chunks must be transformed into certain symbols and expressed outward through some kind of representation. In this regard, the intangible knowledge would become tangible. Thus, representation is to have something standing in for something else, and is the means for representing the things that happened in reality (Echenique 1972; Hesse 1966). Here, representation could be either the act of representing or the product of representing (Greco 1995).

When solving a problem, representations are developed internally and externally by adding information, deleting information as irrelevant, and interpreting information to create means for generating solutions. The means, created internally, is the knowledge, and its structure relating to the task in memory is called *internal representation*. Internal representation is the knowledge in the form of proposition, schema, image, or some sort of isomorph. For solving a problem efficiently, it is critical to build an adequate internal representation at the very beginning of every problem solving process, and then constantly modify the representation to match the problem situation until the solution is finalized. In some cases, an internal representation is a medium representing the external task being confronted. In other cases, it is an abstract structure of the problem. In design cases, the internal representation developed in the mind shall be presented externally by certain means, which is called *external representation*. External representations are the knowledge and structure of the task in the environment used to construct physical objects, symbols, graphs, scripts, programs, and as external rules, constraints, relations or logics embedded in physical configurations (Zhang 1997; Chan 2009, 2011). There are differences in the cognitive procedures applied to solving a well-defined problem versus an ill-defined problem.

For well-defined problems, external representations are usually given, and solving the problem requires mentally constructing one or more internal representations, and finding appropriate rules to execute the tasks required in the external representation. The mutilated checkerboard (MC) problem is a good example (Newell 1965; Wickelgren 1974; Anderson 1980; Korf 1980), which applies a standard 8 by 8 square checkerboard with two diagonally opposite corners removed (see Fig. 2.4). The problem is to cover the 62 remaining squares using 31 dominos, each domino covering two adjacent squares. In this example, the checkerboard is an external representation, and internal representations must be developed to map the external one for processing solutions. Data from an experiment shows that most subjects solve the MC problem by applying dominos, the number of squares, and their geometrical arrangement as their representations (Kaplan and Simon 1990). Since each of the 31 dominos covers two squares, a covering initially seems possible. But, after covering 30 black and white pairs with 30 dominos, the problem situation

Fig. 2.4 The mutilated
checkerboard problem
example

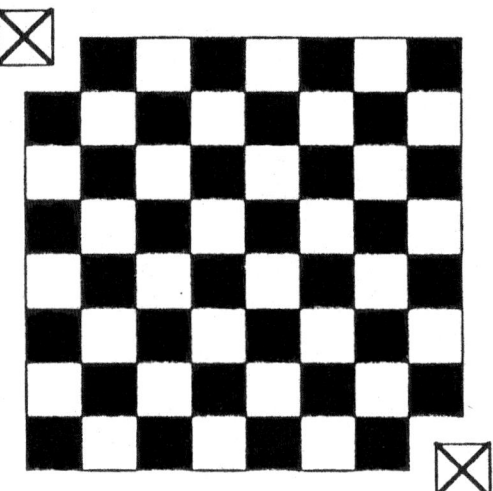

is always left with the impossible situation of having to cover two same-colored squares with one domino. However, switching from the domino representation to partition the squares into two equivalence classes of black squares and white squares would allow subjects to reason about the numbers of squares of each type and to make the crucial inferences needed to prove logically why such a covering ($30 \neq 32$) is impossible (Korf 1980). Thus, finding out the right internal representation or changing representation is critical for solving well-defined problems.

For ill-defined problems faced by writers, painters, computer programmers, lawyers, planners, policy makers and designers, there is no clear external representation given to show the expected final goal state. Problem solvers must construct internal representations first and use external representations to visualize the internal ones to move the process forward (Hayes 1981; Eastman 2001). In design, designers must generate some design concepts in the mind, which are intangible and conceptual in essence, and transform the concepts to internal representations to re-present design concepts, schemes, knowledge chunks, visual images, or mental pictures. Then, the designer would apply some media to implement the concept. The available media for generating external representations include traditional paper-and-pencil mode of sketch drawings, handmade physical models used since the Renaissance Period in 1400, digital models created in digital computers since 1937, and objects generated by digital fabrication. In results, external representations are created, which are visible and tangible, representing the forms to be constructed or to be built. These forms are results of design ideas conceptualized and formulated through a series of mental activities done by the designer. For example, in a building design, the architect must generate some mental image internally while he is designing, and use drawings or models to display the image externally. During the process, the architect must recursively modify both the internal image and external drawing to come up with a product. Since the internal and external representations are both constructed by designers, who have enormous and diversified categories of representations to choose from, ill-defined problems provide more opportunities for designers to develop creative products. As explained by Kosslyn, it is possible for human beings to visualize the objects in question and mentally represent the image to solve it (Kosslyn 1975). If designers make unusual mental images and apply them to solve the design problem, then some creative forms are generated and creativity shows.

In the study of how humans search for representation on solving the MC problem, Kaplan and Simon (1990) showed that cues, hints, heuristics, and prior experience could guide the problem solver to a particular representation, which could then be explored for a solution. Design is the process of making and searching for representation. Designers must select some types of visible external representations to present their invisible internal representation, which could be mentally constructed ideas, concepts, images, symbols, or graphs from experience, to the world. During design, mapping activities between external and internal representations are constantly modified to correspond to the attempted design intentions; dialogs are shaping constantly, and interactions are occurring. In complex problems, without recognizing how any of the two representations are mapping to each other, or if they miss the interactions between the two representations, designers cannot visualize how the design concepts are feasible and how they have been implemented. In some cases, design conflicts happen and designers should look for different representations to resolve the conflict, which might provide opportunities to create innovative solutions. In sum, if a designer has generated a new representation that has not been used before or applied a conventional representation unconventionally to solve the same problem, then the outcome would be creatively different. Thus, representation is a medium used to create and present a design concept, a cognitive mechanism for solving a design problem, and one of the factors that causes creativity. These internal and external representation used in design is the sixth notion of design creativity (DC6).

2.4.6 Design Utilizes Cognitive Strategy

In well-defined problems, there are limited procedures required to solve the problem. For instance, algorithms in mathematics provide a series of steps. Attention paid to the steps results in successful completion of the problem. However, complex tasks in an ill-defined domain cannot be completed by just executing a fixed series of steps; instead, a number of cognitive mechanisms are utilized strategically to deliberately calculate a vast amount of design factors. These cognitive strategies are the use of cognition strategically or methodologically to solve complex problems. A strategy is not a direct procedure or an algorithm. It is a heuristic that supports humans to develop internal procedures to perform higher level operations. For instance, in the field of education, cognitive strategies applied could include the use of guiding procedures in the process of learning, plus practicing basic academic skills simultaneously in a systematic way. Results of utilizing such strategies are to understand the task at hand and learn the procedures to complete the task. A good example of a cognitive strategy developed in teaching reading comprehension is to teach students to generate questions about their reading. Generating questions does not lead to comprehension. But, in generating questions, students have to search the text and combine information, which helps students comprehend what they read (Rosenshine 1997).

Cognitive strategies can also be explained as the cognitive procedures that aid in the performance of specific cognitive tasks. Studies have been done on exploring the nature of the interaction between a problem's intrinsic structure and a problem solver's strategies or behaviors. In design, cognitive strategies are the mental activities of strategically developing a problem structure and a solution structure, which are important in guiding the problem-solving process. It is critical to the designer to have confident control of both the problem structure and the solution structure to maintain a balance between the two. For instance, Akin et al. (1988) investigated the correlation between problem structuring and problem solving, and reported that problem structuring could serve as a mechanism to see how problem parameters were identified by a designer during a design process. This is how designers will handle design problems at the cognitive level. Other strategies used by designers to solve design problems include top-down, bottom-up, inside-out, outside-in, and case-based reasoning strategies.

The top-down strategy is to approach the design first from the macro level, then narrow down to details, whereas bottom-up is to focus on details and individual base elements before taking care of the macro level. The inside-out strategy is to handle design by resolving internal functionalities first, and then consider the overall form afterwards. Outside-in strategy is to start thinking from the external site forces impacting interior functionality and overall form before addressing the interior arrangements. Case-based reasoning is to apply previous experiences, situations, and solutions for handling similar cases as references for solving new problems. Another commonly used design strategy is user-centered method. This method considers the users' behavior, activities, and needs in the building, and the design would be centered on these considerations and arrangements. Then, based on the considered issues, which could either be hypothesized and predicted by the designer or requested from users, a design program is generated. Following the design program, functional arrangements are made to create building forms. Such a set of sequences is a general and conventional strategic operation applied in design studio education.

However, in generating a form that solves design problems, abstract design problems require designers to turn the abstraction into a physical form that meets functional, structural, and material requirements. Designers need methods to accomplish such creations. These design methods include the use of design parti, analogy, metaphor, deformation, as well as other methods not yet reported. These methods have been used by many famous architects, and they are utilized conventionally at early design stages of most design projects, for they are the thinking mechanisms used for design generation. Next, we briefly explore the concepts found within these methods.

A design parti is a basic scheme or main organizing concept of an architectural design formulated in the conceptual design stage. It is an individual module, which could be an abstract concept presented in a simple statement or a fundamental form presented in a basic diagram. As Ching puts it, a design parti is the chief organizing thought or decision behind an architect's design presented in the form of a basic diagram and/or a simple statement (Ching 1995). Designers use a diagram or a statement to formulate an initial concept that might basically resolve most of the design problems. It also has the potential for future growth, and could include the use of one shape or one concept multiple times to create a larger sum. A design parti forms the basis of preliminary design philosophy, which leads to the final solution. Therefore, a parti can be the big idea, the central concept, the essence of the design, or the core element of the project. For example, in urban design, urban fabric is the combination of discrete architecture and traffic network that have been formed and combined by a certain revolution process. During the process, there are basic rules and basic patterns applied to shape the creation. The uses of the rules and patterns are attributed to design methodologies used anytime in the design process, and such rules and patterns are used to implement the parti. Similarly, other than design parti, other methods of analogy, metaphor and deformation could also be used at any stages during the process for form generation.

Analogy is the use of a similar concept to solve a similar problem. The concept could be a solution produced by the designer in previous projects and saved in memory, or a solution done by someone else and applied. Usually, a form created in a design is applied later in several other similar design projects. The previously generated solution is also termed a ***presolution model*** (Foz 1972; Chan 1990, 1993), which is similar to the notion of **case-based reasoning** (Eastman 2001). Analogy is commonly used to generate a solution idea. In design education, students are encouraged to study a number of master cases for design learning. In professional practice, designers would easily recall a number of solutions and test one after the other until the solution is reached (Chan 1993, 2001a).

Metaphor has been recognized as a structure of our conceptual system. Collected from linguistic evidence, Lakoff and Johnson indicated that metaphor is pervasive in everyday life, not just in language but also in thought and action. Using the metaphorical concept of "argument is war" as an example, when we talk about arguments, we think of winning or losing arguments. We see the person we are arguing with as our opponent and we attack their positions. We gain or lose our ground by using strategies. As such, it is clear that we use the notion of war to structure what we do and how we understand what we are doing when we argue (Lakoff and Johnson 2003). Therefore, human thinking processes are metaphorical, which means that the human conceptual system is metaphorically structured and defined. Our ordinary conceptual system is fundamentally metaphorical in nature. Design is a part of thinking and we inevitably apply metaphor to think about design verbally and through imagery as well.

The essential nature of metaphor is to understand and experience one kind of thing in terms of another (Lakoff 1987, 1993). In design, it is the use of one thing to represent something else. The concept of something for another thing could be a substitution of one idea, concept, word, or phrase with a shape, geometry, or object for the purpose of transforming an abstract idea into a physical form. The substitute "form" or "shape" could be a symbol or a sign used to represent a design notion or consideration. For example, the World Expo 70, held in Osaka, used air filled structures for most pavilions. The Japanese pavilion was composed of five circular

Fig. 2.5 Aerial view of Osaka World Expo 1970 site (Source: Davis-Brody, http://www.columbia.edu/cu/gsapp/BT/DOMES/OSAKA/0489-84.jpg. Accessed 10 Oct 2013)

exhibitions halls. These five circles symbolized the five leaves of sakura (see Fig. 2.5), the national flower of Japan. This application of using five circles to represent sakura, which represents the nation, indicates that the constructed artifacts symbolized the nation. This also exemplifies the method of using metaphor for creating a design. Another example is the famous gate that sits in the sea off the shore of Miyajima Island near Hiroshima, Japan. It was built as the main approach, which is the concept of gate, to its Itsukushima Shrine (see Fig. 2.6). This example, similar to **Eero Saarinen**'s big arch design in St. Louis as the gate to the western frontier (see Fig. 2.7) is another example of using an object (arch) that implies a metaphorical form (symbol of a gate) to implement a function (monument). This explains the visual use of metaphor.

Another example of using metaphor approached from a metaphorical object point of view is found in the famous fish designed by Frank Gehry. His notion of fish was to create a representation of the origin of our universe. He said that "Three hundred million years before man was fish" (see http://en.wikipedia.org/wiki/Frank_Gehry). Thus, in looking back at the history, he created a standing glass fish in the Minneapolis Sculpture Garden (1986, see Fig. 2.8) to make a statement. For a few years, he repeatedly used the form of fish in several of his proposed designs and built buildings, including the Richard B. Fisher Center for the Performing Arts in New York's Hudson Valley (2003) and the Port Olympic in Barcelona Spain (1992, see Fig. 2.9). He also generated a number of household items of lamps and sculptures based on the motif of fish.

On the other hand, metaphor could also be used more conceptually than visually. For instance, the California Academy of Science in San Francisco, designed by **Renzo**

Fig. 2.6   Gate at Itsukushima Shrine in Miyajima, Japan (Source: Bo-deh/Wikimedia Commons, http://commons.wikimedia.org/wiki/File:Miyajima_Itsukushima_Shrine_Portal.jpg.   Accessed 10 Oct 2013)

Fig. 2.7   The gateway arch in St. Louis (Source: Matt Kozlowski/Wikimedia Commons, http://en.wikipedia.org/wiki/File:Gateway_Arch_edit2.jpg. Accessed 10 Oct 2013)

Piano, provides a good example (see Fig. 2.10). The roof of the design is a metaphor for the entire project, seen by Piano as topography (Pearson 2009). The idea was to cut a piece of the site, which is a part of the Golden Gate Park, and push it up 35 ft and then put whatever was needed underneath. The green roof is an extension of the park and serves as a thermal buffer for the spaces below. It is clear in this design, the object that represents the metaphorical concept of roof is the Golden Gate Park's topography

Fig. 2.8 The Glass Fish by Frank Gehry in Minneapolis, 1986 (Source: Photo was taken by author)

to shelter the function of the Academy of Science. Other than using metaphor as a big concept in design, metaphor could also be used as a philosophical statement, as proposed by Mies van der Rohe in his modern architecture. He proposed the influential motto of "less is more" to symbolize the economic consideration of spatial arrangement, structural layout, and functional organization. Effects of this phase became a design classic on reducing spatial dimension to the minimal, eliminating unnecessary materials, and fully utilizing extreme simplicity to accomplish beauty.

The last cognitive strategy applied by designers is the method of deformation. Deformation is the application of external force or stress to change the shape, form, or dimensions of an object, which could be utilized at any time in the design process for the purposes of creating new forms. For example, **Peter Eisenman** uses

Fig. 2.9 The Fish by Frank Gehry in the Port Olympic, Barcelona, Spain, 1992 (Source: Till Niermann/Wikimedia Commons, http://commons.wikimedia.org/wiki/File:Barcelona_Gehry_fish. jpg. Accessed 10 Oct 2013

transformation as a step-by-step procedure by which one element is substituted for another, a volume is divided into planes, parallel planes turned to a grid, a grid is rotated and scaled, or even to twist, fold, or bend objects. In a video showing the Guardiola House design in Spain (see Fig. 2.11), he explained through animation the methods of utilizing movement of subtracting, intersecting, and uniting volumes for creating a novel and fragmental form.[16] These design processes can be seen as exercises of deconstructive architecture, which relate to the idea of generating a form with certain linguistic context associated. Such methods of linguistic association and form deformation are popular in the areas of digital architecture and rapid prototyping. Even though the deformation method has no direct relation to cognitive processes, it is a thinking phenomenon resulting from applying a set of particular procedures yielding particular forms, which should be identified as a phenomenon of cognitive strategy for form generation.

These examples explain briefly the use of certain strategies and methods in thinking for generating a design project that solves embedded design problems. The major difference between design strategy and design method is that design strategy is to use a particular method adopted in an early stage to develop an overall plan, or a scenario, or a goal sequence. Such generated schematic plan, scenario, or goal plan would be used to control the design sequences from the beginning and persistently utilized until the design is completed (Chan 1990, 2008). Design methods relate to the procedural

---

[16]The design concept by Peter Eisenman had been recorded in an animation video, which could be found in the following Web page: http://www.youtube.com/watch?v=JAH97LcQ2Cs

Fig. 2.10 The green roof of the California Academy of Science in San Francisco (Sources: Wolfman SF/Wikimedia Commons, http://commons.wikimedia.org/wiki/File:California_Academy_of_Sciences_pano.jpg, and Wikimedia public domain, http://en.wikipedia.org/wiki/File:CalifAcadSciRoof_0820.JPG. Accessed 19, July 2014)

Fig. 2.11 The Guardiola House by Peter Eisenman in Spain (Source: Peter Eisenman, A+U, 89:01, No. 220, Jan 1989, p. 21, Fig. 15)

operations for implementing functional, structural, or constructional requirements. A design method could be used at any time in the process, and many methods could be used in a design for the sake of creating forms. Thus, a design method is a part of a design strategy. If a designer could creatively apply a strategy, be it an analogy, a parti, a metaphor, or a deformation to create a new form, such thinking should be recognized as a creative one. The major point is not the use of the method, but the thinking that strategically activates the application of the method. This application of unique design strategy is the seventh notion of design creativity (DC7).

2.4.7 Design Uses Certain Reasoning

How do human beings make judgments and decisions in thinking? In life, human beings use information presented to them as observation data, generate some hypotheses or premises as the case, apply experience recalled from memory as reference, and utilize logical operations to set up inference for decision making. The commonly used logical operations, described by philosophers, are **deductive** and **inductive reasoning**. Deductive reasoning is to find specific fact from general data, find cause to effect, or start with a general principle and deduce that it applies to a specific situation through which a conclusion can be drawn, or start from reasons and look for consequences (Magnani 2009, p. 10). It does derive from the consequences of the assumed proposition. For instance, when deriving B from A, if the assumptions in A are true, a valid deduction would guarantee a true conclusion of B. Inductive reasoning is to draw from specific to general conclusions, from data to come up with general principles, or to generalize from a sample to a general property, and reasoning from data to a causal hypothesis (Aliseda 2006, p. 33). For example, if inferring B from A, we might have good reason to believe the conclusion of B drawn from the premise of A, but the conclusion of B is not always guaranteed to be true. Inductive reasoning is a commonly used type of reasoning in physics and philology.

These two logical methods of deduction and induction are most popularly discussed theories on human reasoning addressed before the mid-nineteenth century (Fischer 2001, p. 365). They are based on valid premises and therefore yield valid conclusions, which are seen as formal logic reasoning. However, in some cases, some data in the observation set might not be appropriately inferred or deducted, thus the third reasoning mode of **abductive reasoning** was introduced or re-discovered[17] by **Charles Sanders Peirce** (1839–1914) in the late nineteenth century.[18] At first, Peirce emphasized abduction as an evidencing process with a

---

[17] Aristotle (384–322 BC) dealt with deduction and induction but did not analyze abduction in detail; thus, abduction remained absent from the history of logic until it was re-discovered by Peirce (Fischer 2001, p. 365).

[18] In fact, the type of inference called abduction had been studied by Aristotelian syllogistics. But, Peirce had further interpreted abduction essentially as an inferential creative process of generating new hypotheses in broad philosophical and semiotic perspective (Magnani 2009, p. 9). It later is used by scholars to explain the reasoning in scientific discovery.

syllogistic interpretation of rule, case, and result. The famous example used by Peirce was a bag of beans. All the beans in the bag were black (rule), and these beans are from this bag (case), so these beans are black (result). Later on, he compared the relationship of abduction to a guessing instinct. To him, abductive reasoning was guessing through a person's experience to infer the best explanations. He wrote, "the abductive suggestion comes to us like a flash. It is an act of insight, although of extremely fallible insight" (Peirce 1997, p. 242). Abduction is inferring causes from effect. It represents an explanatory principle that even though logically it is invalid, it may still be confirmed inductively. For example, the road is wet because it has rained. The rain is the cause inferred from the consequential effect. In Pierce's perspective, all inferential thinking is to discover something we do not know and thus enlarge our knowledge by considering something we do know. The discovery is done by ways of guessing explanations or by logical leaps of the mind. In scientific discovery, abductive reasoning is seen as the operation of adopting explanatory hypotheses.

The purpose of using abductive reasoning is not to conclude whether something is true or false, but to posit what could possibly be true. The conclusions drawn in this mode are based on probabilities and mostly used within fields of science and research. While abductive reasoning possesses all the elements of true premises and conclusions in formal reasoning, it also consists of the element of probabilities, and thus is classified as a type of informal logic reasoning. As Fischer indicated, philosophically our knowledge of the world might not necessarily come from our logically true inferences, but all our inferences are invented hypotheses, the adequacy of which cannot be proved within logic, but only pragmatically. Therefore, making hypotheses is another form of "think rationally and infer irrationally." As such, knowing represents inferring, inferring means rule-governed interpreting, and interpreting is an act that proves adequate of our knowledge (Fischer 2001). Similarly, scholars use hypothetical reasoning to describe such a form of thinking and knowing, which is giving an excuse or reason for something that has not happened or could possibly happen.

In design, design reasoning is sometimes different from formal logic reasoning, which concentrates on the true or false of premises and conclusions with resulting abstract forms. There is no true or false in aesthetic value or in judging the amount of beauty in architectural forms. Design reasoning leans more towards abductive reasoning, which provides more hypotheses of what may be (March 1984; Martin 2009; Cross 2011). Lorenzo Magnani has described clearly that "abduction is the process of inferring certain facts and/or laws and hypotheses that render some sentences plausible, that explain (and also sometimes discover) some (eventually new) phenomenon or observation; it is the process of reasoning in which explanatory hypotheses are formed and evaluated" (Magnani 2009, p. 8). As such, design activity involves the processes of making hypothetical statements and evaluations from the beginning to the end of the course of a design.

In fact, design reasoning is not totally a formal logical operation with specific algorithms in place, but a certain way of informal reasoning guiding the mode of conduct. It should be seen as all the mental operators that move design actions. Rittel explained that reasoning consists of more or less orderly trains of thought, which include deliberating, pondering, arguing, and occasional logical inferences. But, design reasoning is a process of argumentation based on certain reasons to make decisions on making moves (Rittel 1988). Designers are actively formulating hypotheses inductively from observing certain events, looking for new data points, challenging accepted explanations, inferring possible new forms and new functionalities, and contemplating consequences (Martin 2009).

On the other hand, design requires a lot of effort on handling graphics. Thus, **spatial reasoning** is a part of design cognition, specifically spatial cognition. Spatial reasoning requires mental operations to visualize spatial patterns, and human rationales to manipulate the perceived patterns for accomplishing certain conceptual intentions over a time-ordered sequence of spatial transformations to generate results. Spatial reasoning is a special form of thinking in pictures, versus a form of thinking in words. Both formats of pictures and words are internal representations of the tasks that designers are facing or the problems that designers are solving. The methods used in spatial reasoning are to think about the problem deeply and express the understanding of the problem pictorially, or to relate learned experiences to a given problem situation through identifications, or to connect perceived information with certain concepts and project into the problem situation through association. The pictures applied could be symbols, sketches, or doodles that certain information deems necessary.

All these concepts of design reasoning explain the fundamental operators that motivate and drive the design process ahead. In summary, if a designer applied certain unique reasons that led to the use of certain design strategies yielding an unpredictable result of a novel product, then the conjecture, inference, and hypothetical reasoning could be the driving source of creativity in thinking. It is possible that the reasoning can strategically activate some unique design strategies. Thus, the reasoning used to strategically activate unique design strategies is the eighth notion of design creativity (DC8).

2.4.8   Design Applies Repetition for Design Generation

The last cognitive factor applied in the design processes is **repetition**. Repetition is to repeat the same thinking or action subconsciously or consciously. It is a part of human cognition utilized in language, writing, learning, and design thinking. Details of the nature of repetition and the resulting phenomenon in design can be found in a number of publications (Ross 1933; Ching 1979; Goodridge 1998; Mithen 2005; Chan 2012). This section emphasizes the aspect of repetition as a part of human cognition and its impact in design.

In everyday life, people often repeat the same life routine subconsciously. We repeatedly drink the same brand of beverage, navigate the same route to the same destination, apply the same schedule every day to run errands, go to the same grocery store for shopping, and use the same methods solving similar problems. After the same procedures on a particular task are exercised for a long time, the knowledge of executing them are turned into automatic skills, such as how we learned to ride a bicycle. Therefore, repetition causes automatic execution of skills that becomes a default cognitive procedure for accomplishing tasks. Research done in various fields has shown that repetition is the cognitive strategy used in rhetoric to affect or coordinate attitudes, especially when terms are used repeatedly in question-begging-form to persuade belief (Boisvert 2011); in literature to emphasize a notion and to generate emotional inspiration (Boisvert 2011); in music to generate a piece of music by repeating a fixed rhyme, beat, or melody (Yeston 1976); in psychology to improve learning through **rehearsal** (Waugh and Norman 1965; Atkinson and Shiffrin 1968); and is a human cognitive operation applied every day.

In the fields of design, repetition is one of the cognitive mechanisms used by applying a basic and simple feature as a module with a set of rules repeatedly to create a pattern. Reasons of causing repetition in design could either be the intentional plan or well-practiced procedural knowledge applied recursively to create characteristic patterns that generate the effect of rhythm. As defined, if a regularity of changes in a regularity of measures causes the effect of movement upon our minds, then rhythm occurs (Ross 1933). Rhythm has been seen, in arts and performing arts, as patterned recurrence, repetition, or movement in actions or in artifacts. It is the regular, harmonious recurrence of a specific element, often a single particular entity of line, shape, form, color, light, shadow, and sound. If a designer chooses an entity from these elements, creates a composition, and repeats the composition with a movement or in time, then an order to the whole design is generated. Such an order is the character of rhythmic phenomena. In fact, rhythm generates some regularity, simplicity, balance, and hierarchical order of composition in the design product that possesses a nature of consistency that is well perceived and easily comprehended by beholders.

A good example in design of the rhythm created by repetition is found in work done by Alvar Aalto, the famous Finland architect. Aalto would use the same shape of volume proportionally seven times and rotate the module slightly along the contour of a hill (see Fig. 2.12), use the same shape of balcony seven times on the façade (Fig. 2.13), the same configuration of curved beams four times with skylight to create a patterned shadow on the wall (Fig. 2.14), and the same curved roof shape four times (Fig. 2.15) in his designs. Similar methods can be found in his other building projects. Results of using the same form at least four times generated a wonderful visual order with consistency. These examples demonstrate that Aalto is one of many architects who skillfully applies repetition, which creates the beautiful effect of rhythm.

The same idea of using repetition methodologically (Chan 2012) to generate rhythm went under experimentation by a young designer, Jasmine Brown, in her research. She selected her third year design studio project as the base (see Fig. 2.16). She applied the same design program on the same site, but revised the design to test whether rhythm did occur through repetition. She repeated functional units with regular change of proportion, exposed structural elements with regular intervals in space, and unified skylight units on roof regularly for the purpose of satisfying functional, structural, and energy requirements. Results of the experiments, as shown in

Fig. 2.12 The Shiraz Art Museum by Aalto, 1969–1970 (Source: © Alvar Aalto Museum with permission)

Fig. 2.13 The Paimio Sanatorium design by Aalto (Source: Leon Liao/Wikimedia Commons, http://commons.wikimedia.org/wiki/File:Paimio_Sanatorium2.jpg. Accessed 10 July 2014)

Fig. 2.14 Interior of the Riola Church by Aalto (Source: http://snyfarvu.farmingdale.edu/~straaw/design2/project3/alvaraalto.html. Accessed 10 Oct 2013)

Fig. 2.15 The roof of the Riola Church in Riola di Vergato, Italy by Aalto, 1966, 1969, 1975–1980 (Source: Maija Holma © Alvar Aalto Museum with permission)

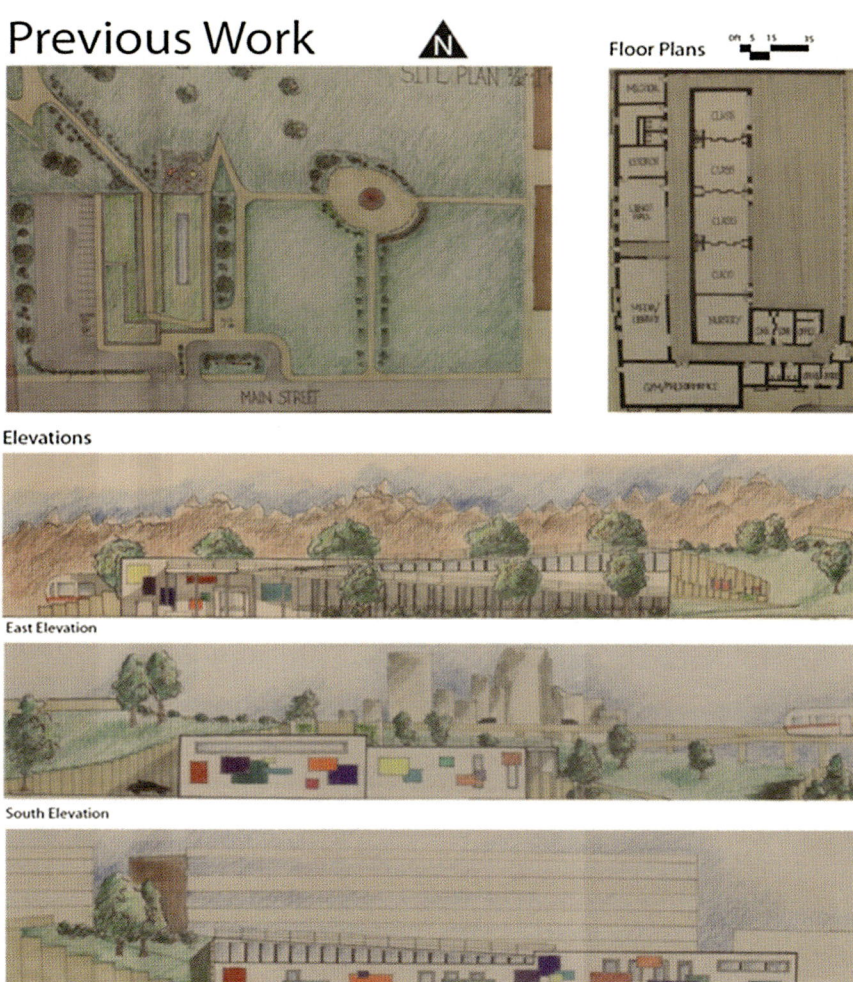

Fig. 2.16 Jasmine Brown's previous design (Source: Renderings by Jasmine Brown, fall 2009)

Fig. 2.17, have systematic spatial orders occurred and rhythmic visual effects revealed. She was pleased with the quality of the revised design.

Figure 2.18 is the interior rendering that has the rhythmic natural light casting rhythmic shadow patterns on the floor to create a pleasant visual image. Such a pleasant image would provide a comfortable perception to the users of the environment. Even though she cannot show that repetition is a part of a cognitive strategy that occurs in design, the results of purposefully utilizing reoccurrence in design did testify to the causal-effect of repetition for rhythm. Some designers consciously use the same features in a design project to generate rhythm, and some would repeat the

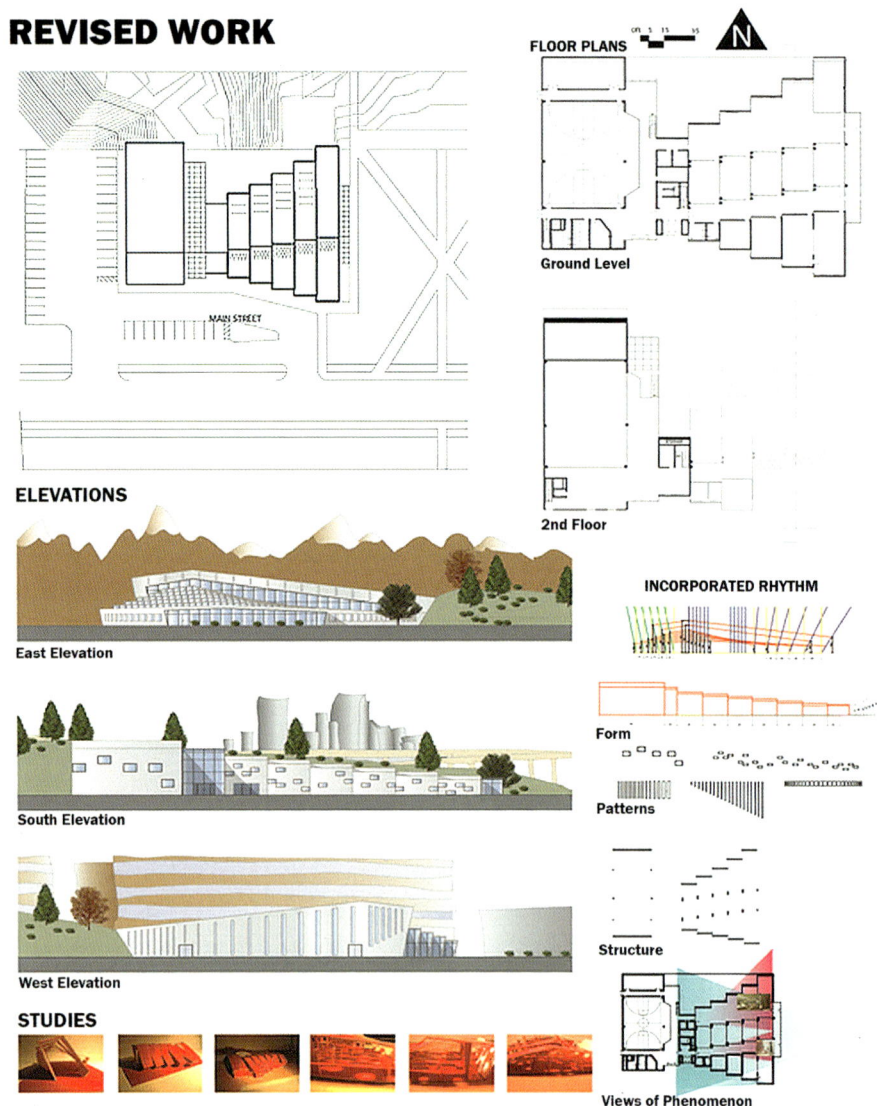

Fig. 2.17 Revised results of applying repetition (Source: Design by Jasmine Brown, spring 2012)

same features (or sub-solutions from previous design) across design projects to signify an individual style, which will be discussed in the following chapters. For instance, Frank Lloyd Wright reused his elevation grammar in his Prairie House design. Thus, rhythm is the cognitive function of repetition, and repetition is the driving force that causes rhythm and style. Since rhythm is ingrained in the human conscience of repetition, it should be a key component of design applied universally. As a summary, design repetition is the last but the most important notion that serves as the definition of individual style.

Fig. 2.18 Interior rendering of the revised design with rhythm (Source: Rendering by Jasmine Brown, spring 2012)

2.5   Overview and Operational Definition of Design Cognition

All these described cognitive mechanisms and strategies used in design could be summarized to clearly explain *what is design cognition?* and *what are the cognitive factors causing style and creativity?* Cognition, in general, is the process of perceiving, receiving, analyzing, storing, retrieving, and utilizing information. Information represents knowledge. Intelligence is a way to manage knowledge. The mental operations used to utilize intelligence include cognition. Various operations compose our thinking and design thinking as well. However, design thinking is different from other thinking processes of solving accounting, financial, statistic, software engineering, or medical problems. Solving design problems requires special knowledge with special strategies to operate the methods, plus special logic reasoning with special procedures to process domain-specific knowledge for creating three-dimensional tectonics to fulfill architectural functions. Skillful designers have certain ways of solving design problems that creates beautiful forms, and which involves the use of strategy, methodology, reasoning, logic, and representation in the entire thinking process to complete the design (Chan 1990, 2001a, 2008).

What does the picture of design cognition look like? Design cognition could be portrayed by including all cognitive activities that may occur and cognitive mechanisms applied in the design process to show a describable picture. The picture could be illustrated by major sequential activities that happened in the mind—similar to the items listed in Table 2.3—that a design is problem solving, with the problem understood and the tasks structured first, goals generated and constraints organized, then use of cognitive strategies of analogy, metaphor, previous cases, or association to generate a scenario (or parti). These major activities comprise the initial stage of conceptual design. Further, the following activities would recursively occur: looking for appropriate representation, using reasoning to operate constraints for accomplishing goals, applying strategies (of analogy, metaphor, deformation, cases, and association) to generate elements as solutions, reflecting the problem structure and solution structure to keep the solution on track, and operating repetition to create rhythm and style. All these activities of association, problem structuring, representation, reasoning, goal sequences, and constraint developments may be operated conventionally to solve the design problems. In some cases, they are uniquely utilized to generate a creative solution. As explained previously, as long as these mechanisms are operated in a novel way that generates a novel form, creativity occurs. However, it is repetition that causes style.

In most cases, designers use the logic of induction to find general conclusions from particular data, or from general data to find particular principles (deduction), or even to use educated guesses to explain things in the data (abduction) for making design decisions. Other than these patterns of reasoning, humans also make hypotheses for a case, and find from experience of heuristic to solve the design problem. As soon as the found experience has appropriate associations linking to the problem task, then it is applied immediately. This pop-up-and-apply phenomenon is the aspect of intuition, which also has some connections with divergent thinking styles. It has been proposed that design problem solving should involve a divergent as opposed to a convergent thinking process (Hatch 1988). In theory, divergent thinking usually happens in a free-flow fashion, and more ideas are generated through building up more associations with knowledge chunks in memory. It is because more links provide more potential solutions than less links, and chances of yielding unexpected surprising results are higher. In terms of logical operation, after the divergent thinking is completed, concepts and information will be structured and arranged by convergent thinking, which is to follow a set of procedures for achieving a solution. This is the fashion of logical operation in design thinking.

What is the operational definition of design cognition? Design activities consist of the operation of reasoning to achieve goals for satisfying constraints, using associations to retrieve knowledge, applying representations to communicate, and utilizing strategies to create form. Thus, the essence of thinking relates to the processing of information in our mind, and design cognition could be operationally defined as *the processes of logic and reasoning on operating design knowledge through a specific representation for conceptualizing, planning, and implementing certain intentions to make artifacts. Different representations used would need different logic procedures*.[19] Design representation is the medium used to communicate design, which could be sketch drawing, flow chart diagram, digital model, physical model, or even scripting. Studies in design cognition should concentrate on how designers use graphic thinking, how they execute reasoning, what representations have been used, and on exploring the patterns of operating logics for generating solutions in the design thinking process.

2.6   Other Affecting Design Thinking Factors

Reasoning and logic operations have connections with the design methods and strategies described earlier. But, different design media require different cognitive procedures that definitely yield different reasoning phenomena. For instance, design processed in the conventional paper-and-pencil mode versus the use of a digital system, such as Building Information Modeling (BIM), might have different design methods used in different phases of schematic design, design development, and constructional design stages. The conventional pencil-and-paper method is used to conceptualize design ideas. After ideas are developed, sketch drawings can be scanned and imported into the digital system to serve as visual guides for three-dimensional digital modeling. Alternatively, concepts could also be drawn two-dimensionally in a drawing system and imported to the BIM system for 3D modeling, which is related to the **top-down** or **conceptual driven** method. On the other hand, if a design is approached from the digital system, a grid system would usually be generated in the system and used as reference, desired details of structural components, corresponding walls, and standard architectonic elements would be added to the grid patterns with bottom up or data driven methods, or by applying a massing tool to create representative masses and put details afterwards as with top-down conceptual driven methods. However, design approached from digital systems is not popularly accepted in design education.

Regardless of whether conceptual driven (top-down) or data driven (bottom-up) thinking methods are applied in the BIM system, the flexibility of the software system cannot match the information processing speed of human brain, and the needed multiple functions and long procedures for execution in the digital system would consequently affect the human design vision and delay the progress of thinking. Thus, the various media utilized impacts design thinking. This is due to the fact that during the problem solving process, the mind creates similar mental representations to match up the external representation of the media used at hand for communication and design dialog. If the media used were sketch drawings, then mental pictures would exist in the mind (Chan 1997). If the media were physical models or digital models, then similar

---

[19] An *operational definition* is a definition of a measure that is detailed and concise, and is used to identify one or more specific observable conditions or events.

abstracted three-dimensional images would be generated to match the external models for continuing the design dialogues. And, if the task at hand were mathematic calculations, such as is applied in parametric modeling, then the internal representation would be numeric symbols (Chan 2009). In this situation, without having a monitor displaying executions, it is difficult to mentally predict design results. Yet, in order to understand the process of thinking and to explore the impact of design media to thinking results, more in-depth studies on design representations are needed. In Chap. 7, the effect of representation on creativity will be further explained.

2.7 Study Methods

Studies in exploring design thinking have applied various methods of conducting experiments to observe design activities, interviewing designers to collect needed information, concentrating on one designer's works and investigating the patterns and behaviors displayed as case study, asking designers to think aloud while working and recording the design processes for data collection, or mixing these methods for the purposes of accomplishing research objectives. These methods can be categorized by the following types.

2.7.1 Controlled Experiment Type

Controlled experiments have been used to observe certain aspects of design behavior. Thomas and Carroll studied various tasks, i.e. restaurant design, letter writing, and software design etc., to explore the cognitive processes (Thomas and Carroll 1979). By analyzing videotaped interactions between designer and client, and the subject's introspection, they showed that design problems were structured in terms of subproblems, which were dynamically produced during design. This was proven by the evidence that the overall attack on a design problem was organized into relatively smaller and simpler cycles. This finding could be interpreted as partial solutions generated locally by trial-and-error (generate-and-test) search. They reported that right representations suitable for the problem in question helped in solving the problem. They also pointed out that the goal structure the designer brought to the design situation would drastically alter the design activity and the design product. Thus, a crucial aspect in design is the specification of design goals.

Similar methods by running controlled experiments for observing responding activities has been done by Carroll et al. (1980) to study the representation and presentation in design problem solving. Their experiments were to ask 81 subjects to work on two design problems that were isomorphic in structure—an office layout and a manufacturing schedule—and to observe the design behavior. Their findings indicated that graphic representation helped solve the well-defined problem (manufacturing schedule) as well as the ill-defined problem (the office layout).

Moreover, certain presentations encouraged the utilization of graphic representation and therefore enhanced problem-solving ability. The controlled observation methods allowed experimenters to study a particular aspect by using special experimental tasks. However, the subject matter in these experimental tasks were not suitable for studying the overall cognitive behavior in architectural design processes. For instance, the office layout, the restaurant interior design, the letter writing, the software design, and the manufacturing schedule used in their experiments were not architectural design tasks and were not able to yield evidence explaining architectural problem-solving behavior. As indicated by Carroll et al., their study could not detect the subjects' goal structure because the design problems used in their experiments were too artificial and the subjects were just manipulating the arrangements of design blocks (Carroll et al. 1980).

2.7.2 Interview Type

Interview method used interviews with questionnaires, and asked designers to recall their design activities and report back to the interviewers (Krauss and Myer 1970; Darke 1979; Lawson 1994; Cross and Clayburn Cross 1996). This method has been used by Krauss and Myer (1970) to study the preliminary design process of a nursery school, which was a real project. They used charts to reconstruct and to record the architect's design activities. Such a method mainly relied on observations and designer's retrospections from reviewing field documents for data collection. They reported that designers carried out two activities that were considered as the essence of design. The first one was to make forms (solutions) relating to the pertinent data and analysis. The second one was to reevaluate the problem and possible solutions (forms), emphasizing one set of criteria (constraint) after another. These activities were performed in a continuous cycle throughout the whole design. This aspect was analogous to the generate-and-test concept. Another observation they made was that the order in which designers changed constraints could affect the final solution, and the design decision revealed a designer's characteristic judgments.

In studying built architecture design, **Darke** was the first one who reported this method in 1979 as the retrospective process to interview architects on their housing designs. Her study intended to find if during the design period the architectural designers have in mind an image or expectations about users (Darke 1979). She used the technique of interviewing the architects of five housing projects, and tape recorded the interviews. She relied on observations of sketched and written outputs, and on asking designers to recall their own processes, a retrospection and introspection method, for data collection. Darke's analysis of interviews showed that architects use a few simple objectives (an architectural term which refers to the intended to be achieved issues, or global constraints) to reach an initial concept (also called a design scenario). This initial concept further generated a visual image. Darke did not explain how an initial concept generated a visual image; she indicated that this visual image might come early in the design process, or it might appear after a certain amount of preliminary analysis, which was a conjecture or conceptualization of a possible solution. She called the initial concept and the objective the primary generator. She pointed out that, in most cases, the design concept was arrived at before the requirements had been worked out in detail. Once the concept was generated, it gave rise to a conjecture and an analysis followed. Thus design progressed in a generate-conjecture-analysis cyclic fashion, which differed from the analysis-synthesis model developed in the early 1960s.

From her study, Darke also reported that such an interview might not get perfectly accurate data, and there were problems. For example, architects might not be able to get the correct and original memories back, might interpret the activities from a different perspective, and might not be able to explain details of some graphic output (Darke 1979). Similarly, **Ericsson and Simon** (1996) indicated that subjects might provide incomplete reports due to the mental processes of forgetfulness or selective reporting of thoughts. **Durkin** (1994) posited that interviews and questionnaires provided potentially erroneous or distorted and therefore inaccurate reports. In some cases a person's recollection of thinking might be totally different from the actual thoughts made at the time.

Other interview examples are the studies by Cross (2011) of Gordon Murray's famous Formula One race car design, and on Kenneth Grange, a product designer. In his interviews with Murray, he had enough information to clearly explain the winning aerodynamic feature to lower the car for gaining speed, the innovative devices of a hydro-pneumatic suspension and the generated "stop pit" in the racing processes, and the working methods of managing a team applied by the designer to achieve winning records. Similarly, in his interview with Grange on the prototype of the sewing machine and the nose of the high speed train, Cross explained that the designer's works are not based on styling or re-styling of a product, but from a fundamental reassessment of the purpose, function and use of the product. From interviewing designers and collecting enough data, he could explain what these two designers had learned from failure and the working methods utilized by them to generate good products. The methods used by Cross are asking designers questions and request for retrospection on a few products to study the character of each design, which also is a type of case study.

2.7.3 Case Study Type

A case study is a careful analysis of a person, group, event, or a phenomenon on factors in relation to context. It can be used to explore causation for finding underlying principles. The method on conducting a case study is to generate hypotheses, then provide a systematic way of looking at events, collecting needed data, analyzing related information, and reporting the results to test the hypotheses. The case study has been used as a teaching method at the Harvard Business School through interviewing leading practitioners of business and writing detailed accounts of what these managers were doing, to serve as textbooks (Barnes et al. 1994; Dul and Hak 2008).

In Psychology, it is used as a descriptive research tool to provide detailed descriptions of rare events and specific conditions (Christensen 1994). One interesting approach is the critical case study, which has strategic importance in relation to a general problem that applies to a larger population. For instance, if just one observation does not fit with the proposition, it is considered not valid generally and must therefore be either revised or rejected. Thus, for a scientific proposition, if it is false, then observation or experiment will at some point demonstrate its falsehood.

In medicine and psychology, case study is an in-depth study of one person on the subject's life to find patterns and causes for behavior. In management, it is conducted over an extended period of time to explore the progression of an event, or used to study historical information of events or organizations. For gathering data, case study typically combines data collection methods such as archives, interviews through structured survey-type questions or open-ended questions, questionnaires, documents, and direct observations. The evidence collected may be qualitative (e.g. texts) and quantitative (e.g. numbers), or both. Because of the quantitative nature, models can be generated and tested through case studies, and new variables identified in case studies can be formalized in models (Bennett and Elman 2006).

In applications, case studies can be: exploratory to serve as a pilot study before developing further hypotheses, explanatory to investigate causal effect and test theory (Pinfield 1986; Anderson 1983), descriptively to generate a descriptive theory and provide description test (Kidder 1982), collectively to study a group of individuals, or generate theory (Gersick 1988; Harris and Sutton 1986). These studies are usually expected to be generalized to many others. But, such studies are sometimes subjective and not able to cover a larger population. In order to use it more effectively, scientific methods of gathering data and a well planned research agenda should be fully constructed. For instance, Eisenhardt (1989) outlined a rigorous set of eight procedures with associated criteria for using case studies to sequentially build up a ground theory. These steps include getting started (to define research question), selecting cases, crafting instruments and protocols (for data collection), entering the field, analyzing data, shaping hypotheses, enfolding literature, and reaching closure. They do set up a roadmap for building theory. Most empirical studies lead from theory to data. However, the accumulation of knowledge involves a continual cycling between theory and data, thus, the use of case studies does allow a researcher to go from data to theory (Eisenhardt 1989).

As a summary, the case study method allows us: to understand the way new ideas arise and come to fruition in an innovative product (Candy and Edmonds 1996); to sequentially observe the true and false side of the proposition for testing the hypotheses (Flyvbjerg 2006); to study the reasoning involved in justifying a decision chain from a given design brief to a given solution (Galle 1996); and to develop tools and guidelines for improving team design practice in product designs (Valkenburg and Dorst 1998). In art and architecture, case studies have been used to explain a designer's design method that generates some characteristics in design products. For instance, Frank Lloyd Wright used similar grid systems to fit the module of five casement windows in his prairie houses (Chan 1992, 2000), and Alvar Aalto used repetition to create the phenomenon of rhythm (Chan 2012). In this book, the causing factors of creativity will be explained through case studies in later chapters.

2.7.4 Think Aloud Type

Among all the methods used and reported, the most suitable one for closely studying thinking is **protocol analysis**, which involves videotaping the verbalization of thoughts while designing in real time as a case and analyzing the case. The method is to ask subjects to "think aloud" while working on a given task in an experiment, with the collected data to be systematically analyzed to find insight on cognitive processes in solving problems. The earliest documented analysis was done by Watson (1920) to illustrate the characteristics of cognitive process in problem solving. De Groot (1978) used protocol analysis, based on the concept of thinking as a hierarchically organized linear series of operations, to study chess players selecting moves in games. Even though de Groot's protocols were recorded by hand and were incomplete, the data provided enough information to compose a rather detailed illustration of problem solving processes of novice and expert chess players. After tape and video recorders were developed in the 1950s, their utility for instantly recording sounds made the thinking data transparent. Particularly, a video recorder could also record drawing while designers talked. Thus, concurrent verbal reports were recognized as major data source on cognitive processes in specific tasks (Anderson 1987).

Studies on applying "think aloud" and "protocol analysis" techniques had been done in various fields, including: accounting to discover human decision making (Belkaoui 1989); architecture to analyze design behaviors (Eastman 1970; Akin 1979; Chan 1990; Lloyd and Scott 1994); artificial intelligence to show problem solving strategies (Conati and Vanlehn 2000); computer simulation to find out search methods applied in problem solving (Newell and Simon 1956); decision making to explore reasoning (Montgomery and Svenson 1989); education to understand human learning methods (Chi et al. 1989); human-computer interaction to describe user group interaction (Howard 1997); second-language learning to study knowledge formulation (Faerch and Kasper 1987); software engineering research to reveal engineers' behaviors and strategies on the usage of methodologies (Guindon 1990; Davies and Castell 1992; Hughes and Parkes 2003); industrial design (Dorst 1995; Valkenburg and Dorst 1998); mechanical engineering design (Lloyd et al. 1996; Atman et al. 1999); electronic engineering design (McNeill et al. 1998); and others in exploring the character of problem space (Goel and Pirolli 1992).

In testing architectural thinking, studies have been done in running design experiments to identify the mental operators and representations applied in the design processes (Eastman 1970; Akin 1978; Chan 1989). When exploring cognitive behavior, various cognitive models have been developed to best fit the experimental data (Akin 1986; Chan 1990). While analyzing design activities on parti, design strategies, and graphic thinking, protocol analyses are used as the study tool to describe characters of design activity (Foz 1972; Chan 1990; Goldschmidt 1991; Valkenburg and Dorst

1998) and patterns of behavior (Chan 1993; Suwa et al. 1998). In fact, through analyzing the protocol data, researchers can effectively simulate designers' cognitive processes through computer programs. Detailed explanations on methods and procedures of using protocol analysis for studying architectural design have been provided (Chan 1990, 2003) and discussed (Chan 2008).

Regarding the data efficiency on studying cognitive tasks, we can learn about what we are thinking by verbally describing what is going through our mind while performing the tasks. Such verbal description is called *verbal protocol*. However, to learn the mental tasks of designing, which requires the development of mental images, is more complicated than other problem solving tasks. Because the activities of design require sketching in designing, there are concerns regarding the limitations of using the think aloud method to collect data for studying design thinking. The following three items are the major concerns:

1. Gaps in verbal protocol data: Human beings think faster than they can talk and draw. People might not be able to quickly catch the marvels that flash in their mind before they are gone.
2. Cognitive actions cannot be verbalized: Some actions are automatic skills that are not that easy to describe through language; particularly, some drawings might have multiple meanings.
3. Cognitive overload: It might consume more energy to process information while four tasks of talking, drawing, perceiving, and processing information are occurring simultaneously.

Although there are deficiencies in and criticism of this method, some researchers have addressed the issues of existing gaps in verbal data, unable to be verbalized cognition actions, and cognitive overload (Akin 1979; Eastman 2001; Cross 2001). Regardless, this method is still justified for capturing first-hand data for cognitive studies. For instance, a workshop on "analyzing design activity" held in 1994 at the Delft University of Technology to study the virtues of protocol analysis, demonstrated the value of its data analysis. In that workshop, the same sets of videotapes and transcribed protocols were sent to a group of researchers to perform the analysis in any form they chose (Dorst 1995). The workshop yielded many interesting results based on the diverse array of approaches researchers used. Their findings varied, all arising from the same data set, from the study of social activity in a group design process (Cross and Cross 1995) and episodic knowledge used in the design activity (Visser 1995) to verbal and visual coding applied in the design process (Akin and Lin 1995), among others. This demonstrates that verbal data are a rich data mine that has potential for various explorations. In fact, a single set of data generated 20 different papers selectively published in the journal *Design Studies*, which showed clearly that protocol data is a rich set of original data bank and protocol analysis is a rigorous research tool for empirical studies on design activity.

Even though the protocol data cannot catch 100 % of the thinking phenomenon and there are limitations, this method has been improved by filling the gap in the data through retrospective post-task interviews and questionnaires (Ericsson and Simon 1980, 1996; Chan 2008). Methods of coding and decoding protocol data have also been carefully studied and developed to ensure the reliability of data analysis (Chan 1990, 2008; Gero and McNeill 1998). In fact, any verbalization produced by a subject while problem solving does clearly represent the contents of the subject's working memory (Ericsson and Simon 1996). Thus, it is regarded as the best method for studying and understanding human thinking, and is used to study individual style in this book. Interestingly, the term *protocol analysis* has also been used in computer science, but it relates to the concept of employing proper software and hardware tools to operate the package of data for transiting a network's media, and thus specifically called *network protocol analysis*.

2.7.5   Other Study Types

Other than the discussed methods, virtual reality is another means for studying design thinking in processes. For example, in 1997, a **Virtual Architectural Design Tool** (VADeT) was created in an immersive virtual environment (CAVE). This VADeT system had a number of metaphorical icons (see Fig. 2.19) serving as design tools for generating, modifying, and editing three-dimensional objects of architectural elements (see Fig. 2.20), and other tools for defining dimensions, materiality, and colors of the objects. By using these tools in the system, users could create a design and observe how designers design in virtual space. Experiments having architecture students use the system to design a kitchen were conducted right after the system was built (see Figs. 2.21 and 2.22). Through

Fig. 2.19 The icons used in the VADeT system (Source: "Collaborative Design in Virtual Environments", 2011, pp. 34–35, "Virtual Representation and Perception in Virtual Environments" by CS Chan, Fig. 1, p. 34; with permission from Springer Science + Business Media B.V.)

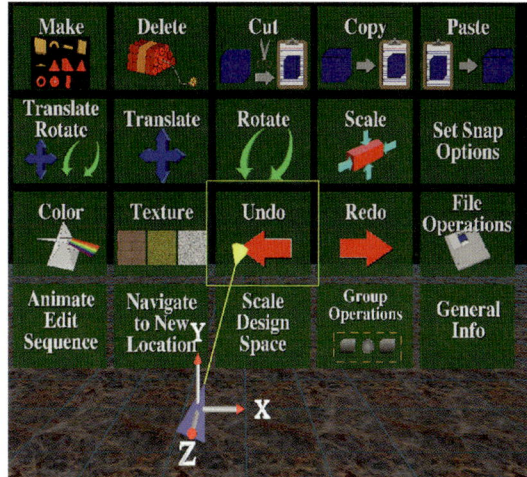

Fig. 2.20 The menu selections used in VADeT (Source: "Collaborative Design in Virtual Environments", 2011, pp. 34–35, "Virtual Representation and Perception in Virtual Environments" by CS Chan, Fig. 1, p. 34; with permission from Springer Science + Business Media B.V.)

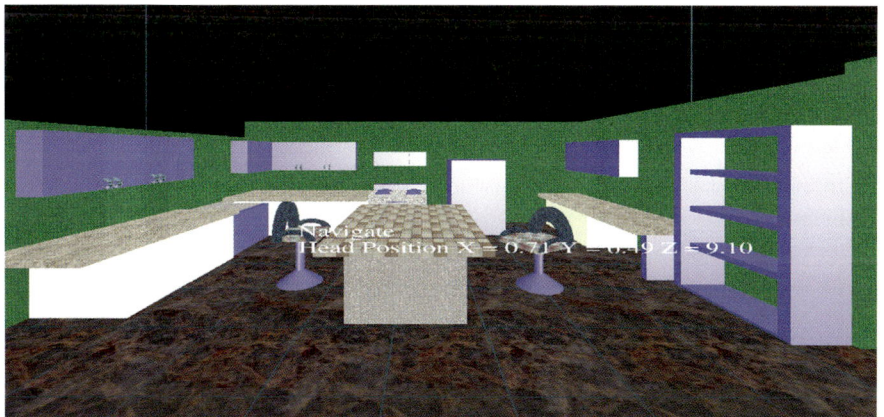

Fig. 2.21 A design example A done in the VADeT system (Source: "Collaborative Design in Virtual Environments", 2011, pp. 34–35, "Virtual Representation and Perception in Virtual Environments" by CS Chan, Fig. 3, p. 35; with permission from Springer Science + Business Media B.V.)

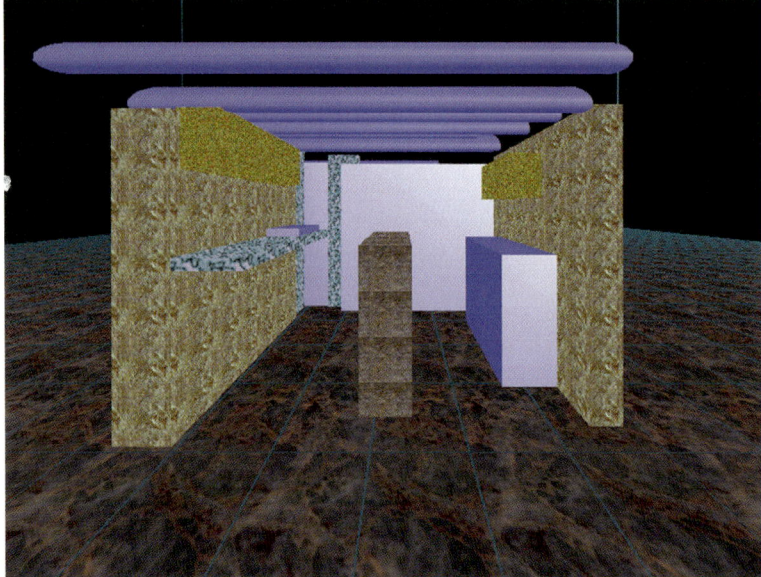

Fig. 2.22 A design example B done in the VADeT system (Source: "Collaborative Design in Virtual Environments", 2011, pp. 34–35, "Virtual Representation and Perception in Virtual Environments" by CS Chan, Fig. 3, p. 35; with permission from Springer Science + Business Media B.V.)

compiling the protocol analysis on the collected verbal data, it was found that the overwhelming sense of immersion and projection in the virtual reality environment altered design behavior and thinking routines. Several interesting findings were discovered in the experiments.

1. In conventional design, designers heavily rely on using scale to get accurate measurements for spatial layouts. In the full scale environment and in this system, there was no physical scale available in the virtual environment; designers used their bodies as scale references.
2. Subjects focused a great deal of attention on the proportions of each object, their spatial relationships with adjacent objects, and their locations in the space that fulfilled functional links and visual connections with other objects in the scene.
3. Design processes in the virtual environment were almost purely visual, with considerable attention devoted to the sizing, texturing, and coloring of the details of the objects, and less on reasoning and logical problem solving.
4. No alternative design solutions were created or considered in the experiments and the entire design process was linear progression. This might have been due to the large work-load consumed in the environment.

In short, the subjects' design processes seemed more intuitive rather than deliberate. It is possible that their perception was overwhelmed by 3D images and their thinking processes were driven mainly by geometric (visual) thinking instead of

Table 2.4 Summary of cognitive actions causing design creativity (DC) and individual style (IS)

ID:	Cognition causing design creativity (DC) and individual style (IS)
DC1:	Knowledge repertoire or pre-solutions.
DC2:	Association of diversified knowledge information.
DC3:	Unique goal sequences, utilization of external & internal constraints.
DC4:	Quantity of extrinsic environmental constraints.
DC5:	Problem structuring and restructuring from what-if reflection.
DC6:	Internal and external representations.
DC7:	Design strategy used in design.
DC8:	Reasoning used to strategically activate design strategies.
IS1:	Repetition of forms or of any DC variables in processes defines style.

conventional logical reasoning (Chan et al. 1999; Chan 2001b). It could also be due to the fact that the full scale of the immersive virtual environment created a very strong sense of presence (Chan and Weng 2005). Therefore, designers focused their attention on the objects created and ignored overall functional layouts. As such, design cognition, under this new context, may be adjusted automatically to accommodate the new sensations created by this exotic visual world. This example explains the influences of design representation, which might have changed the design strategies used, to thinking patterns. It also correlates with the concept that design could be characterized as a construction of representations (Visser 2006).

In this chapter, all cognitive factors that constitute design cognition operating in the design process are explained. Factors causing design creativity and individual style are highlighted and summarized in Table 2.4. For the items as explained in Sects. 2.4.3, 2.4.4, and 2.4.7 of: (1) having more of extrinsic constraints would decrease the chances of creativity (DC4), (2) the shaping and changing of problem structures would cause creativity (DC5), and (3) reasoning used that affects creativity (DC8), they do need a longer period of observation time to trace their causal-effects. Thus, they are highlighted in the table as a reference.

These included cognitive factors are mechanisms used to process cognition, and each of them has different levels of significance on the formulation of design products. Some cognitive factors are more influential and dominant than others. In the discussion provided in this chapter, the major cognitive factor or mechanism dominating or influencing design is design representation, either internal or external. The secondary factor is the design strategy used. Domain knowledge (e.g., environmental issues, comprehension of spatial arts, and principle of compositions etc.) and design intention (e.g., goals and constraints used to construct the problem structure) are the third and fourth factors contributing to the completion of a solution, regardless whether the solution is a creative one or not.

Future studies in this area should concentrate on exploring the relationships between the level of significance of these cognitive factors and design creativity. A measurement scale should be further developed to evaluate their levels of significance to design outcome. Other studies could also be conducted to explore the new thinking patterns that might emerge for adjusting the cognitive capacity to cope with the evolution of information technology and digital architecture. For instance, the design of the virtual scene and representation of avatars in the Internet-based virtual world has an impact to the issues of reasoning in virtual reality versus reasoning in reality.

It is clear that design thinking logic and reasoning in the digital architecture area would have been changed cognitively due to the change of design representations. In the following chapters, how design cognition affects designers' individual style will be introduced by conducting experiments for protocol analyses and by case studies to provide evidence for explaining the causes of creativity. Thus, concepts defined in this chapter are intended to provide readers with enough information for conducting future studies and a feasible starting point for chapters to follow.

References

Akin O (1978) How do architects design? In: Latombe JC (ed) Artificial intelligence and pattern recognition in computer aided design. North-Holland, New York, pp 65–104

Akin O (1979) An exploration of the design process. Des Methods Theor 13(3/4):115–119

Akin O (1986) Psychology of architectural design. Pion, London

Akin Ö (ed) (1997) Descriptive models of design. Des Stud (special issue) 18:4

Akin O, Lin C (1995) Design protocol data and novel design decisions. Des Stud 16:211–236

Akin O, Dave B, Pithavadian S (1988) Heuristic generation of layouts (HeGeL). In: Gero J (ed) Artificial intelligence in engineering design. Computational Mechanics Publications, Southampton, pp 413–444

Alexander C (1963) The determination of components for an Indian village. In: Jones JC, Thornley DG (eds) Conference on design methods. Pergamon Press, Oxford, pp 83–114

Alexander C (1964) Notes on the synthesis of form. Harvard University Press, Cambridge, MA

Alexander C (1971) State of art in design methodology: interview with C. Alexander. DMG Newsletter, pp 3–7

Alexander C, Ishikawa S, Silverstein M et al (1977) A pattern language, towns, buildings, constructions. Oxford University Press, New York

Aliseda A (2006) Abductive reasoning: logical investigations into discovery and explanation, vol 330. Springer, The Netherlands

American Heritage Dictionary (2013) http://ahdictionary.com/. Accessed 5 Mar 2014

Anderson JR (1980) Cognitive psychology and its implications. W. H. Freeman, San Francisco

Anderson P (1983) Decision making by objection and the Cuban missile crisis. Adm Sci Q 28:201–222

Anderson JR (1987) Methodologies for studying human knowledge. Behav Brain Sci 10:467–505

Anderson JR, Bower GH (1973) Human associative memory. V. H. Winston & Sons, New York

Archer B (1965) Systematic method for designers. In: Cross N (ed) Developments in design methodology. Wiley, New York, pp 57–82

Archer B (1970) An overview of the structure of the design process. In: Moore GT (ed) Emerging methods in environmental design and planning. MIT Press, Cambridge, MA, pp 285–307

Archer B (1979) Whatever became of design methodology? Des Stud 1(1):17–18

Atkinson RC, Shiffrin RM (1968) Human memory: a proposed system and its control processes. In: Spence KW, Spence JT (eds) The psychology of learning and motivation, vol 2. Academic, New York, pp 89–195

Atman C, Chimka J, Bursic KM, Nachtmann HL (1999) A comparison of freshman and senior engineering design processes. Des Stud 20(2):131–152

Baker GE, Dugger JC (1986) Helping students develop problem solving skills. Technol Teach 45(4):10–13

Bardach E (1981) Problems of problem definition in policy analysis. Res Publ Pol Anal Manag 1:161–171

Bardach E (2000) A practical guides for policy analysis. The eightfold path to more effective problem solving. Chatham House Publishers, New York

Barnes L, Christensen C, Hansen A (1994) Teaching and the case method: text, cases, and readings. Harvard Business School Press, Boston

Bayazit N (2004) Investigating design: a review of forty years of design research. Des Issues 20(1):16–29

Belkaoui A (1989) Human information processing in accounting. Quorum Books, New York

Bennett A, Elman C (2006) Qualitative research: recent developments in case study methods. Annu Rev Polit Sci 9:455–476

Blank MS, Foss DJ (1978) Semantic facilitation and lexical access during sentence processing. Mem Cognit 6:644–652

Boisvert DR (2011) Charles Leslie Stevenson. In: Zalta EN (ed) The Stanford Encyclopedia of Philosophy. http://plato.stanford.edu/archives/sum2011/entries/stevenson/. Accessed 4 May 2014

Bower GH (1970) Organizational factors in memory. Cogn Psychol 1:18–46

Broadbent G (1969) Design method in architecture. In: Broadbent G, Ward A (eds) Design methods in architecture. George Wittenborn, New York, pp 15–21

Buschmann F, Meunier R, Rohnert H, Sommerlad P, Stal M (1996) Pattern-oriented software architecture: a system of patterns. Wiley, New York

Candy L, Edmonds E (1996) Creative design of the Lotus bicycle: implications for knowledge support systems research. Des Stud 17(1):71–90

Carroll JM, Thomas JC, Malhotra A (1980) Presentation and representation in design problem-solving. Br J Psychol 71:143–153

Casti J (1998) The Cambridge Quintet: a work of scientific speculation. Addison-Wesley, Reading

Chan CS (1989) Cognition in design process. In Proceedings of the 11th Annual Conference of the Cognitive Science Society. Lawrence Erlbaum, Hillsdale, pp 291–298

Chan CS (1990) Cognitive processes in architectural design problem solving. Des Stud 11(2):60–80

Chan CS (1992) Exploring individual style through Wright's design. J Architect Plann Res 9(3):207–238

Chan CS (1993) How an individual style is generated? Environ Plann B 20(4):391–423

Chan CS (1997) Mental image and internal representation. J Architect Plann Res 14(1):52–77

Chan CS (2000) Can style be measured? Des Stud 21(3):277–291

Chan CS (2001a) An examination of the forces that generate a style. Des Stud 22(4):319–346

Chan CS (2001b) Design in a full-scale immersive environment. In: Proceedings of the 2da. Conferencia Venezolana sobre Aplicaciones de Computadoras en Arquitectura, pp 36–53

Chan CS (2003) Thoughts of Herbert A. Simon – on cognitive science in design. In: Chiu ML (ed) CAAD talks 2: dimensions of design computation. Garden City Press, Taipei, pp 34–43

Chan CS (2007) Does color have weaker impact on human cognition than material? CAAD future 07. Springer, Amsterdam, pp 373–384

Chan CS (2008) Design cognition: cognitive science in design. China Architecture & Building Press, Beijing

Chan CS (2009) The impact of design representation to design thinking. New Architect 3:88–90

Chan CS (2011) Design representation and perception in virtual environments. In: Wang XY, Tsai J (eds) Collaborative design virtual environments. Springer, Amsterdam, pp 29–40

Chan CS (2012) Phenomenology of rhythm in design. J Front Architect Res 1(3):253–258

Chan CS, Weng C (2005) How real is the sense of presence in a virtual environment? In: Bhatt A (ed) Proceedings of the 10th international conference on computer aided architectural design research in Asia. TVB School of Habitat Studies, New Delhi, pp 188–197

Chan CS, Hill L, Cruz-Neira C (1999) Can design be done in full-scale representation? In: Proceedings of the 4th design thinking research symposium – design representation, vol II. MIT, Boston, pp 139–148

Chase WG, Simon HA (1973) Perception in chess. Cogn Psychol 4:55–81

Chi MTH, Hassok M, Lewis MW, Reimann P, Glaser R (1989) Self-explanations: how students study and use examples in learning to solve problems. Cognit Sci 13:145–182

Ching FDK (1979) Architecture: form, space & order. Van Nostrand Reinhold, New York

Ching FDK (1995) A visual dictionary of architecture. Van Nostrand Reinhold, New York

Christensen LB (1994) Experimental methodology, 6th edn. Simon & Schuster, Needham Heights

Collins AM, Loftus EF (1975) Spreading activation theory of semantic processing. Psychol Rev 82:407–428

Collins AM, Quillian MR (1969) Retrieval time from semantic memory. J Verbal Learn Verbal Behav 8:240–248

Conati C, Vanlehn K (2000) Toward computer-based support of meta-cognitive skills: a computational framework to coach self-explanation. Int J Artif Intell Educ 11:398–415

Cross N (1984) Developments in design methodology. Wiley, New York

Cross N (1999) Natural intelligence in design. Des Stud 20(1):25–39

Cross N (2000) Designerly ways of knowing: design discipline versus design science. In: Picazzaro S, Arruda A, De Morales D (eds) Design plus research, proceedings of the Politenico di Milano conference, pp 43–48

Cross N (2001) Design cognition: results from protocol and other empirical studies of deign activity. In: Eastman E, McCracken M, Newstetter W (eds) Design knowing and learning: cognition in design education. Elsevier, Amsterdam, pp 79–103

Cross N (2011) Design thinking: understanding how designers think and work. Berg, Oxford

Cross N, Clayburn Cross A (1996) Winning by design: the methods of Gordon Murray, racing car designer. Des Stud 17(1):91–107

Cross N, Cross AC (1995) Observations of teamwork and social processes in design. Des Stud 16:143–170

Cross N, Edmonds E (eds) (2003) Expertise in design. University of Technology, Australia Creativity and Cognition Press, Sydney

Cross N, Dorst K, Roosenburg N (1992) Research in design thinking. Delft University Press, Delft

Cross N, Christiaans H, Dorst K (eds) (1996) Analysing design activity. Wiley, Chichester

Darke J (1979) The primary generator and the design process. Des Stud 1(1):36–44

Davies S, Castell A (1992) Contextualizing design: narratives and rationalization in empirical studies of software design. Des Stud 13(4):379–392

Davis SF, Palladino JJ (2002) Psychology, 3rd edn. Prentice-Hall, Upper Saddle River

De Groot AD (1978) Thought and choice in chess. Mouton, The Hague

Dery D (1984) Problem definition in policy analysis. University Press of Kansas, Lawrence

Dorst K (1995) Analyzing design activity: new directions in protocol analysis. Des Stud 16(2):139–142

Dorst K, Cross N (2001) Creativity in the design process: co-evolution of problem-solution. Des Stud 22(5):425–437

Dul J, Hak T (2008) Case study methodology in business research. Butterworth-Heinemann, Oxford

Dunn WN (2004) Public policy analysis: an introduction. Prentice Hall, Upper Saddle River

Durkin J (1994) Expert systems: design and development. Prentice-Hall, Upper Saddle River

Eastman C (1970) On the analysis of intuitive design processes. In: Moore GT (ed) Emerging methods in environmental design and planning. MIT Press, Cambridge, MA, pp 21–37

Eastman C (2001) New directions in design cognition: studies of representation and recall. In: Eastman C, McCracken M, Newseller W (eds) Design knowing and learning. Elsevier, Amsterdam, pp 147–198

Ebbinghaus H (1913) A contribution to experimental psychology. Teachers College, Columbia University, New York

Eberhard JP (1970) We ought to know the difference. In: Moore GT (ed) Emerging methods in environmental design and planning. MIT Press, Cambridge, MA, pp 363–367

Echenique M (1972) Models: a discussion. In: Martin L, March L (eds) Urban space and structures. Cambridge University Press, Cambridge, pp 164–174

Eisenhardt K (1989) Building theories from case study research. Acad Manag Rev 14(4):532–550

Ericsson KA, Simon HA (1980) Verbal reports as data. Psychol Rev 87:215–251

Ericsson KA, Simon HA (1996) Protocol analysis: verbal reports as data. MIT Press, Cambridge, MA, pp 48–62

Faerch C, Kasper G (1987) Introspection in second language research. Multilingual Matters, Clevedon

Fischer HR (2001) Abductive reasoning as a way of worldmaking. Found Sci 6(4):361–381

Flyvbjerg B (2006) Five misunderstandings about case study research. Qual Inq 12(2):219–245

Fowler M (2002) Patterns of enterprise application architecture. Addison-Wesley, Boston

Foz ATK (1972) Observations on designer behavior in the parti. Master's thesis, Massachusetts Institute of Technology

Frith C (2007) Making up the mind: how the brain creates our mental world. Blackwell Publishing, London

Galle P (1996) Design rationalization and the logic of design: a case study. Des Stud 17(3):253–275

Gamma E, Helm R, Johnson R, Vlissides J (1994) Design patterns: elements of reusable object-oriented software. Addison-Wesley, Boston

Gero JS (ed) (2004) Design computing and cognition'04. Kluwer Academic Publishers, Dordrecht

Gero JS (ed) (2006) Design computing and cognition'06. Springer, Dordrecht

Gero JS (ed) (2010) Design computing and cognition'10. Springer, Dordrecht

Gero JS, Goel AK (eds) (2008) Design computing and cognition'08. Springer, Dordrecht

Gero JS, McNeill T (1998) An approach to the analysis of design protocols. Des Stud 19(1):21–61

Gersick C (1988) Time and transition in work teams: toward a new model of group development. Acad Manage J 31:9–41

Gluck MA, Mercado E, Myers CE (2007) Learning and memory: from brain to behavior. Worth, New York

Goel V, Pirolli P (1992) The structure of design problem spaces. Cognit Sci 16:395–429

Goldschmidt G (1991) The dialectics of sketching. Creativ Res 4(2):123–143

Goldschmidt G, Porter W (eds) (2004) Design representation. Springer, London

Gombrich EH (1960) Art and illusion: a study in the psychology of pictorial representation. Pantheon, New York

Goodridge J (1998) Rhythm and timing of movement in performance drama, dance and ceremony. Jessica Kingsley, London

Greco A (1995) The concept of representation in psychology. Cogn Syst 4(2):247–256

Guindon R (1990) Designing the design process: exploiting opportunistic thoughts. Hum Comput Interact 5:305–344

Harley TA (1995) The psychology of language: from data to theory. Psychology Press, East Sussex, pp 175–205

Harris S, Sutton R (1986) Functions of parting ceremonies in dying organizations. Acad Manage J 29:5–30

Hartt F (1994) History of Italian Renaissance. Abrams, New York

Hatch L (1988) Problem solving approach. In: Kemp WH, Schwaller AE (eds) Instructional strategies for technology education. Glencoe, Mission Hills

Hayes JR (1981) The complete problem solver. The Franklin Institute, Philadelphia, pp 51–69

Heath T (1984) Method in architecture. Wiley, New York

Hesse M (1966) Models and analogies in science. University of Notre Dame Press, Indiana

Holl S (2006) Questions of perception: phenomenology of architecture. William Stout Publishers, San Francisco

Howard S (1997) Trade-off decision making in user interface design. Behav Inform Technol 16(2):98–109

Hughes J, Parkes S (2003) Trends in the use of verbal protocol analysis in software engineering research. Behav Inform Technol 22:127–140

Jones JC (1963) A method of systematic design. In: Jones JC, Thornley DG (eds) Conference on design methods. Pergamon Press, Oxford, pp 53–73

Jones JC (1970) Design methods: seeds of human futures. Wiley, London

Jones JC (1977) How my thoughts about design methods have changed during the years. Des Methods Theor 11:1

Kaplan C, Simon HA (1990) In search of insight. Cogn Psychol 22:374–419

Kidder T (1982) Soul of a new machine. Avon, New York

Koffka K (1935) Principles of gestalt psychology. Harcourt Brace, New York

Kohler W (1930) Grouping in visual perception. In: Murchison C (ed) Psychologies of 1930. Clark University Press, Worcester, pp 143–147

Korf RE (1980) Toward a model of representational changes. Artif Intell 14:41–78

Kosslyn SM (1975) Information representation in visual images. Cogn Psychol 7:341–370

Kosslyn SM, Pomerantz JR (1977) Imagery, propositions, and the form of internal representations. Cogn Psychol 9:52–76

Krauss RI, Myer JR (1970) Design: a case history. In: Moore GT (ed) Emerging methods in environmental design and planning. MIT Press, Cambridge, MA, pp 11–20

Lakoff G (1987) Women, fire and dangerous things: what categories reveal about the mind. University of Chicago Press, Chicago

Lakoff G (1993) The contemporary theory of metaphor. In: Ortony A (ed) Metaphor and thought. Cambridge University Press, New York, pp 202–251

Lakoff G, Johnson M (2003) Metaphors we live by. University of Chicago Press, London

Lang J, Burnette C, Moleski W, Vachon D (1974) Emerging issues in architecture. In: Lang J, Burnette C, Moleski W, Vachon D (eds) Designing for human behavior: architecture and the behavioral sciences. Dowden, Hutchinson & Ross, Inc., Stroudsburg, pp 3–14

Lawson B (1994) Design in mind. Butterworth-Heinemann, Oxford

Lawson B (2006) How designers think, the design process demystified. Elsevier, Amsterdam

Lloyd P, Christiaans H (eds) (2001) Designing in context. Delft University Press, Delft

Lloyd P, Scott P (1994) Discovering the design problem. Des Stud 15(2):125–140

Lloyd P, Lawson B, Scott P (1996) Can concurrent verbalization reveal design cognition? Des Stud 16(2):237–259

Logan B, Smithers T (1993) Creativity and design as exploration. In: Gero JS, Maher ML (eds) Modelling creativity and knowledge-based creative design. Lawrence Erlbaum Associates, Hillsdale, pp 139–175

Magnani L (2009) Abductive cognition. The epistemological and eco-cognitive dimensions of hypothetical reasoning. Springer, Berlin/Heidelberg

Maher ML, Poon J (1996) Modeling design exploration as co-evolution. Comput Aided Civ Infrastruct Eng 11(3):195–209

March LJ (1984) The logic of design. In: Cross N (ed) Developments in design methodology. Wiley, New York, pp 265–276

Martin RL (2009) The design of business: why design thinking is the next competitive advantage. Harvard Business Press, Boston

McClelland J, Rumelhart D et al (1986) Parallel distributed processing, vol II. MIT Press, Cambridge, MA

McDonnell J, Lloyd P (eds) (2009) About: designing – analysing design meetings. Taylor & Francis, London

McNeill T, Gero J et al (1998) Understanding conceptual electronic design using protocol analysis. Res Eng Des 10(3):129–140

Minsky M, Papert S (1972) Research at the laboratory in vision, language, and other problems of intelligence. Artificial Intelligence Progress Report, Memo, 252

Mithen S (2005) The singing Neanderthals: the origins of music, language, mind and body. Weidenfeld & Nicolson, London

Montgomery H, Svenson O (1989) Process and structure in human decision making. Wiley, Chichester

Newell A (1965) Limitations of the current stock of ideas about problem solving. In: Kent A, Taulbee OE (eds) Electronic information handling. Spartan Books, Washington, DC, pp 195–208

Newell A (1969) Heuristic programming: ill-structured problems. In: Aronofsky J (ed) Progress in operations research. Wiley, New York, pp 360–414

Newell A (1990) Unified theories of cognition. Harvard University Press, Cambridge, MA

Newell A, Simon HA (1956) The logic theory machine – a complex information processing system. IRE Trans Inf Theory 2(3):61–79

Newell A, Simon HA (1972) Human problem solving. Prentice-Hall, Englewood Cliffs

Newell A, Shaw JC, Simon HA (1958) Elements of a theory of human problem solving. Psychol Rev 85:151–166

Norman D (2002) The design of everyday things. Basic Books, New York

Paivio A (1971) Imagery and verbal processes. Holt, Rinehart and Winston, New York

Paivio A (1986) Mental representations: a dual coding approach. Oxford University Press, New York

Pearson C (2009) California academy of science. Archit Rec 2009–01:58–69

Peirce CS (1997) Pragmatism as a principle and method of right thinking: the 1903 Harvard lectures on pragmatism. SUNY Press, Albany

Piaget J (1953) The origin of intelligence in the child new fetter lane. Routledge & Kegan Paul, New York

Piaget J (1967) The psychology of intelligence. Routledge & Kegan Paul, London

Pinfield L (1986) A field evaluation of perspectives on organizational decision making. Adm Sci Q 31:365–388

Posner MI (1973) Cognition: an introduction. Scott Foresman, Glenview

Pylyshyn ZW (1973) What the mind's eye tells the mind's brain: a critique of mental imagery. Psychol Bull 80:1–24

Reed SK (1974) Structural descriptions and the limitations of visual images. Mem Cogn 2:329–336

Reitman WR (1964) Heuristic decision procedures, open constraints, and the structure of ill-defined problems. In: Shelley MW, Bryan GL (eds) Human judgments and optimality. Wiley, New York, pp 282–315

Rittel H (1988) The reasoning of designers. Arbeitspapier A-88-4. Institut für Grundlagen der Planung. Universität Stuttgart, Stuttgart

Rittel H, Webber M (1973) Dilemmas in a general theory of planning. Policy Sci 4(2):155–169

Rodgers P (ed) (2012) Articulating design thinking. Libri Publishing, Faringdon

Rosch E (1978) Principles of categorization. In: Rosch E, Lloyds B (eds) Cognition and categorization. Laurence Erlbaum Associates, Hillsdale, pp 189–206

Rosenhead J (1996) What's the problem? An introduction to problem structuring methods. Interfaces 26(6):117–131

Rosenshine B (1997) Advances in research on instruction. In: Lloyd JW, Kameenui EJ, Chard D (eds) Issues in educating students with disabilities. Lawrence Erlbaum Associates, Mahwah, pp 197–220

Ross DW (1933) A theory of pure design; harmony, balance, rhythm. Peter Smith, New York

Rowe P (1987) Design thinking. MIT Press, Cambridge, MA

Rumelhart DE, Ortony A (1977) The representation of knowledge in memory. In: Anderson RC, Sprio RJ, Montague WE (eds) Schooling and the acquisition of knowledge. Lawrence Erlbaum Associates, Hillsdale, pp 99–135

Schon DA (1983) The reflective practitioner. Temple-Smith, London

Schön DA (1987) Educating the reflective practitioner: toward a new design for teaching and learning in the professions. Jossey-Bass, San Francisco

Simon HA (1969) Sciences of the artificial. MIT Press, Cambridge, MA

Simon HA (1973) The structured of ill-structured problem. Artif Intell 4:181–201

Simon HA, Newell A, Shaw JC (1962) The processes of creative thinking. In: Gruber HE, Terrell G, Wertheimer M (eds) Contemporary approaches to creative thinking. Lieber-Atherton, Inc, New York, pp 63–119

Stewart S (ed) (2011) Interpreting design thinking. Des Stud (special issue) 32:6

Stiny G, March L (1981) Design machines. Environ Plann Plann Des 8:245–255

Suwa M, Purcell T, Gero J (1998) Macroscopic analysis of design processes based on a scheme for coding designers' cognitive actions. Des Stud 19:455–483

Teuber ML (1973) New aspects of Paul Klee's Bauhaus style. In: Teuber ML (ed) Paul Klee: paintings and watercolors from the Bauhaus years, 1921–1931. Des Moines Art Center, Des Moines, Iowa, pp 6–17

Thevenot C, Oakhill J (2006) Representations and strategies for solving dynamic and static arithmetic word problems: the role of working memory capacities. Eur J Cogn Psychol 18:756–775

Thomas JC, Carroll JM (1979) The psychological study of design. Des Stud 1:5–11

Tidwell J (2005) Designing interfaces: patterns for effective interaction design. O'Reilly Media, Inc., Sebastopol

Turing AA (1950) Computing machinery and intelligence. Mind LIX(236):433–460

Valkenburg R, Dorst K (1998) The reflective practice of design teams. Des Stud 19(3):249–271

Vesely A (2007) Problem delimitation in public policy analysis. Central Eur J Publ Pol 1(1):80–100

Visser W (1995) Use of episodic knowledge and information in design problem solving. Des Stud 16:171–187

Visser W (2006) Designing as construction of representations: a dynamic viewpoint in cognitive design research. Hum Comput Interact 21:103–152

Wade J (1977) Architecture, problems, and purposes. Wiley, New York

Watson JB (1920) Is thinking merely the action of language mechanisms? Br J Psychol 11:87–104

Waugh NC, Norman DA (1965) Primary memory. Psychol Rev 72:89–104

Wertheimer M (1923) Untersuchungen zur Lehre von der Gestalt II. In: Psycologische Forschung, 4, 301–350. Translation, entitled laws of organization in perceptual forms, published in Ellis, W. (1999) A source book of gestalt psychology. Routledge, London, pp 71–88

Wickelgren WA (1974) How to solve problems. Freeman, San Francisco

Wimmer GE, Shohamy D (2012) Preference by association: how memory mechanisms in the hippocampus bias decisions. Science 338(6104):270–273

Winograd TW (1975) Frame representations and the declarative-procedural controversy. In: Bobrow DG, Collins A (eds) Representation and understanding: studies in cognitive science. Academic, New York, pp 185–200

Woolley RN, Pidd M (1981) Problem structuring – a literature review. J Oper Res Soc 32(3):197–206

Yeston M (1976) The stratification of musical rhythm. Yale University Press, New Heaven

Zhang J (1997) The nature of external representations in problem solving. Cognit Sci 21(2):179–217

Part II
Style

Chapter 3
Development of Studies in Style

"**Art**" is a field that is highly abstract, conceptual, metaphysical, and difficult to measure. It is a generational result influenced by psychological and cultural phenomena, and usually is probed by philosophers through aesthetics. **Aesthetics**, on the other hand, is a way of studying the arts and is a branch of philosophy—a philosophy of beauty—that provides a theory of the beautiful and of the **fine arts**. So, the study of art usually comes along with the study of aesthetics. However, as the study of art became more complex and because of the profound nature of art, scholars and philosophers developed a notion of "style" and used it to help analyze beauty and explore methods of creation, and to distinguish differences between individual artists, groups, and schools.

Anthropological studies in the 1960s of aesthetics showed how art evolved and functioned in different cultural settings and how changes in customs occurring in social evolution influenced art in societies. Gradually, studies of the arts were examined in the fields of sociology, psychology, and archeology, rather than only history and philosophy; scholars in social studies began focusing on the sociological phenomena occurring in the creation processes. Recently, scholars in the field of industrial design have been exploring style in the automobile industry to help establish identifiable and recognizable images of different products to represent a signature image of a company.

Thus, style is a norm developed for categorizing art and studying artistic artifacts. But, the trend of theoretical development in fine arts has shown that although style is used as a means for studying art to some degree, the conceptual current of art, still, is dominating and influencing the notions utilized in the studies of style. As such, the academic study of style has not yet come into its own as a unique direction or approach. This is because style is regarded as the product of the historical development of art evolution. However, in my opinion, style should be treated as the product of human cognition, which will be explained in the following chapters. In order to get a full comprehension of the notion of style, the historical background of how the concept of style evolved from the concept of art is described in the following sections.

© Springer International Publishing Switzerland 2015
C.-S. Chan, *Style and Creativity in Design*, Studies in Applied Philosophy,
Epistemology and Rational Ethics 17, DOI 10.1007/978-3-319-14017-9_3

3.1 Historical Development of Art Theories

In early cultures, cave paintings were regarded as attempts to represent or record nature through irrational rendering of monstrous and grotesque images, which have no association with artistic theory. For example, at the Neolithic period dating back 5,000–10,000 years ago, art was mainly symbolic. During that period, large petroglyphs were imprinted on the surface of rocks. These petroglyphs were symbolic, and mostly silhouettes—two dimensional representations of the outlines of objects—with mostly animals and humans involved in hunting and fishing activities, which were the settlers' main occupation. The picture on the left of Fig. 3.1a is a petroglyph showing an animal in a more than 30,000 year old cave painting in the Chauvet Cave, France; the right hand picture is a rock painting of hunting dated from about 8000 BC in Val Camonica, Italy.

As shown in the pictures, primitive art was merely a symbol of nature (Childe 1962, p. 164) and a means of representation. A clear example can be seen in the desert areas of Australia, where some pictures painted on rocks were only used for signs and signals. These figures are primitive, with only lines to show artificial objects. Figure 3.2 shows two clay paintings done by native Australians in Ayers Rock. Kept in two caves, recognized as their supreme libraries, these paintings were used for documentary purposes to make statements about certain facts and beliefs of that area. This phenomenon of such recording is similar to the use of making knots in a rope or string as a means of representation for archiving events in ancient Chinese history, and the Knotted-String Records by the Incas Empire to record numerical, statistical, and narrative information (Urton 2003).

In the Mesopotamia period (3500–331 BC), natural and formal elements (i.e., advanced geometric figures) were unified in art. Natural elements and figures were integrated into certain meaningful forms. But, in ancient Egypt (3200–1340 BC), art

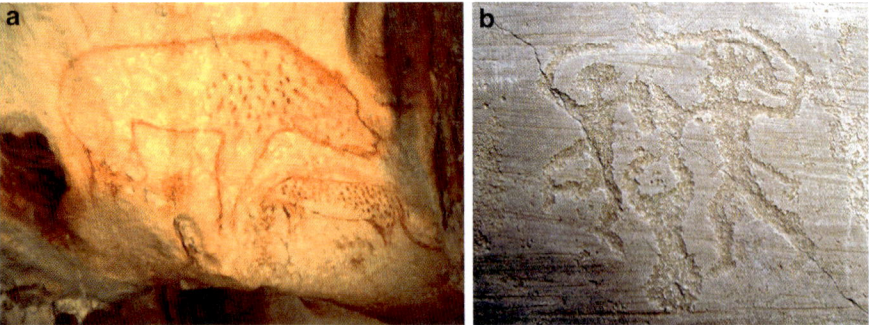

Fig. 3.1 (**a**) Examples of the Neolithic art in petroglyph (Sources: (**a**) Is taken from: Carla Hufstedler/Wikimedia Commons, http://en.wikipedia.org/wiki/Cave_painting. Accessed 10 Oct 2013. (**b**) Is taken from: Luca Giarelli/Wikimedia Commons, http://en.wikipedia.org/wiki/File:Scena_di_duello_R6_-_Foppe_-_Nadro_(Foto_Luca_Giarelli).jpg. Accessed 10 Oct 2013)

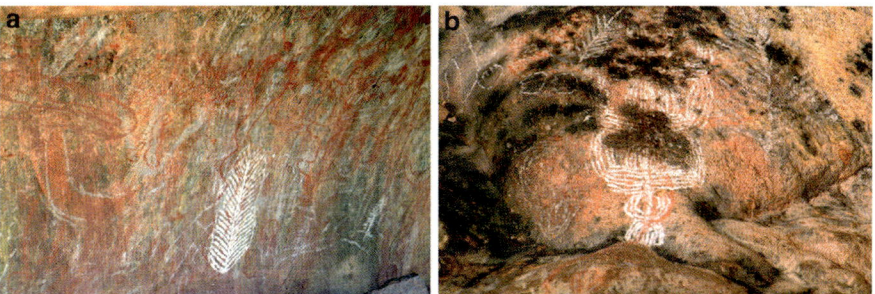

Fig. 3.2 Examples of the art painted in caves, Ayers Rock, Australia (Source: Photos taken by the author)

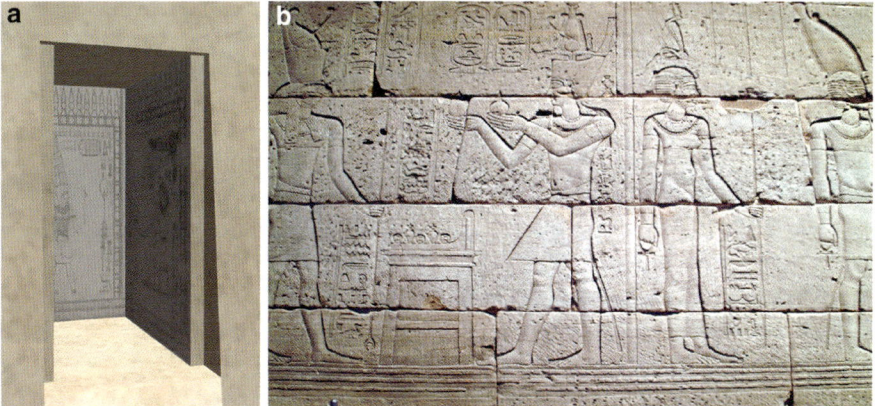

Fig. 3.3 The interior of the Mortuary Temple of Queen Hatshepsut and the Exterior wall picture of Temple of Dendur, Egypt, displayed at Metropolitan Museum of Art in New York City (Source: Digital model created at Iowa State University)

was developed for the dead. Egyptians built pyramids and decorated tombs with pictures of the gods (see Fig. 3.3), or pictures of events and activities that the deceased participated in when they were alive and wished to be involved in doing for eternity. The picture on the right of Fig. 3.3 is the engraved picture on the external wall of the Temple of Dendur, displayed in the Metropolitan Museum of Art in New York City. Later, "**true art**" came into being, which is recognized only in highly developed cultures, and in which artists would combine their knowledge of nature forms with some inferences to bring beauty to their imagination and put into works (Schapiro 1962, pp. 281–282). Greek culture, regarded as highly developed, gradually matured and was, therefore, recognized as creating true art. Philosophers such as Plato and Aristotle in Athens explained the nature of human knowledge, of which art is a part.

3.1.1 Theory of Art by Plato

The earliest theory of art was developed by **Plato** (427–347 BC), whose notion on art was the earliest concept in written record that has great intrinsic value. His theory was that art was "imitation of truth, of nature, or of physical things in reality" (Cavarnos 1973), which in turn imitated the "Form" that comes from nature and from life. This notion, explained extensively in the *Republic* (Jowett 1967), was that art imitated the objects and events of ordinary life. Architecture as well as sculpture, painting, dance, music, literature, and drama were categorized as a part of fine arts. Yet, the art of building, to Plato, had a dimension of aesthetic value beyond the imitation of nature, because it utilized the greatest number of measures and instruments and yielded a high level of accuracy, making it more rigorous than other arts.

This notion of art in architecture was depicted in his Dialogues with Socrates on *Philebus* (Jowett 1967, vol. III, p. 566, 56b). Furthermore, Plato cited in the *Republic* that architecture was a kind of specialized knowledge, and he viewed architecture as an excellent instance of an art that is based on knowledge other than just imitation. He explained that "the science of house-building is a kind of knowledge which is defined and distinguished from other kinds and is therefore termed architecture (Jowett 1967, vol. II, *Republic Book IV*, p. 292, 438d). Together with sculpture and poetry, architecture is an art that may express beauty, harmony or rhythm, or the opposites of these (Jowett 1967, vol. II, *Republic Book III*, p. 249, 401a-c). For example, beauty and harmony is reflected in the Greek Parthenon (Fig. 3.4) and Erechtheum on the Acropolis of Athens, which were viewed by Plato as the architectural achievements of his age (Cavarnos 1973, p. 32). Significant examples of forms that were influenced by Plato's theories are marble engravings used in the capital of the five column orders in ancient Greek and Roman temples. These marble engravings are sculptures of natural leaves found in plants.

Influenced by Plato's concepts, the notion that art comes from imitation of nature dominated western aesthetic theory for centuries, from Greek, Rome, Romanesque,

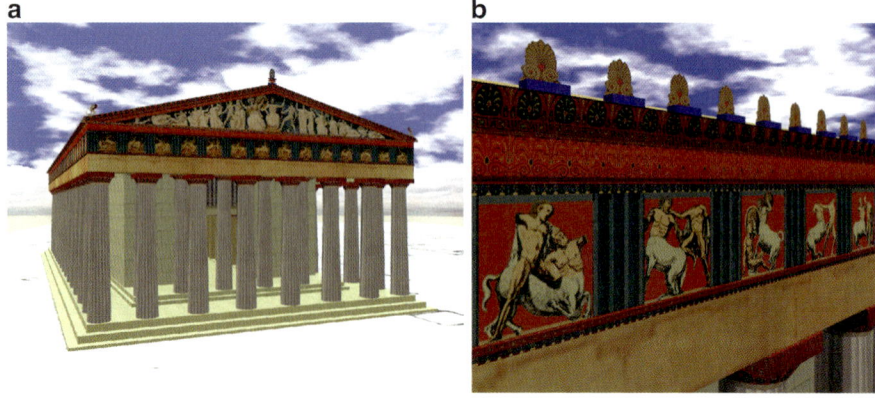

Fig. 3.4 Exterior of the Parthenon and details of some sculptures on frieze (Source: Digital model created at Iowa State University)

Gothic, up to the Renaissance period. During these periods, notions of art did not change direction and theory. In medieval periods, art was focused on conveying religious beliefs and attitudes. Art concepts and theories up to the period of the Renaissance had no major breakthroughs. In the early Renaissance, **Giorgio Vasari** (July 30, 1511–June 27, 1574), the famous Italian painter and architect, explained that painting is just the imitation of all the living things of nature with colors and designs just as they are in nature. In the eighteenth century, a new idea developed that art was not merely the imitation of reality, but also was an act of imagination. The central idea of treating art as the act of imagination was that art must express an artist's emotion through his or her own imagination and sensitivity. This was expounded by Sir **Joshua Reynolds** (1723–1792) in 1786 under the title of "Discourse 13," which is a part of the "Discourses on Art" series addressed from 1769 to 1790 to his students (Reynolds 1997, pp. 229–244). Since then, aesthetics has been cultivated as an autonomous study.

3.1.2 Theory of Art by Reynolds

Reynolds was the most important and influential of eighteenth century English painters. As the founder and first elected president of the Royal Academy of Arts, established in 1768, his 15 Discourses, delivered to the Academy students, remains a persuasive advocacy of eighteenth century aesthetics. He explained his perspectives of depicting an ideal in nature instead of representation of detail, on the importance of imagination over imitation, and on the need for good taste and sensibility (Weitz 1970, p. 4). Since then, art has been considered as the act of expression that leads to the artist's emotional projection and the beholder's emotional stimulation.

    Reynolds' notions of art in architecture differed from Plato's theory of imitation. He indicated that architecture is not just an imitative art. Architecture should apply itself directly to the imagination, without the intervention of any kind of imitation (Reynolds 1997, p. 241). This explains the fundamental theory of art in general and in architecture as a discipline in the eighteenth century. However, Reynolds' theories were highly abstract and speculative as a whole, devoted largely to questions on whether beauty is an objective property or a subjective human feeling. But, most artists and philosophers had begun to discover laws of beauty in the universe.

3.2 Modern Art Theories and Movements

Beginning with the nineteenth century, many aesthetic theories moved to the use of scientific methods to explore the nature of art and what it really was. These new theories, instead of deducing rational from religious and metaphysical assumptions, were more empirical and drew conclusions from the observations made on creative works, inner correspondence, art products, and their related phenomena. These scientific methods applied for exploring art included the use of psychology that provided a general conception of human basic processes and functions, and concepts of sociology and anthropology that showed how art functioned in different cultures and how social evolution triggered custom changes to change culture through time. They formulated and tested hypotheses about recurrent patterns in art works or inner correspondence, on predicting and controlling outcomes through statistical measurement and estimating probabilities from experiments to obtain extensive reliabilities. Thus, aesthetics were more scientifically analyzed, and these became the modern methods of studying aesthetics.

Consequentially, notions of aesthetic theories were affected in the late nineteenth and early twentieth centuries by influences on the changing trend in art movements due to changes in modern culture, which also changed artists' methods of art creation. Important major influential art movements are selectively listed in the following and displayed together with representative photos to provide visual reference for comparisons.

1. **Impressionism**: The history of modern art begins with impressionism, an art movement founded in Paris in 1863 when **Edouard Manet** (1832–1883) exhibited his painting "Dejeuner sur l'herbe," or "Luncheon on the Grass" (see Fig. 3.5), at the Salon des Refuses (or "Exhibition of Rejects"). In April of 1874, a group of painters exhibited their influential art, and the term "**Impressionism**" was coined by journalist **Louis Leroy**. Impressionists were

Fig. 3.5 The Luncheon on the Grass painted by Edouard Manet (Source: Wikimedia public domain, http://en.wikipedia.org/wiki/Edouard_Manet. Accessed 10 Oct 2013)

interested in the actual visual experience and the effect of light and movement on the appearance of objects. The impressionism concept spread beyond the art world to music and literature, but less in architecture.

2. **Arts and Crafts Movement** and **Art Nouveau**: When the industrial revolution began in the mid nineteenth century, there were concerns about the cheap mass production of industry that would drive art away from everyday life. Advocates turned away from the use of machines and towards handicraft. They tried to produce inexpensive and handsome products by hand. The **Arts and Crafts Movement** had its beginnings in the late nineteenth century in England, pioneered by **William Morris** (1834–1896). Morris and his associates produced hand crafted metalwork, jewelry, wallpaper, textiles, furniture, and books. Their ideas spread to other countries and became identified with other international interests in design, particularly **Art Nouveau**. Art Nouveau, it could be said, was a continuation of the Arts and Crafts Movement. But, this movement advocated nature as the true source of all good design. The characteristics of this movement included the use of the sinuous curved lines from grasses, lilies, vines, peacock feathers, butterflies, and insects, with an asymmetrical arrangement of forms and patterns. Figure 3.6 is the wallpaper designed by **John Henry Dearle** in 1897. The influences of Art Nouveau to design are significant. Famous examples are the **Vienna Secession, which** proposed to separate arts from tradition in Austria and the **Bauhaus Design School** in Germany. Figure 3.7 left is the Secession's building built in 1897 in Austria, and a wall painting by **Gustav Klimt** is on the right.

Fig. 3.6 Wall paper designed by John Henry Dearle, 1897 (Source: Wikimedia public domain, http://en.wikipedia. org/wiki/Arts_and_Crafts_ Movement. Accessed 10 Oct 2013)

Fig. 3.7 Secession building in Vienna built in 1897 (**a**, *left*), wall painting by Gustav Klimt (**b**, *right*) (Sources: **a**, *Left*: photo taken by the author. **b**, *Right*: Wikimedia public domain, http://commons.wikimedia.org/wiki/File:Gustav_Klimt_014.jpg. Accessed 10 Oct 2013)

3. **Cubism**: The name was first used by **Henri Vauxcelles** in 1908. It was one of the most influential visual art styles in the twentieth century, and revolutionized painting and sculpture and inspired movements in music and literature. Created by **Pablo Picasso** and **Georges Braque**, Cubism actually provided a liberating technique of slicing a subject apart and joining it together in a wholly new view of objects in space. In cubism, objects were broken up, analyzed and re-assembled in an abstracted form and presented the subject from a multitude of viewpoints to represent the subject in a greater context. The picture on the left of Fig. 3.8 is a painting of a woman and a guitar by Braque in 1913. The objects of woman and guitar are sliced into pieces and shown as a whole from multiple angles. The chaotic shadows and background created a three-dimensional illusion. The picture on the right of Fig. 3.8 is the guitarist painted by Picasso in 1910.

4. **Formalism**: This theory, developed around 1920, maintains that certain combinations of lines, colors, shapes, volumes, and subjects would evoke a unique response to such combinations. Thus, art is an instance of significant "form" and anything that has no such form is not art. **De Stijl** is one example, founded in 1917 in the Netherlands. This movement used simple compositions in vertical and horizontal directions, with black and white as the primary colors, and asymmetrical rectangular forms, see the work by Piet Mondrian (1872–1944) on Fig. 3.9a. It influenced painting, sculpture, architecture (see Fig. 3.9b), furniture, and decorative arts. **Mies van der Rohe** was a modern architect who was influenced by this movement the most.

Fig. 3.8 Woman with a guitar by Braque (1913) on (**a**, *left*) and the guitarist by Picasso (1910) on (**b**, *right*) (Sources: © 2014 Artists Rights Society (ARS), New York/ADAGP, Paris, and © 2014 Estate of Pablo Picasso/Artists Rights Society (ARS), New York)

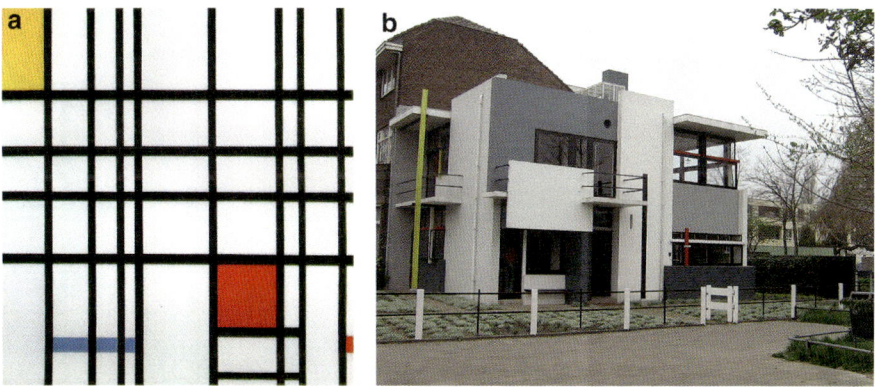

Fig. 3.9 (**a**, *left*) An example of De Stijl movement by Piet Mondrian (Source: Wikimedia public domain, http://en.wikipedia.org/wiki/File:Mondrian_CompRYB.jpg. Accessed 10 Oct 2013). (**b**, *right*) Schroder House designed by Gerrit Rietveld (1924) (Source: Wikimedia public domain, http://commons.wikimedia.org/wiki/File:RietveldSchroederhuis.jpg. Accessed 10 Oct 2013)

5. **Dada and surrealism**: Dada[1] was an informal international movement that rejected war and expressed their rejection of reason and logic by embracing chaos and irrationality. Many Dadaists believed that the reason and logic of capitalist society had led countries into war. Dada was not art but "anti-art," a movement that created surrealism. Surrealism is a cultural movement in painting and poetry that began in 1924 by the French poet **Andre Breton**. The surrealists claimed to create forms and images not by reasons, but by unthinking impulse and blind feeling, or sometimes by accident. It used elements of surprise, unexpected juxtapositions and illogical connections to create a magical world in painting and poetry that was more beautiful than the real one. Figure 3.10 is the famous 1931 painting by **Salvador Dali** of sagging, limp watches, as if time is melting.

6. **Abstract-Expressionism**: Prevailing around 1930–1960, Abstract-Expressionism had a strong influence on painting. No rules existed in making a picture, to allow for pursuit of freedom in the making and expression of art. Artists would base their creation on their personal subjective experiences.

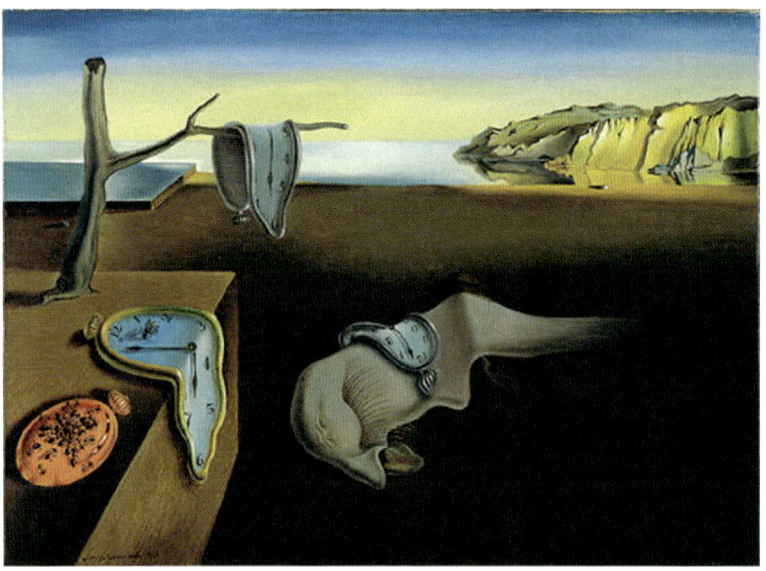

Fig. 3.10 The persistence of memory by Salvador Dali (1931) (Source: © Salvador Dali, Fundacio Gala-Salvador Dali, Artists Rights Society (ARS), New York 2014)

---

[1] **Dada**: The word "dada" has a legend. In 1916, a group of artists wanted a name for their new movement. By stabbing a French-German dictionary with a paper knife, they picked the name from where the knife point landed. In French, dada is a child's word for "hobby-horse"; *c'est mon dada* means *it's my hobby*.

7. **Emotionalist**: This movement proposed the embodiment and projection of emotion into artworks to characterize them as art.
8. **Intuitionist**: Intuitionists define art as coming from a specific creative, cognitive, and spiritual act that has images and intuitions involved in the expression.

Of course, there were many movements in the field of fine arts that diversified the media applied for expression and the artists' purposes of expression. In fact, all the art movements since the nineteenth century have been edited and compiled into a long list, available on-line.[2] The brief historical developments of art theories, starting from the ancient time to the modern age and explained in this chapter, provide just a schematic outline of art notions. This short outline represents a visual record of a rich source of ideas and understandings about how the world presents to us, which is valuable for referencing social and cultural development throughout history as countless artworks displayed in galleries and museums. However, art is not only created by artists, but also experienced by beholders through perception. Thus, researchers have started to explore scientifically (Pepperell 2012) and neurologically (Ramachandran and Hirstein 1999) how art is essentially perceived by viewers rather than how art is created by artists.

The scientific approach to understand the perception of art is to research the nature of visual experience and the process of perceiving art. Conventionally speaking, the world has objects and events with properties that exist independently of how we see them. The human visual system creates an internal representation, or a model of the objects or events that represent the external world. It is with this model that we construct our experience, and it might not be exactly identical to the real world (Koenderink 2011). **Chris Frith** explained how the brain creates our mental world: "When I look at a tree in the garden, I don't have the tree in my mind. What I have in my mind is a model (or representation) of the tree constructed by my brain" (Frith 2007, p. 170). Thus, humans perceive things in the world and construct a model of the perception for recognition. If the perceived object is not fully recognizable in the early perceptual stage, or the feature of visual information derived from the world is not intrinsic, then our brain has to fetch from the available data and cognitively augment what is seen in the later perceptual stage (Humphreys and Riddoch 1987; Farah 2004). This cognitive process could be explained by paintings of Cubism created by Braque and Picasso (see pictures in Fig. 3.8). They painted in a way that provided viewers with numerous clues about the identity of the subjects, but also denied viewers full recognition. Thus viewers had to go through a series of perception and interpretation to satisfy the "visual indeterminacy" (Pepperell 2012).

In another study on how viewers cognitively respond to art, it was proposed that beauty was in the eye of the beholder and there were universal rules underlying all artistic experience (Ramachandran and Hirstein 1999). Such universal rules are the common denominator underlying all types of arts. Certain rules are deployed by artists consciously or unconsciously to titillate the visual areas of the brain.

---

[2] Art movements that are treaded as tendencies or styles in art from the nineteenth century to twenty-first century have been listed and linked to the group of artists online, URL: http://en.wikipedia.org/wiki/Art_movement

How beholders appreciate art is based on three components—the logic of art, the rationale used to have the particular form in that art, and the neuron activity that occurred in the brain. This book will address particularly on the logic of art. The logic of art includes the ability of the artist to abstract the essential features of an image and discard redundant information, which is what our vision has evolved in pattern recognition. The first rule of all art is a representation of caricature. The second rule is the ability to group visual information into a pattern, which is a part of our neuron system that is capable of combining signals together to form a big picture (Singer and Gray 1995; Crick and Koch 1998). Such a feature-ground grouping is not just to group information, but to segment information out from noisy background to make it meaningful. The third rule is the ability to isolate a single visual modality before the signal in the modality is amplified. Isolating a single area allows one to direct attention more effectively to this one source of information. The fourth rule is to sort out and extract features before grouping, which is to discard redundant information and extract contrast from regions to allocate attention. The fifth rule is symmetry, which is fully recognized cognitively and is a built-in aesthetic preference. The sixth rule is the generic viewpoint principle, or the foreground versus background concept in vision, explained in Gestalt psychology (see footnote 1 in Chap. 2). Our vision favors generic vantage points in perception and dislikes suspicious coincidences of view (Barlow 1986). The seventh rule is perceptual problem solving, which means a puzzle picture may paradoxically be more alluring than one in which the message is obvious. The last rule is metaphor, in which it is possible that discovering hidden similarities between dissimilar entities is one of the fundamentals of visual pattern recognition, and when it happens, a signal is sent to our neuron system for information processing. All these rules have connections to perception in the human vision system and in art works (Ramachandran and Hirstein 1999). Thus, these rules are proposed as artistic experience existing as inherent elements in human cognition. Therefore, they are applied automatically by artists for creation and by beholders for experience.

Along with the concept development of art, the notion of style is also historically affiliated with art, starting from ancient Greece. Style, as a part of art, exists only after "**true art**" is established (see Sect. on 3.1), and **true art** was formed originally in Greek culture and further developed afterwards. Therefore, concepts of style were developed from the Greek culture through Roman, Renaissance, Rococo, to modern art. Only after the Renaissance period were notions of style addressed by scholars, and since, have matured to some degree. Research methods were gradually developed in different fields to meet different purposes of studies. In the following sections, the history of style, the changes in methods of study, and studies conducted in various fields are briefly described.

3.3 Historical Development of Studies in Style

The word *style* has its origins in Greek, passing into Latin. The Greek original means a wooden pile, a stone pillar, and a metallic graver for writing and drawing. The Latin word *stilus* or *stylus* signifies an iron instrument similar to a pencil in size

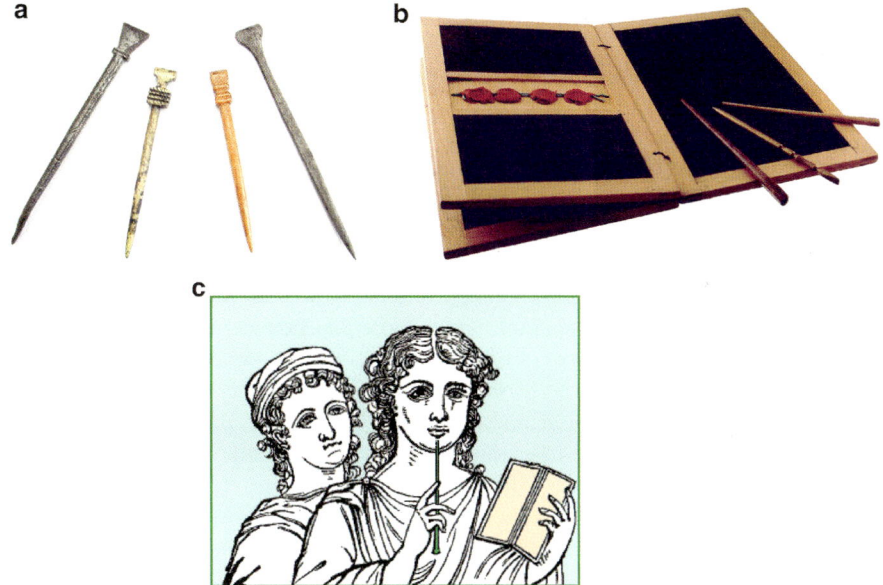

Fig. 3.11 (**a**) Stilus in medieval age. (**b**) Stilus and wax tablet. (**c**) The use of stilus & tablet (Sources: (**a**) Numisantica (http://www.numisantica.com), http://commons.wikimedia.org/wiki/File:Stylus.jpg. Accessed 26 July 2014. (**b**) Wikimedia public domain, http://en.wikipedia.org/wiki/Wax_tablet#mediaviewer/File:Wachstafel_rem.jpg. Accessed 17 July 2014. (**c**) Public domain, http://penelope.uchicago.edu/Thayer/E/Roman/Texts/secondary/SMIGRA*/Stilus.html. Accessed 10 Oct 2013)

and shape used for writing on waxed tablets, which is sharpened at one end for writing and flat at the other end to serve as an eraser for smoothing the wax on the tablets (Smith's Dictionary of Greek and Roman Antiquities 1875 edition, s.v. STYLUS, p. 1071). Figure 3.11a shows examples of four medieval writing utensils. Figure 3.11b is a set of wax tablet with three styli, and Fig. 3.11c is a woodcut that shows an engraving of a man and a woman. The man holds a sharp metal stilus in his right hand and a folding wooden frame with wax writing tablets inside in his left hand. The general idea of the word *stilus* conveys an object tapering like an architectural column; later, it translated into *style* in English.

Other than signifying a writing instrument, the literal and original meaning of *style* was a characteristic way of making letters and a characteristic manner of dressing thought in words. Thus, it has a strong connection with language and writing, and the earliest concept of style was Greek writings on rhetoric.[3] To them, style was the quality of language correctness, clarity of language, and suitability and shortness of expression. It was the art of using language effectively and persuasively. Particularly, word order, clarity, and rhythm have been considered extensively in poetry and prose writing. In the Roman Period, public speech, characterized by a

---

[3] See Rhetoric, https://en.wikipedia.org/wiki/Rhetoric#History

studied or elevated style of oration, became an important part of public life. Orators needed to be knowledgeable about all areas of human life and culture. Marcus Tullius Cicero (106–43 BC), the best known ancient orator, established the outlines of Latin eloquence and style, asking his orator followers to learn Greek rhetoric, focused on Roman ethics, linguistics, philosophy, and politics. He emphasized the importance of all forms of representation in emotion, humor, stylistic range, irony and digression rather than pure logic reasoning in oratory. At that time, style became the way of presentation, which also relates to the presentation in arts.

After the Roman period, Europe moved into the **Middle Ages**, which extends from the end of the Roman Empire to approximately the 1500s. In the Middle Ages period, the church became the single great force that bound Europe together. The church touched almost everyone's life through several important events. Churches baptized a person at his birth, performed the wedding ceremony at his marriage, and conducted the burial services at his death. Life philosophy, painting, and writing were all centered around the Christian religion in the Middle Ages, which was a constraint in art. Another constraint was the feudal status. In the Middle Ages, Europe was divided into many kingdoms that were split into hundreds of vassals ruled by independent rulers (duke, prince, baron, or count) with their own fiefs. Figure 3.12 shows a fortified town in Mont Saint-Michel, France. These noblemen

Fig. 3.12 The fortified town and abbey of Mont Saint-Michel, France (Source: Wikimedia public domain, http://commons.wikimedia.org/wiki/File:France-Mont-Saint-Michel-1900_bordercropped.jpg. Accessed 10 Oct 2013)

Fig. 3.13 The coronation of Charlemagne de France depicted in the fourteenth century, 1375–1379 (Source: Photo Credit: Bridgeman-Giraudon/Art Resource, NY)

ruled their fiefs through a form of government called **feudalism**.[4] In the feudal states, no one could build a castle, collect taxes, regulate trade, or hold important court trails without the ruler's permission. Therefore, under religious and feudalism constraints, art in the Middle Ages was inactive, focused on religious subjects, and lacked diversity, see Fig. 3.13.

Yet, during the late Middle Ages, with the growth of economic activities in Europe, the increase in royal power, the decline of Feudalism, and the political troubles in the church between popes and kings, scholars and artists gradually became less concerned with religious thinking, and became more interested in understanding man and the world around man. This new cultural view was termed "**humanism**."

---

[4] In the Middle Ages, Europe was divided into many kingdoms, ruled by kings. In creating their empire, kings depended on the assistance of loyal noblemen called vassals. A nobleman became a vassal when he pledged his loyalty and promised his service to the king. The king then became a lord to his vassal. Most vassals held important positions in the king's army and served as knights. Many vassals had their own knights, who also pledged services to the king. In return, the kings rewarded their vassals by granting them estates called fiefs, which included the manors on the land, the buildings, the villages of each manor, and the peasants who farmed the manors.

More writers composed prose and poetry not in Latin but in the vernacular languages, which opened a new literary age. Thus, the Renaissance came into being.

In the Renaissance period (fourteenth to seventeenth centuries), scholars found and translated ancient writings that had been lost during the Middle Ages. From the influences of old writings, artists invented new techniques to make their works as beautiful and lifelike as those of the ancient masters. Like the Greeks and Romans, they made man, rather than God, their center of interest. At this period, man began to put importance on the individual. Thus, artists, architects, poets, and writers used brave new forms and techniques to show new ideas. In this humanist period, painters favored portraits, landscapes, and scenes; sculptors reflected a new understanding of human anatomy; writers emphasized individual personality; architects centered on man and his needs. Particularly, architects also paid as much attention to designing houses, palaces, and public buildings as they did to designing churches. They also tried to make the buildings comfortable for the users. It is because of their opposition to the ideological control by the church that ancient Greek and Roman architecture was reconsidered. A new movement of the classical tradition that had inspired many good architectural projects through some 400 years, the architectural vocabulary of Greece and Rome, was the basis of design during this period (Jordan 1969, p. 167).

The Renaissance period was an age of adventure and curiosity; artists were fascinated with the world about them. Man was the center and individuals had opportunities to extend their horizons. Unlike the artists of the Middle Ages, who were unknown craftsmen, Renaissance artists gained respect and attention from emperors, bishops, and the public. In return, they even tried to explore more new ways of expression. For instance, **Giotto di Bondone**, the famous early Renaissance painter and architect in Florence, broke the decorative and stiff style of medieval painting; he made his figures look like real men and women with strong human feelings, see Fig. 3.14. At that time, artists and writers searched for new styles to express the new ideas of their time. The concept of utilizing style in arts had begun.

From Greek to Roman to Renaissance, style has long been concentrated in literature, language, and linguistics, and used to study the systematic method of creation. For example, in literature, style was explained by **Thomas De Quincey** (1893) as a theory of composition, an art of constructing sentences in writing and weaving them into coherent wholes (Scott 1893, p. 104). In the arts, style referred to an inner peculiarity that showed itself in the outward representation through characteristic marks (Wackernagel 1923). For instance, in the eighteenth century, the study of style tended to interpret the meaning of expression delivered by works of art. **Giorgio Vasari** (**1511–1574**), the highly regarded art historian of the Renaissance, began to study the change of style by means of biological metaphor. He believed that style, like human bodies, had a birth, a growth, an aging, and a death (Ackerman 1963, p. 170), which attributed to the life history of the organism in a recurrent cycle of childhood, maturity, and old age that is similar to the phenomenon of the rise, maturity, and decline of the culture as a whole.

This cyclical process has its character in Western art and architecture. For example, the Renaissance is preceded by Romanesque, Gothic, and **Carolingian** styles, and followed by **Baroque** and **Rococo** styles up to modernism. The same series occurred in the Egyptian, Greek, and Roman worlds. This life cycle metaphor

Fig. 3.14 Fresco by Giotto in Cappella dell'Arena, Padua, Italy (Meeting at the Golden Gate) (Source: Photo Credit: Scala/Art Resource, NY)

schema survived into the nineteenth century, when it was refined by **Charles Darwin**'s evolutionism, which sees the process as an unfinished evolution from the most primitive to the most advanced forms. This concept is quite obvious in the western architectural period style evolution. As such, each historical period has its evolution in birth, growth, and vanishing stages. Yet, in oriental architecture, the changes of style are not as significant as in western architecture.

In Chinese architecture, for example, most buildings were residences, temples, and palaces. In residential buildings, courtyard houses of wood structures were recognized as the standard dwelling typology for more than 900 years. Most courtyard houses shared similar appearances due to the rigid regulation of building codes, as specified in the grammar book "**YingZao FaShi**", written by **Li Jie** (Pirazolli-t'Serstevens 1971, p. 60; Glahn 1984, pp. 47–57) in 1100 and reinforced by all emperors before 1911 (Chan and Xiong 2006). Therefore, the homogeneous floor plan spatial layout, two bays three column elevation, and standardized elevation decorations set up a typical style of popular courtyard housing in China, particularly the courtyard houses in the Inner City of Beijing (see Fig. 3.15). Because of the rigid constraints given by the YingZao FaShi, plus the restrictions enforced by the supreme emperors, no single building form could deviate from this plan. Thus, no clear style ever did exist in ancient Chinese architecture.

In the nineteenth century, **Alois Riegl** (1858–1905), one of the major figures in the establishment of art history as a self-sufficient academic discipline, developed the relationship between stylistic development and cultural history. He attempted to chart the entire history of western art through the changes of architectural ornaments as the record of a "context with nature." This context took different forms

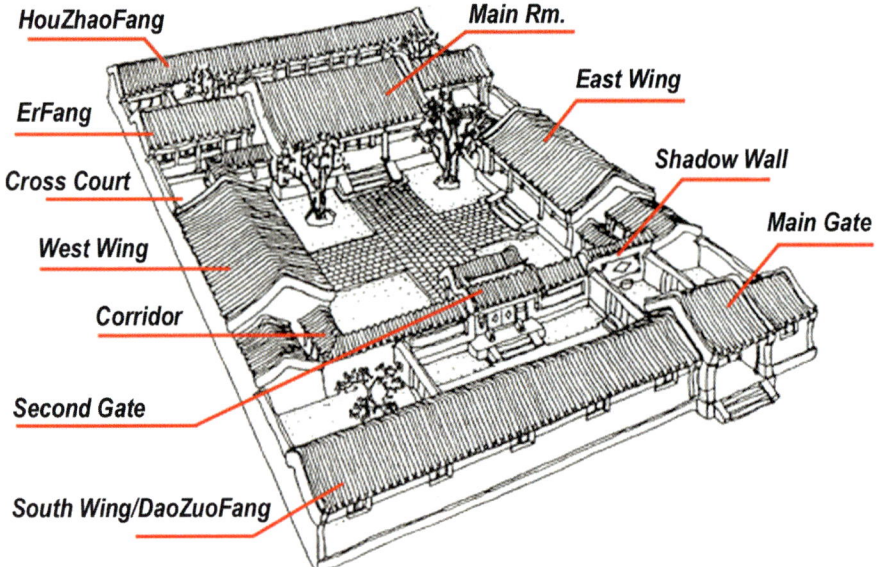

Fig. 3.15 A typical three court courtyard housing in the Inner City of Beijing (Source: Chan and Xiong (2006). Figure 10)

depending on the changing concepts of nature by humans (Riegl 2004). But, one of Riegl's concepts on typifying art history is that the best solution to an artistic problem is the one that best fulfills the artist's aim. To him, art was an active creative process in which new forms arise from the artists' intentions (or will to art) to solve specifically artistic problems. Thus, forms were changed correspondingly with the changes of time and cultural histories. Riegl developed a new perspective to explain the dynamics of style in place of the biological metaphor.

Starting in 1950, when **Meyer Schapiro** (1962) reviewed past notions of studies on style in his article 'Style,' he concluded that there was a need for a modern theory of style. This was a turning point in the field. **James Ackerman** (1963) contributed to this change by urging and providing fresh thinking about style. Following these ideas, **Ernst Gombrich** (1960) and **George Kubler** (1962) made major contributions to the field of style, which are studies of the features used by artists to recognize their artistic intentions. Details of each concept and contribution will be discussed in Chap. 4.

3.4 Other Approaches in Style

Style has also been explained as a purely artistic thing, with fashion that defines a style. For example, in product design, style is something created in an artistic way to show fashion for attracting visual attention. For instance, the Macintosh iPod series in the 2000s is a classic example. It might be possible that the shape of the created form is unique, new, and has not been generated before and is appealing to viewers to buy or to follow. Therefore, style, fashion, and pleasure might go together, and fashion design, among all the fields of design profession, possesses such characters.

Fashion design is a profession started in the nineteenth century by designing clothes that were functional and beautiful. From the user point of view, dress is used to protect the human body (down jacket and T-shirt), to attract attention (evening gown and tuxedo), to show emotion (happy or sad), to show religious beliefs (long red robes by Buddhist monks and long black suits by orthodox Jewish men), to show celebration (wedding dress), and to show identification (hospital gown and military uniform). From the users' point of view, dress is a symbol or a sign, representing a users' identity and the group that they belong to. From the socio-cultural and psychological points of view, the dress of any given period is exactly suited to the actual climate of the time. For instance, younger generations dressed like Elvis Presley or in the Hippy style in the 1950s, 1960s and 1970s. From the designers' point of view, there are many chances to create a new form with new color and materials, and while they tend to be one time creations, they can be used by a large population to shape a culture. As such, a master designer is able to lead, spot, modify, and adapt to consumer trends. Therefore, fashion design has a quick, seasonal, and powerful impact to society.

Analyzed from the design thinking point of view, there are many combinations of materials, colors, styles, and patterns to select from and decide upon. Fashion designers must first consider functionality of a garment, who will wear it, and in what situations it might be worn in order to create appropriate products. In designing a garment, designers usually will sketch their ideas on paper, which is an illustrated drawing of the design to present to their clients. After a design is set on paper, a pattern is made out of hard paper and sewed to serve as a sample pattern. Or, the designer can also make the pattern out of muslin, make necessary adjustments to satisfy the fit and flow of the sewing, and a final paper sample pattern with assigned fabric will be made afterwards. The made sample pattern is used to generate a final product for a customer, or as a template, sent to the assembly line for mass production. Therefore, the designer does not know in the beginning of design what the final clothing will look like, and there are many options to make a satisfactory garment. In this regard, fashion design fits into the category of ill-defined problem solving. But the final product has a distinctive and habitual trend in style that lasts for one or two seasons. So, fashion design is the one that makes fashionable style.

Another approach to fashionable style is to digitally implement the creation of a style through the exploration of design grammars, which is the field of shape grammar. As explored by Stiny, it sometimes is more efficient to design with a clearly defined knowledge of form and pattern than abstract intuition. Shape grammar that carries out rules of generation in certain procedural sequences to gradually create design is a method capable serving the purpose (Stiny 1980). The concept of shape grammar was developed by Stiny and Gips in the 1970s (Stiny and Gips 1972; Lauzzan and Williams 1988). By definition, a shape is a finite arrangement of 2D or 3D nonzero length straight lines. When shapes are combined to form a new shape, they have a certain spatial relation. A shape grammar is defined over a set of shapes, and map shapes into shapes to generate a language of shapes. The notion is based on: (1) a set of shapes; (2) a set of rules representing spatial relations; and (3) an initial shape to start the rule operations. After executing the rule operations, a final shape will be created.

For example, a shape grammar S has four parts: $S = (V_T, V_M, R, I)$.

- V_T and V_M are two finite sets of shapes.
- R is a finite set of shape rules of the form $\alpha \rightarrow \beta$, where α and β are shapes made up in V_T or shapes in V_M. When a rule is applied to a given shape, a resulting shape will have the given shape consisting of the right side shape of the rule substituted the shape of the left side of the rule.
- I is the initial shape.

By the language, a shape is generated from a shape grammar by beginning with the initial shape I, and recursively applying the shape rules in the set R. Shape grammars have the following sequences: (1) find part of the shape in I that is geometrically similar to the left side of a rule; (2) find the geometric transformations of translation, rotation, reflection or mirror, scale, glide-reflection to make the left side of the rule identical to the corresponding part in the shape I, and apply these transformation mechanisms to the right side of the rule; (3) substitute the right side of the rule for the corresponding part of the shape I, then; (4) when no shape rule in the grammar can be applied, a design is completed. Thus, shape grammar is executed by recognizing a particular shape from the left-side and replacing the recognized shape to the right-side shape of the rule. The right-side shapes are either transformed into left-side shapes or new additional shapes. Figure 3.16 is an example of a design generated by implementing two production rules and three rotations.

Shape grammars apply a specific class of production system through the format of unordered and data based rules, instead of sequential instructions of computation, to

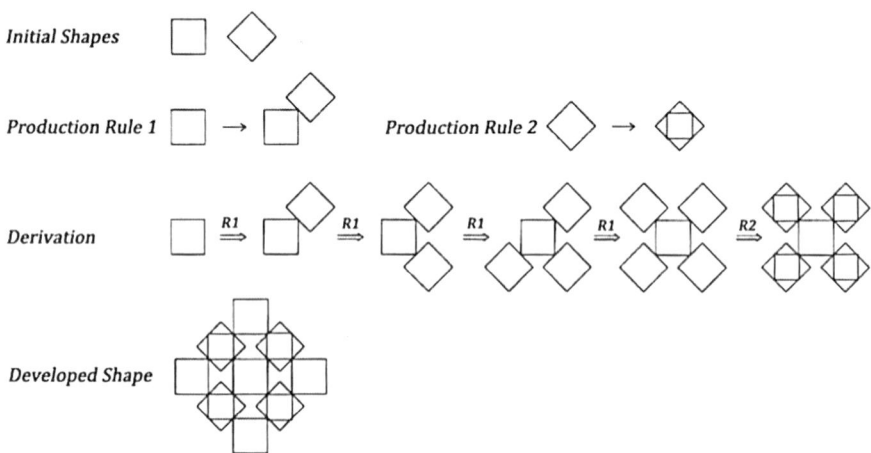

Fig. 3.16 A design by shape grammars

generate shapes or designs. These production systems could represent knowledge in some way about the functionality and the form of a product. From the conceptual design point of view, it is sometimes challenging to convert social-cultural design issues into geometric rules for shape creation. Shape grammar is, however, regarded as an attractive design tool for studying design due to the following reasons: (1) shape grammars can directly model a parametric geometry; (2) it can operate directly on geometry to generate shapes; (3) it can immediately provide emergence of shape. Thus, shape grammars can be used to reveal the clear steps for exploring and explaining the spatial relationship and hierarchy of designs (McCormack et al. 2003).

Many research projects and grammars have been developed since the 1970s, which can be categorized in three major trends of applications and research. These three trends apply different methodologies with similar approaches and procedures. One is the trend of developing original grammars by utilizing a series of schema used to create new and original design. The second group is the analytical grammars, which are for analyzing historical designs through deducing the rules of the grammars used in the original designs. The third and most recent one is the maintaining grammars by identifying the original shapes and appropriately combining them to maintain the group style. These three major trends have been used by scholars to study design solutions and the style of design.

1. Original grammars: This category is used to create a new design and a new style. The approach is to invent the vocabulary of shapes and spatial relations, defining initial shapes and rules to determine the spatial relations, and creating an initial shape to begin with. Then, the combination of these rules becomes grammar rules for the generation of designs. This approach has been applied to explore the space puzzle of modern apartment houses in Seoul (Kyung 2007). It also has been used in engineering to create coffee-makers (Agarwal and Cagan 1998) and to generate cellular automated rule patterns (Crawley et al. 2007).

2. Analytical grammars: This is done through the following steps: distinguishing the vocabulary of shapes and a set of spatial relations, which in common, design a style; defining the shape rules that determine the spatial relation; defining an initial shape to start the shape generation, and; setting up appropriate grammar rules to sequentially generate the overall shape from the initial shape. A number of projects have been conducted with this approach, for instance, the rules used to generate Chinese ice ray lattice designs (Stiny 1977), Palladian villas (Stiny and Mitchell 1978; Mitchell 1990; Sass 2007), Mughul gardens (Stiny and Mitchell 1980), Hepplewhite chairs (Knight 1980), Frank Lloyd Wright's prairie houses (Koning and Eizenberg 1981), Queen Anne houses (Flemming 1987), and traditional Taiwanese houses (Chiou and Krishnamurti 1995) to name just a few.

3. Maintenance grammars: This category is the notion of consistently using the appropriate sets of grammars and the vocabulary of original shapes to maintain and manifest the same style. This method has been utilized for encoding the key elements of a brand into a repeatable language to generate products consistently, as in Buick automobile designs (McCormack et al. 2003). In painting, shape grammar has also been used to study painting rules through computer algorithms to regenerate a painting. Kirsch and Kirsch (1986) analyzed the patterns in

Fig. 3.17 A surrealistic painting by Joan Miro (Resource: Wikimedia public domain, http://commons. wikimedia.org/wiki/ File:Mirop.jpg, Accessed 10 Sep 2014)

Richard Diebenkorn's abstract painting and used parametric shape grammars to develop the grammar for the painting structure. They also scanned the Spanish painter Joan Miro's drawings and manually traced the outlines of the shapes. Then, they developed programs to control the shapes through translation, rotation, and stretch to make various anthropomorphic shapes similar to Milo's style of composition (Kirsch and Kirsch 1988). Figure 3.17 shows a surrealistic painting by Miro for reference.

By using simple geometrical shapes, one can analyze visual styles of complex designs and recreate them. But, the most significant phenomenon of shape grammar is that design principles of a designer can be revealed by analyzing the corpus used across designs. For instance, the fireplace is the generating center of the prairie house designed by Wright. Froebelian-type blocks are added around the fireplace to generate basic compositions. The core unit of the basic composition could be one axis of growth to form asymmetrical plans, or two subordinate axes of growth to establish cruciform plans. Thus, from the study of shape grammar, authors argue that all the prairie house designs share the same principle of organization, and the basic composition is simply an arrangement of interdependent layers of blocks about a fireplace with axes of growth (Koning and Eizenberg 1981). Therefore, shape grammar not only helps to discover or interpret principles behind design thinking, but also to find ways to develop new design within the same style. For instance, a "split" algorithm of dividing a square horizontally or vertically and re-split the previously split rectangles to generate various building plans have been developed for the Palladian design style (Hersey and Freedman 1992).

Apparently, shape grammar and style have strong theoretical connections in research and applications. Grammars can be used to study a designer's designs by analyzing and interpreting the generational principles used in designs, and converting them into production rules to interpret the creation. If the identified production rules can generate and mimic more than four products designed by the same designer, then it is valid to say that these production rules do represent an individual style (Chan 2000) and reveal the designers' methods successfully. However, a designer's thinking is more compli-cated in solving complex design problems. For studying design thinking and cognition, some firsthand data on how designers process design information, how they carry out design activity, how an artificial thing appears, and how the design knowledge is embodied in an artifact all require appropriate methods of inquiry. In sum, shape gram-mars could be used to: (1) analyze a particular style through deducing what exactly the rules that create a style are; (2) predict the generation or revolution of a style over time, which relates to the explanation and what-if analysis; (3) explore all possible variations in a style to revive that style; or even (4) generate a new style.

3.5 Studies in Individual Style

Explanations on different approaches to style in this chapter have covered the aspects of expressions of artistic products, interpretations of artistic creation, and methods used to digitally reproduce stylistic design products. However, this book concentrates more on the style that is created from the individual cognitive thinking level. Individual style is defined through a designer who might apply some cognitive patterns to repeatedly generate certain features that categorize his or her signature style—not just a one time creation. Thus, in fields other than fashion design and shape grammar, studies in design have followed the trend in the late twentieth cen-tury to explore the causes of creation. Theorists advocated that the study of art should not only concentrate on the forms in products, but also study the creative process. The creative process is defined as any process resulting in the production of a work of art (Sparshott 1965). Studies have rarely been conducted concerning the relationships between the creative process and the production of art, or how a style is generated during the creative process (Wollheim 1979). Among those that do exist, there have been several discoveries (Whyte 1961; Sparshott 1965; Weitz 1970; Kubler 1979). Particularly, the use of computers as a means for exploration has been valuable in drawing attention to studying style as a set of detectable pat-terns, which will be further discussed in Chaps. 5 and 6.

References

Ackerman J (1963) Style. In: Ackerman J, Carpenter R (eds) Art and archaeology. Prentice-Hall, Englewood Cliffs, pp 174–186

Agarwal M, Cagan J (1998) A blend of different tastes: the language of coffee makers. Environ Plann B 25(2):205–226

Barlow HB (1986) Why have multiple cortical areas? Vision Res 26(1):81–90

Cavarnos C (1973) Plato's theory of fine art. Astir Publishing Co, Athens

Chan CS (2000) Can style be measured? Des Stud 21(3):277–291

Chan CS, Xiong Y (2006) The features and forces that define, maintain, and endanger Beijing courtyard housing. J Architect Plann Res 24(1):42–64

Childe G (1962) Old world prehistory: Neolithic. In: Tax S (ed) Anthropology today: selections. University of Chicago Press, Chicago

Chiou SC, Krishnamurti R (1995) The grammar of Taiwanese traditional vernacular dwellings. Environ Plann B 22:689–720

Crawley E, Speller T, Whitney D (2007) Using shape grammar to derive cellular automata rule patterns. Complex Syst 17:79–102

Crick F, Koch C (1998) Consciousness and neuroscience. Cereb Cortex 8(2):97–107

De Quincey T (1893) Essays on style, rhetoric, and language. Allyn and Bacon, Boston

Farah M (2004) Visual agnosia. MIT Press, Cambridge, MA

Flemming U (1987) More than the sum of parts: the grammar of Queen Anne houses. Environ Plann B 14:323–350

Frith C (2007) Making up the mind: how the brain creates our mental world. Blackwell Publishing, London

Glahn E (1984) Unfolding the Chinese building standards: research on the Yinzao fashi. In: Steinhardt NS (ed) Chinese traditional architecture. China Institute in America, New York, pp 47–57

Gombrich EH (1960) Art and illusion: a study in the psychology of pictorial representation. Pantheon, New York

Hersey G, Freedman R (1992) Possible Palladian Villas. MIT Press, Cambridge, MA

Humphreys G, Riddoch M (1987) To see but not to see: a case study of visual agnosia. Lawrence Erlbaum Associates, Hillsdale

Jordan RF (1969) A concise history of western architecture. Harcourt Brace Jovanovich Inc., London

Jowett B (1967) The dialogues of Plato, vol I–IV, 4th edn. Clarendon Press, Oxford

Kirsch JL, Kirsch RA (1986) The structure of paintings: formal grammar and design. Environ Plann B 13(2):163–176

Kirsch JL, Kirsch RA (1988) The anatomy of painting style: description with computer rules. Leonardo 21(4):437–444

Knight TW (1980) The generation of Hepplewhite-style chair back designs. Environ Plann B 7:227–238

Koenderink J (2011) Vision as a user interface. In: Rogowitz BE (ed) Human vision and electronic imaging, SPIE-IS&T Electronic Imaging, vol 7865. International Society for Optics and Photonics

Koning H, Eizenberg J (1981) The language of the prairie: Frank Lloyd Wright's prairie houses. Environ Plann B 8(3):295–323

Kubler G (1962) The shape of time. Yale University Press, New Haven

Kubler G (1979) Towards a reductive theory of visual style. In: Lang B (ed) The concept of style. University of Pennsylvania Press, Philadelphia, pp 119–127

Kyung WS (2007) Space puzzle in concrete box: finding design competence that generates the modern apartment house in Seoul. Environ Plann B 34:1071–1084

Lauzzan R, Williams L (1988) A rule system for analysis in the visual arts. Leonardo 21(4):445–452

McCormack JP, Cagan J, Vogel CM (2003) Speaking the Buick language: capturing, understanding, and exploring brand identity with shape grammars. Des Stud 25(1):1–29

Mitchell WJ (1990) The Logic of Architecture. MIT Press, Cambridge, MA

Pepperell R (2012) The perception of art and the science of perception. In: Rogowitz BE, Pappas TN, de Ridder H (eds) Human vision and electronic imaging XVII, SPIE-IS&T Electronic Imaging, vol 8291. International Society for Optics and Photonics

Pirazolli-t'Serstevens M (1971) Living architecture, Chinese. Grosset & Dunlap, New York

Ramachandran VS, Hirstein W (1999) The science of art: a neurological theory of aesthetic experience. J Conscious Stud 6(6–7):15–51

Reynolds J (1997) Discourses on art. Yale University Press, New Haven

Riegl A (2004) Historical grammar of the visual arts. Zone Books, New York

Sass L (2007) A Palladian construction grammar – design reasoning with shape grammars and rapid prototyping. Environ Plann B 34:87–106

Schapiro M (1962) Style. In: Tax S (ed) Anthropology today: selections. University of Chicago Press, Chicago, pp 278–303

Scott F (1893) Essays on style, rhetoric, and language. Allyn and Bacon, Boston

Singer W, Gray CM (1995) Visual feature integration and the temporal correlation hypothesis. Annu Rev Neurosci 18:555–586

Smith W (1875) A dictionary of Greek and Roman antiquities. John Murray, London. http://penelope.uchicago.edu/Thayer/E/Roman/Texts/secondary/SMIGRA/home.html. Accessed 15 Feb 2013

Sparshott F (1965) The structure of aesthetics. University of Toronto Press, Toronto

Stiny G (1977) Ice-ray: a note on the generation of the Chinese lattice designs. Environ Plann B 4:89–98

Stiny G (1980) Introduction of shape and shape grammars. Environ Plann B 7:343–351

Stiny G, Gips J (1972) Shape grammars and the generative specification of painting and sculpture. In: Freiman CV (ed) Proceedings of IFIP Congress 71. North-Holland, Amsterdam, pp 1460–1465

Stiny G, Mitchell W (1978) The Palladian grammar. Environ Plann B 5(1):5–18

Stiny G, Mitchell W (1980) The grammar of paradise: on the generation of Mughal gardens. Environ Plann B 7(2):209–226

Urton G (2003) Signs of the Inda Khipu: binary coding in the Andean Knotted-String records. University of Texas Press, Austin

Wackernagel W (1923) Poetics, rhetoric, and the theory of style. In: Cooper L (ed) Theories of style. Macmillan, New York, pp 1–22

Weitz M (1970) Problems in aesthetics. Macmillan, London

Whyte LL (1961) A scientific view of the creative energy of Man. In: Philipson M (ed) Aesthetics today. World Publishing, Cleveland, pp 349–374

Wollheim R (1979) Pictorial style: two views. In: Lang B (ed) The concept of style. University of Pennsylvania Press, Philadelphia, pp 129–145

Chapter 4
Style as Identities in Design Products

Architecture, together with painting, sculpture, music, and poetry, has long been classified in the domain of "**Arts**" or "**Fine Arts**" (Greene 1940) and studied within this fine arts domain. For example, architecture has been extensively studied by architectural historians to understand historical developments, and by theorists to explore individual characteristics (Pothorn 1982). Thus, architectural style has been described as the collective characteristics of buildings where structure, unity and expressiveness are combined in an identifiable form related to a particular period or region, sometimes to an individual designer or school of design (Smithies 1981, p. 25). All these notions and approaches attempt to examine the prominent features in forms and to systematically illustrate the underlying cultural, social, economical, and technical relationships among different regions, periods, and designers that are associated with these features.

Historically, a style is identified by recognizable features (forms) that appear in certain products created by one person (individual style—e.g. van Gogh style), by a group of persons (group style—e.g. Prairie style), across some geographical areas (regional style—e.g. Chicago Jazz music), or through a period of time (period style—e.g. Renaissance style) that is labeled by a noun to signify its existence. Thus, the study of prominent features existing in art works and buildings that can be used to identify a style is the foundation of the "**theory of style**." For example, the world in **Vincent William van Gogh**'s paintings was represented as a vortex of lines (Gombrich 1960, p. 241) and his painting could be recognized by very bold brush strokes with dots to show shapes (see the series of paintings in Fig. 4.1). Whether the painting subject is a portrait, a vase of flowers, a red vineyard, or a wheat field, the type of brushwork is typical of **van Gogh**, which can be utilized to identify the work as clearly as a signature. Thus, the brush strokes are one symbol used to recognize his **individual style** of painting.

Regarding the phenomenon of a group style, for instance, **Prairie Style** was used originally to identify Frank Lloyd Wright's designs from 1901 to 1910. His Prairie House Style had stimulated a number of followers who applied a number of similar methods in their designs, which formed a **group style**. For period style, a good

© Springer International Publishing Switzerland 2015
C.-S. Chan, *Style and Creativity in Design*, Studies in Applied Philosophy,
Epistemology and Rational Ethics 17, DOI 10.1007/978-3-319-14017-9_4

Fig. 4.1 A self-portrait of van Gogh, flowers and a vase, a vineyard, a wheat field, and a starry night (Source: Wikipedia public domain, URL: http://en.wikipedia.org/wiki/Vangogh. Accessed 10 Oct 2013)

example is the Renaissance style that was born in Italy and was primarily royal and mercantile, especially north of the Alps (Jordan 1969, p. 167). But, it had prevailed for several hundred years across Europe and formed a major design trend that was associated with an historical period and named after it. Even though the period of Renaissance style had three stages of **Early**, **High**, and **Late Renaissance**; characteristics of Renaissance buildings, in general, included extensive use of classical columns and pilasters, symmetrical arrangement of windows and doors, triangular pediments, square lintels, arches, domes, and niches with sculptures. For instance, the combination of heavy Corinthian and Ionic columns in sequence on the two stories in the cloister facade of **Santa Maria della Pace** in Italy, designed by **Donato Bramante** around 1,500, serves as a good example (Beazley 1988) – that is, **Ionic** column below, **Corinthian** above, each with its correct base, capital and entablature, with piers carrying arches (see Fig. 4.2). The Scuola Grande di San Marco in Venice, Italy is another example sharing the same characteristics in design (see Fig. 4.3). The other example of St. Peter's Basilica in the Vatican is the greatest creation of Renaissance architecture in the High Renaissance period. The completion of this church was contributed consecutively by a great number of architects, but its resulting form maintained similar characteristics recognizable as Renaissance style in totality (see Fig. 4.4).

In the study of style, style is used to differentiate changes in art. Art historians and critics study the formation and evolution of art. They use style as a benchmark

Fig. 4.2 Cloister facade of **Santa Maria della Pace** in Italy (Photo by John Maves)

Fig. 4.3 Scuola Grande di San Marco in Venice, Italy (Source: Picture was adapted from Giovanni Dall'Orto/Wikimedia Commons, http://en.wikipedia.org/wiki/File:Venezia_-_Ospedale_-_Foto_G._Dall%27Orto,_2_lug_2006_-_03.jpg. Accessed 10 Oct 2013)

Fig. 4.4 St. Peter Basilica in Vatican, Italy (Source: Picture was adapted from Wikipedia public domain,   URL: http://en.wikipedia.org/wiki/Image:Petersdom_von_Engelsburg_gesehen.jpg. Accessed 10 Oct 2013)

to date and to locate original works and as a means to trace the relationships among groups (Schapiro 1962). Historians of culture or philosophers of history study forms and qualities shared by all the art of a culture during a particular span of time. Thus, art historians and critics create classes such as Impressionism, Baroque art, or Chicago Blue Jazz on the assumption that a certain complex of elements common to a group of work is sufficiently stable, distinct, and relevant to justify characterizing it as a style (Ackerman 1963). Style in this sense is represented by the constant forms (Schapiro 1962, p. 278) or stable elements in works (Ackerman 1963). This notion has been supported through a number of studies. For example, Newton indicated that style is the outward manifestation of the artist's temperament (1957, p. 467). Schapiro argued that "By style is meant the constant form—and sometimes the constant elements, qualities, and expression – in the art of an individual or a group" (1962, p. 278). Ackerman explained that certain characteristics often appear in other products of the same artist(s), era or locale (1963, p. 164). A distinguishable ensemble of such characteristics is called a style.

For these scholars, styles have been the categories created to distinguish one period in social history from another, under the condition that these periods developed visually coherent artistic products that could be discerned from the products of preceding or succeeding periods (Schwarting 1984). Thus, the word *style* has been applied to culture or to material items and used by anthropologists to characterize civilizations or artifacts (Kroeber 1957, 1963). For instance, Greek culture had a fairly definite idea about the appearance of a temple, and builders conformed to the default of marble wide-span beam and tall column, rectangular structure with two-pitch roof. Furthermore, the Doric column used in most early temples is an example. In this case, identifying characteristics have nothing to do with the personal manner, but rather to the cultural background. Because most groups' styles are rooted in a common background (culture), they are often labeled according to the date or place of its origin.

In other studies, the relationships (syntax) between recognizable forms to establish an orderly hierarchy among persons, groups, regions, or periods to build up structures for the history of any arts was examined (Jencks 1977, p. 80). Some studies focused on either identifying significant forms that manifested a style (Alexander 1963) or reviewing their developing background (Pothorn 1982), and further tracing their context and associations as they interacted with other forms (Kroeber 1957; Schapiro 1962; Smithies 1981). Other studies, which differ from identifying forms and syntax, concentrated on the way of doing things (Sparshott 1965). In other words, a pattern of doing things the same way defined a style. For instance, jazz music has a strong but flexible rhythmic understructure with solo and ensemble improvisations on basic tunes and chord patterns that characterize it as jazz. **Ernst Gombrich** states that style is any distinguishable way in which an act is performed or an artifact is made (Gombrich 1968), but also is coming from "**choices**" made from alternatives. To him, the history of taste and fashion is the history of preferences, of various acts of choice between given alternatives (Gombrich 1960).

Beyond examining the choices that influence style, a few studies have also investigated the factors of thinking patterns that cause a style (Simon 1975)—especially, the individual style in architectural design (Chan 1992, 1993). These studies initiate an interesting question about how to recognize an individual style from different design products. Particularly, can a style be recognized and measured? The answer to this question is based on the premise hypothesized in this chapter that "a style can be treated as an entity." This concept is similar to how a color can be operationally defined in information science. In fact, color exists everywhere in the real world, which is very attractive, abstract, and subtle. It is not only challenging to verbally describe the minor changes in tone, the quality of saturation, and the quantity of intensity, but also is difficult, in the old manner, to describe or measure two identical colors exactly by words to get their identical hue, saturation, and brightness. In the field of information technology, however, there is a hexadecimal number that specifies a color to be displayed on Internet Web pages, and there are four major common color models (HSB, RGB, CMYK, L*a*b) available for printing and rendering.[1] The properties of color can be mathematically defined using one of the color models. Although all models yield different numeric numbers for one color, as demonstrated in PhotoShop, they represent the same color for monitor display and

---

[1] In color rendering and printing, a few color models are developed. The four common ones applied in most computer software are (1) hue, saturation, and brightness (HSB); (2) red, green, and blue (RGB); (3) cyan, magenta, yellow, and black (CMYK); and (4) CIE L*a*b, which L is for lightness and a and b are the color-opponent dimensions of x, y in the color space.

printing. For instance, one shade of light gray has HSB value of 0°, 0 %, 61 %; RGB of 155, 155, 155; CMYK of 42, 29, 29, 7 %; L*a*b of 72, 0, 0; and a hexadecimal value of 9b9b9b. This manner of representing colors by numbers develops a new convention.

Similarly, a style can be treated as an object or entity that is identified by repetitious forms, **features**, and **syntax**.[2] Features, in this book, are treated as constant elements that appear in art products, whereas the geometric relationships between the repeated forms define syntax. In these regards, a matrix or formula could be developed to signify a style. As long as these elements (features), syntax, and matrix are recognized, identified, and constructed, a style is established and defined. Thus, styles can be treated as special entities that can be quantified, recorded, and measured.

Unlike the way in which styling is applied in industrial design to create design language for generating stylistic forms (Tovey 1997), or for developing a formal model to communicate stylistic concepts through the computer for form generation (Chen and Owen 1997), this chapter will develop a descriptive model for scientifically analyzing style, based on the hypothesis of treating style as an object. The purpose is to establish **algorithms** to explain how style can be defined, compared, identified, and measured. This will provide an application tool capable of comprehending a particular style in architectural design, painting, and sculpture.

4.1   Features Identification

A style is recognized by means of perception across products. Perception is to recognize, be aware of, or understand the message revealed in an artistic product. The artistic products can be actions in dancing and drama, patterns in music, or physical design objects of sculpture, painting, furniture, and building in fine arts. Thus, design objects are products that have a set of **features** that appear. The term *feature* covers many meanings of patterns (detail treatments), physical forms (materials and treatments), or characteristics (textures and colors). Other than the meaning used in structural linguistics (Hampton and Dubois 1993), it is applied in this chapter to cover the forms of design products. Furthermore, the meaning of *feature* can also cover the functional and geometrical relationships between forms.

In painting, for instance, features shown in impressionists' painting are visible short and broken brush strokes of pure, unintended and unmixed pigments, open and simplified composition, emphasizing light, light colors, ordinary subject matter, innovative visual angles, and the surfaces of the paintings are textured with thick paint. These features specifying impressionism are found in **Claude Monet's** (the left picture in Fig. 4.5), **Paul Cezanne's** (the right picture in Fig. 4.5), and

---

[2]The geometric relationships between the repeated features (forms) and other elements define syntax, which has not been focused in this book.

Fig. 4.5 Claude Monet's "Impression, Sunrise"; and Paul Cezanne's "Les Grandes Baigneuses" (Sources: Picture of (**a**) was adapted from Wikipedia public domain, URL: http://en.wikipedia.org/wiki/File:Claude_Monet,_Impression,_soleil_levant,_1872.jpg. Picture of (**b**) was adapted from Wikipedia public domain, URL: http://en.wikipedia.org/wiki/File:Paul_C%C3%A9zanne_047.jpg. Accessed 10 Oct 2013)

van Gogh's paintings (see Fig. 4.1), and even used by **Eugene Delacroix** and **Pierre-Auguste Renoir**. These painters have similar strokes that make them part of the category of artists working in France in the nineteenth century. Thus, if a set of features (or forms) occurs repetitiously in a number of art products, a style emerges. If a feature in an object is originally created by a designer, it serves as one of the signatures of the designer's style. A designer's signature feature must be a "**prominent feature**" that has properties to attract audience or user attention. The attracting properties could be color, scale, material, texture, composition, motion, sound, or a combination of these.

Therefore, any "**signature feature**" to be considered as stylistic should have the following properties: (1) this feature is a form or composition distinguished by some particular shape, configuration, or proportion; (2) it has some special contextual relationship with other features; (3) it is generated originally by a designer through creative processes of initial inventing; (4) it also is adapted or copied by a designer from other sources with specific functionalities attached; but (5) it must be a member of a set of prominent features repeatedly applied by the designer or by a group of designers. In sum, a representative stylistic feature must be a prominent signature feature that has been constantly reappears in many of the creator's design products.

For instance, common features appearing in the "**Prairie House Style**" include a low hip roof, a band of casement windows, continuous bands of sill, extended terraces with low parapet and coping, watertable, corner blocks, planting urns, massive brick chimney, continuous wall between sill and watertable, overhanging eaves, and symmetric side facade (Chan 1992, 1994). Some features can be founded and seen in other designs, but most were created and put together uniquely by **Frank Lloyd Wright** to fit special functions, and repeatedly appeared in many design products. In this case, this set of common features signifies one of Wright's

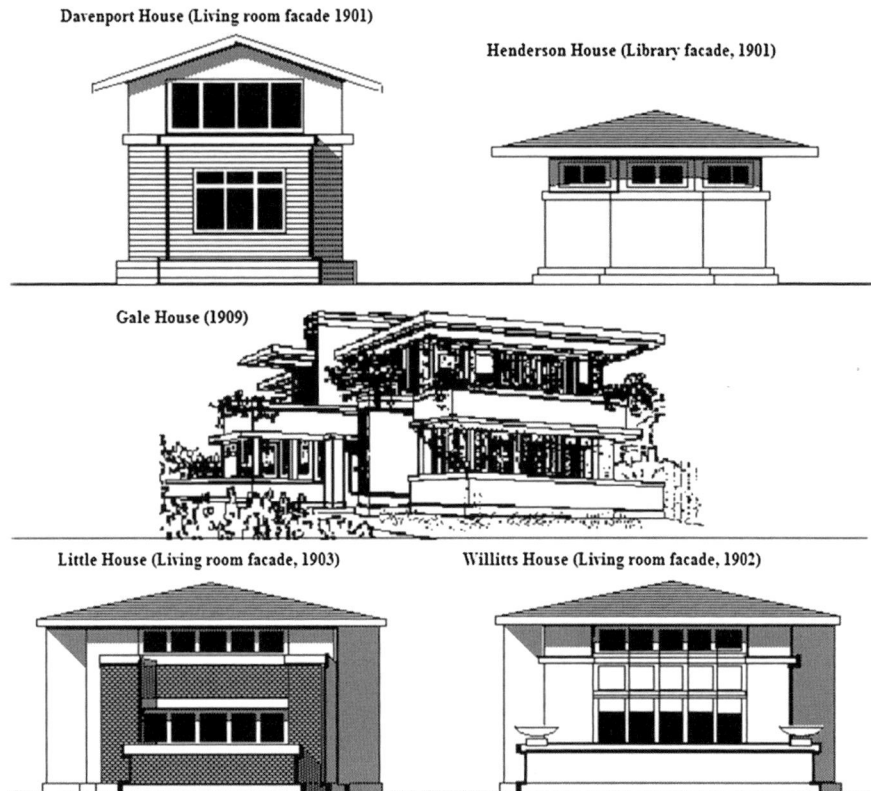

Fig. 4.6 Features in Wright's designs (Source: Reprinted from Chan (2001). Figure 1, p. 321, Copyright (2001), with permission from Elsevier)

individual styles—the **Prairie Style** (see Fig. 4.6). Studied from a broader point of view, these same features can also be seen as an ensemble in the Prairie Style houses designed not only by Wright from 1901 to 1910 but also by his students and followers. Thus, this set of common features do signify a **group style**. In any case, such features found in a group of houses are called the "**critical common features**" that define a stylistic and critical set for determining a style.

4.2   Style Identification

After a set of common features is identified, all objects that possess the same feature set should have the same style. In some cases, some objects have more members of the set than others. For instance, ten features appearing in one design will more strongly suggest a style than having five features present. Thus, the number of features appearing in an object affects how easily a style of that object can be recognized, which suggests the notion of **perceptibility**. In other words, greater or fewer numbers of features existing in an object will alter the perception of a particular style, which signifies the degree of "**style within a style**" class.

Similarly, 10 common features appearing in 20 design objects will more strongly suggest the style shared by the group than 5 common features do. The number of the set of common features indicates the notion of expressiveness. Thus, the number of features appearing in the group of objects affects the strength of the style shared by the group. Greater or fewer members in the set appearing in a group of objects will affect the expression of styles, which denotes the degree of style "between" style classes. These cognitive phenomena of "**between style**" and "**within style**" relating to the concept of the degree of style have been demonstrated by psychological experiments used to explore the *operational definition of style* and the *measurement of style*, as explained in the following sections.

4.3   Operational Definition of Style: Experiment 1

As explained earlier, all art works, as long as they possess features that were drawn from the same set of critical common features, should have similar appearance and could be categorized as the same style. For example, in Fig. 4.7, object i and j share the same set of common features of A, B, C, D, and E, which could be a subset of another larger set of common features. But, in this situation, it already suggests the existence of style X. Thus, **a style is operationally defined by the constantly appearing critical common features in art works of a person, group, period, or region**. In other words, a style will be perceived by the set of critical common features appearing among objects. (For convenience, the term **critical common features** will be shortened to **common features** in this chapter.) A psychological experiment was conducted to test the hypothetical theory for verifying the definition, particularly for individual style. In that experiment, subjects were first-year students from

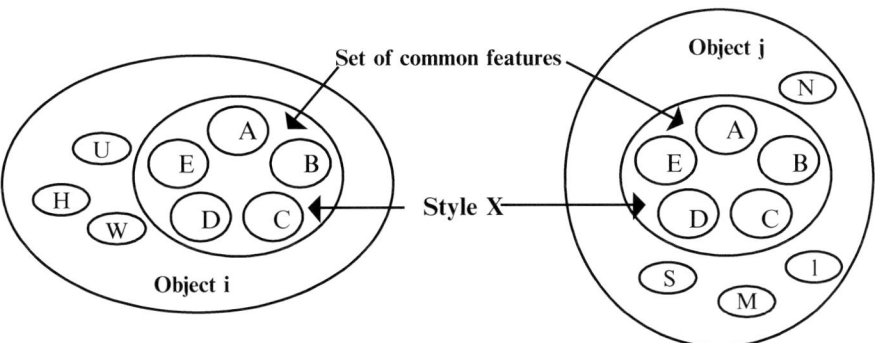

Fig. 4.7 Objects i and j of style X share the same set of features

the College of Humanities and Social Sciences at Carnegie Mellon University who were taking their first psychology course during the time the experiment was conducted. Thirty-one students from fields other than architecture participated in the experiment for course credit. Because they were coming from fields other than architecture, they would not be familiar with all the buildings presented to them and this prevented them from using prior knowledge to distinguish known style, buildings, or architects.

4.3.1   Experimental Materials

Materials presented were pictures of buildings designed by famous architects signifying three major styles. Each style was represented by 11 buildings designed by two architects. The three styles were the **Prairie Style** of **Frank Lloyd Wright** and **Vernon Watson**, the **Modern Style** of **Richard Meier** and **Michael Graves** (also called **New York 5** Style), and the **Vernacular Style** of **Charles Moore** and **Robert Venturi**.[3] Wright, Meier, and Moore were the three main architects chosen for study and each had ten buildings included. These residential buildings were selected to show common features in the pictures. For example, the number of common features appearing in the stimuli ranged from nine to ten for Wright, which included low hip roof, a band of casement windows, continuous bands of sill, extended terraces with low parapet and coping, watertable, corner blocks, planting urns, massive brick chimney, continuous wall between sill and watertable, overhanging eaves, and symmetric side façade, shown in the ten buildings in Fig. 4.8. The photo on the very lower right corner of Fig. 4.8 is the design by Watson.

Photos in Fig. 4.9 show the designs by **Richard Meier** that have features ranging from five to six, including full-height glass wall with mullions expressing the interior structural grid, round columns, circular staircase, horizontal band of parapet, walls of vertical wooden siding painted white, pipe handrail on patio, overhang circular platform, protruded solid round staircase, protruded staircase with solid handrail, protruded staircase with pipe handrail, and overhand staircase with pipe handrail, etc. The photo on the lower right corner of Fig. 4.9 is the design by **Michael Graves**.

Figure 4.10 shows the photos that have three to five features in the buildings designed by **Moore** and **Venturi**, including double-pitch roof, single pitch roof, vertical redwood siding, protruded small units with single-pitch roof, shingle roofing, left and right declining composition, white stucco surfacing, and full opening. The photo on the lower right corner of Fig. 4.10 is the design by **Venturi** in 1970. These lists of features were carefully developed by extensively

---

[3] Designs by **Charles Moore** and **Robert Venturi**, in general, represent vernacular style in the States. Using photographs of their works published in various architectural periodicals (Bloomer and Moore 1977; Allen 1980; Johnson 1986), 11 buildings were selected as examples and experimental stimulus for explanatory purposes. See Fig. 4.9.

Fig. 4.8 Prairie style of Frank Lloyd Wright and Vernon Watson (Sources: Hitchcock (1942) include: **1-Heller House**, Figure 43; **2-Fricke House**, Figure 64; **3-Heurtley House**, Figure 71; **4-Willits House**, Figure 73; **5-Little House**, Figure 89; **6-Little House**, Figure 91; **7-Barton House**, Figure 90; **9-Martin House**, Figure 102; **10-Robie House**, Figure 164; all these photos are: © 2014 Frank Lloyd Wright Foundation, Scottsdale, AZ/Artists Rights Society (ARS), NY. Photo **8-Dana House** taken from Manson (1958), Figure 82, reproduced with permission of Wiley & Sons, Inc. Photo **11-Rockwell House** taken from Sprague (1976), Figure 74, reprinted with permission)

labeling: (1) all namable features, and (2) primitive architectural elements on each picture, verified by three jurors who were faculty members at Carnegie Mellon University. Finally, the set of critical common features was identified as the features appearing on at least three pictures. These pictures were mounted on a 4 by 6 in. white index card labeled with an ID number (see the numbers in front of the building names on Figs. 4.8, 4.9, and 4.10) for the experiments. In this experiment, subjects were required to sort out 33 cards of buildings shown in Figs. 4.8, 4.9, and 4.10 into piles, each pile representing a style decided by subjects. Subjects were not told the exact number of cards nor the number of styles, but were instructed that they could place any number of cards in each pile. Results of the experiment yielded interesting cognitive phenomena of categorization, similarity, and mapping of identifying features.

Fig. 4.9 Pictures of modern style by Richard Meier and Michael Graves (Sources: Photos used for the experiments as shown in 3–9 were adapted from the following resources: Museum of Modern Art (1975):**12-Smith House** (p. 116); **13-Smith House** (p. 117); **14-Saltzman House** (p. 130); **15-Saltzman House** (p. 131). Photos taken from Meier (1984) include: **16-House in Pound Ridge** (p. 46); **17-House in Westbury** (p. 54); **18-Shamberg House** (p. 67); **19-Shamberg House** (p. 67); **21-Douglas House** (p. 76). Photo of **20-Douglas House** taken from Meier (1976), all these photos are ©Ezra Stoller/Esto. Photo of **22-Snydreman House** taken from Dunster (1979). Fig. 10, permission by Michael Graves & Associates)

Fig. 4.10 Vernacular style by Charles Moore and Robert Venturi (Sources: Photos taken from Allen (1980) for **23-Bonham House** (p. 17); **24-Bonham House** (p. 9); **26-Sea Ranch Swim Club, I** (p. 40); **27-Johnson House** (p. 47); **28-Santa Barbara Faculty Club** (p. 66); **30-Koizim House** (p. 79); **31-Burns House** (p. 90); **32-Swan House** (p. 105). Photos of 23–28 were photograph by Morley Baer. ©2014 by the Morley Baer Photography Trust, Santa Fe. Used by permission – All reproduction rights reserved. Photo of **30-Koizim House** is ©Ezra Stoller/Esto. Photos for **25-Sea Ranch Condo** and **29-Sea Ranch Spec House II** taken from Littlejohn (1984), in Moore Tour section. Photo of **32-Swan House** used by permission from Norman McGrath. Photo of **33-Trubeck House** taken from Moos (1987) (*top*))

4.3.1.1   Experimental Results A: Feature Set Identification

Data collected from Experiment 1 as listed in Table 4.1 show that more **common features** shown in pictures of one style allowed viewers to recognize them as the same category of style and they tended to put the pictures into the same pile. For instance, common features in the Prairie style picture ranged from 8 to 11, and 55 piles are sorted by 31 subjects. Among them, 16 subjects put these 10 pictures of **Wright**'s and 1 of **Watson**'s into 1 pile. For the Modern style pictures, there were five to six common features appearing and 49 piles were arranged by subjects. Within these, 17 subjects put the 10 building photos of **Meier** and 1 of **Graves** into 1 pile. But, for Vernacular style with merely three to five common features appearing in the building pictures, none of the 31 subjects put the **Moore**'s and **Venturi**'s into 1 pile; but 78 piles instead (comparing the 55 for Prairie and 49 for the Modern). Amazingly, 16 cases of misplacement happened in the Vernacular style, versus two cases in the Modern style and only one case in the Prairie style. This demonstrated the phenomenon that a style was recognized by common features appearing in

Table 4.1 Results of sorting three styles

Name of style	Number of features	Number of piles	Average pile per subject	Number of misplaced piles	Number of single-style piles
Prairie	8–11	55	1.77	1 to Vernacular	16
Modern	5–6	49	1.58	2 to Vernacular	17
Vernacular	3–5	78	2.52	15 to Modern & 1 to Prairie	0
Wright	8–11	51	1.65	1 to Moore	16
Meier	5–6	48	1.55	2 to Moore	16
Moore	3–5	75	2.42	15 to Meier	0

objects, and more features appearing in objects would strongly hold the significance of the style and reduce the chance of mixing perceptions.

Following the theory, more features in the Prairie Style would let viewers recognize the style easier with fewer piles and would result in more piles for the Modern Style. Yet, data has shown a discrepancy, which might be caused by the extraneous variable of the character of feature contents. In comparing the Prairie Style features displayed in black and white photos with detail conveyed in grey, the features of Modern Style are the results of new design culture with large scale, big dimensions, and significant shapes in white color which are prominent and visually appealing. Therefore, the strong features of the Modern Style, built around 1970, would be more culturally familiar to and more fashionably preferred by college students than the old or weak features shown in the Prairie Style built around 1910. To young viewers, familiar and strong forms are more recognized, perceived, and identified than the old and weak ones. This observation might explain how the variation of feature contents might balance out the variability of the number of features on holding the style together.

4.3.1.2 Experimental Results B: Subset of Style

Experiment 1 has shown that each style was characterized by a distinct set of common features, but not all the features appear in each building. Sets of features can vary, permitting the definition of subsets nested in styles. Thus, features appearing in the buildings also constitute subsets. For example, Table 4.2 lists the total set of common features (numbered 1 through 8) and five subsets (ranged from A to E) that appear in the Vernacular Style represented by buildings designed by **Moore** and **Venturi**. The relationships of set and subset are also graphically shown through a Venn diagram in Fig. 4.11. In this diagram, a set of common features can stand alone, representing one style (or one building), or be subordinated to another set (or another building), representing a larger group of styles. For instance, subset D{1,3,4} stood by itself and also has been part of subset C to form a larger group, and subset E{1,4,5} represents **Venturi**'s style.

Table 4.2 Features apparent in the stimuli of the vernacular style

Building name	Feature ID number	Subset	
28: S. B. Club, 1968	2,4,6,7,8	A	**Common set of features**
30: Koizim, 1971	2,4,6,7,8	A	1. Double pitch roof
31: Burns, 1972	2,4,6,7,8	A	2. Single pitch roof
24: Bonham, 1962	2,3,4	B	3. Vertical redwood siding
25: Ranch I, 1965	2,3,4	B	4. Protruded small units with single pitch roof
26: Ranch, 1966	2,3,4	B	5. Shingle roofing
27: Johnson, 1966	1,3,4,5	C	6. Left and right declining composition
29: Ranch II, 1969	1,3,4,5	C	7. White stucco surfacing
32: Swan, 1976	1,3,4,5	C	8. Full opening
23: Bonham, 1962	1,3,4	D	
33: Trubeck, 1970	1,4,5	E	

Source: reprinted from Chan (2000) Figure 6, p. 287, Copyright (2000), with permission from Elsevier

Fig. 4.11 Subsets in the stimuli of the vernacular style presented by a Venn diagram (Source: Copyright Pion Limited, London. Figure 1, p. 226, reproduced with permission. Chan (1994), Websites: www.pion.co.uk, www.envplan.com)

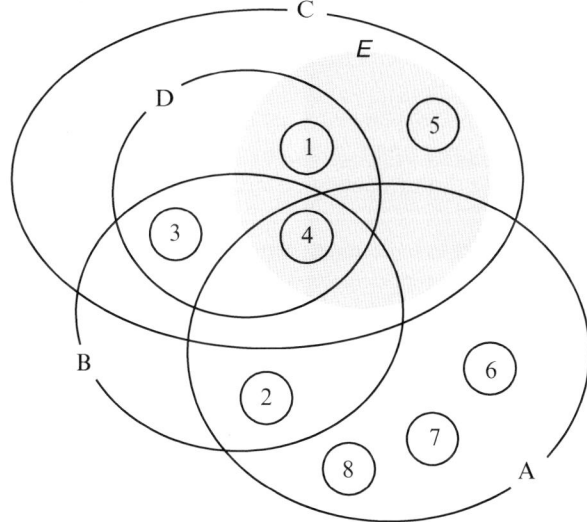

Although the subjects did not know the architects and had not seen the pictures before they entered the lab, they had no difficulty in sorting out the pictures into piles. They did not place pictures in a random manner; rather, the pictures were clustered into certain fixed groups. In **Moore**'s pictures, for example, there were three buildings (cards with numbers 28, 30, 31 in Fig. 4.10) having the set of (2,4,6,7,8), and 27 out of 31 subjects (87 %) identified it and grouped these cards together. Among them, 10 subjects put these 3 cards into a single pile to represent 1 style, whereas 12 subjects mixed them as a group with **Meier**'s cards, because the 3 buildings looked similar to **Meier**'s buildings. This result suggests that the subjects identify styles by distinguishing the same sets of "common features." In other words, common features, as listed in Table 4.2, were used as clues by the subjects for sorting and categorizing pictures. According to **George Lakoff** and **Mark Johnson**, a "categorization" in psychology is "a natural way of identifying a kind of object or experience by highlighting certain properties, downplaying others, and hiding still others" (1980, pp. 163–164). In this regard, more features in an art work would hold the symbol stronger and make it easy for the style to be identified. Subjects may have used different features or methods to categorize the pictures other than the list provided on Table 4.2 for identifying Vernacular, Modern, or Prairie styles. For clarification, interviews with the subjects were conducted after the experiment to determine the features used by them. The interviews verified that the subjects did use the technique of comparing buildings by similar features and these features were indeed the ones included on the list, which has yielded an interesting phenomenon of the notion of recognition by similarity that relates to one of the cognitive abilities on **categorization**. This categorization also explains the reason that photos 28, 30, and 31 by **Moore** shared certain similarities with features of **Meier**'s style, which caused misplacement.

4.3.1.3 Experiment Results C: Categorization and Similarity

The word *category* means a set of entities or objects that are considered equivalent and are grouped together based on some criterion or rules. Therefore, to categorize is to render discriminately different things equivalent, to group objects and events into classes, and to respond to them in terms of their class membership rather than their uniqueness (Brunner et al. 1956). Studies of categorization have included natural-kind categories (Rosch 1973; Rosch and Mervis 1975; Rosch et al. 1976; Smith and Medin 1981) and man-made objects (Reed and Friedman 1973) with prototype model (Posner and Keele 1968) and the feature-frequency model (Hayes-Roth and Hayes-Roth 1977). The prototype model suggests that a category is centered on a representative prototype, whereas the feature-frequency model emphasizes a match of individual features and combinations of features. Another major approach to categorization is to compare the similarities between objects (Smith 1989). Similarity is a fundamental principle by which individuals create concepts, **form generalizations**, and classify entities.[4] The method has been described in terms of geometric analysis of similarities (Shepard 1974) or by a hierarchical clustering scheme (Johnson 1967). A more general model of similarity is "**feature matching**."

In **feature matching**, objects are represented as collections of features, and similarity is treated as a combination of the measures of common and distinctive

---

[4] **Form generalization**: In psychology, a form represents the entity of a notion. Form generalization is the formation of a general concept which is an act or process of recognizing or perceiving similarity or relation between different stimuli, as between words, colors, sounds, lights, concepts, or feelings, to make a common knowledge.

features existing in objects (Smith 1989; Tversky 1977). Thus, the similarity between objects x and y is monotonically related to:

$$S(X,Y) = \theta * f(x \cap y) - \alpha * f(x-y) - \beta * f(y-x), \quad \theta, \alpha, \beta \geq 0$$

In this model, X and Y are two objects of art works, while x and y are the sets of features in X and Y, respectively. S(X, Y) is the similarity between the objects X and Y. (x ∩ y) is the common set of features in both objects. (x − y) is the set of features in X but not in Y. If $(\theta = 1, \alpha = \beta = 0)$ is true, similarity between objects of X and Y is determined only by their common features. On the other hand, if $(\theta = 0, \alpha = \beta = 1)$ is true, the dissimilarity is determined by their distinctive features only (Tversky 1977). This model provides weighted difference of the measures of their common and distinctive features. The scale of the function f (i.e., θ, α, β), determined by the intensity of the geometric form, frequency of appearance, viewers' familiarity with the form, and the amount of informational content it consists of, can be used as an index to reflect the salience or prominence of the various features. The parameters in the model can be assigned any number by the priority of the intensity of geometry, frequency of appearance, familiarity, and the quantity of information contents to show strength.

For instance, if X is the object selected to compare the **similarity** with Y, then features in X should be weighted (α) more heavily than the features in Y (β). If the common features are larger in proportion than the distinct features, then θ should be weighted more than α and β. However, studies (Rosch and Mervis 1975; Tversky 1977) have shown that the relative weight assigned to the common and the distinctive features may differ in various tasks. If subjects were asked to compare differences, they would give more attention to the distinctive features than to the common features. In this style experiment, subjects were not asked to single out any particular building as a referent for comparison during the picture cards' sorting process. Every building was treated as equally important. In perceiving architectural images for style recognition, there is no assigned subject of comparison with the referent, all features are hypothetically treated as equally important and salient to balance out the visual bias and different focuses of attention; therefore, the parameters have equal weight ($\theta = \alpha = \beta = 1$) and the model can be rewritten as:

$$S(X,Y) = f(x \cap y) - f(x-y) - f(y-x)$$

Where S(X, Y) is the similarity between objects X and Y, and f (x ∩ y) is the function of the quantity which is a positive integer denoting the number of common features in X and Y, f(x − y) is the number of distinctive features in X but not in Y. Applied to the field of architecture, the value S(X, Y) of the similarity between two objects, or buildings; is dominated by the quantity of common features and distinctive features in X and Y. This value could range from positive (highly similar) to negative (highly dissimilar). An increase in the common features increases similarity and decreases difference, whereas an increase in the distinctive features decreases similarity and increases difference. For example, three objects—A, B,

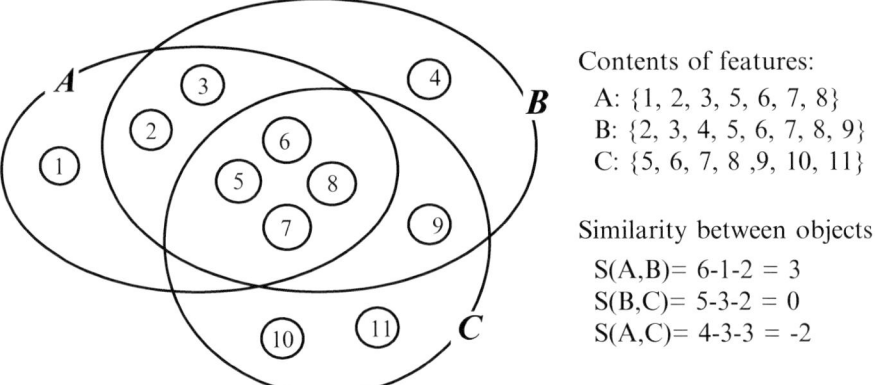

Contents of features:
A: {1, 2, 3, 5, 6, 7, 8}
B: {2, 3, 4, 5, 6, 7, 8, 9}
C: {5, 6, 7, 8 ,9, 10, 11}

Similarity between objects
S(A,B)= 6-1-2 = 3
S(B,C)= 5-3-2 = 0
S(A,C)= 4-3-3 = -2

Fig. 4.12 Venn diagram of three objects *A*, *B*, and *C* with three sets of features

and C—can be symbolized by having three sets of features, and these sets of features can be Venn diagrammed as in Fig. 4.12. Numbers on the lists of sets represent features. The similarity scores between AB, BC, and AC are 3, 0, and −2, respectively. If 0 stands for the breaking point, then positive 3 of S(A, B) is greater than S(A, C) of −2, and A looks more similar to B than to C.

Charles Moore's style, in general, represents vernacular style, and is characterized by the set of features as listed in Table 4.2, but not all the features appear in each building.[5] This is easily explained by **Moore's** collaboration with other designers whose own styles affected the final product.[6] The list of common features consists of four subsets symbolized by letters from A to D that were identified from these ten buildings (see the top ten items on Table 4.2). The Venn diagram on Fig. 4.13 explains their relationship. The similarity between the Santa Barbara Faculty Club (1968) and Ranch (1966) is: S(Club, Ranch)$=2-3-1=-2$; whereas S(Club, Burns)$=5-0-0=5$. This indicates that there are variations of magnitudes of objects within a style. Some objects look more similar than others and chances of showing up in the same pile are higher. Yet, if there were more common features appearing in objects, then the value of similarity would be steadily presenting across objects and its representative style would be stronger.

The value of S(X, Y) is also assumed to be a function of the number of times that x and y appear in the same pile. The greater the similarity values, the greater the probability that two buildings will be placed on one pile, otherwise the probability

[5]The frequency of appearance varies among individual styles. In **Wright's** Prairie Style, for instance, a large set of common features was constantly apparent in most of his designs. Thus, **Wright's** style was more constant than **Charles Moore's** style.

[6]Two personal interviews with Professor **Charles Moore** were conducted in Austin, Texas on 15 and 16 of November, 1991. This piece of information was obtained from these personal conversations.

Fig. 4.13 Venn diagram representing subsets of features in Charles Moore's 10 buildings

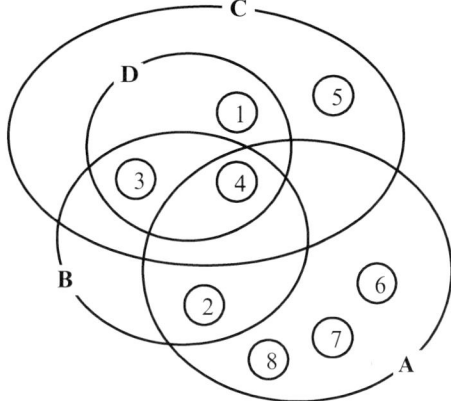

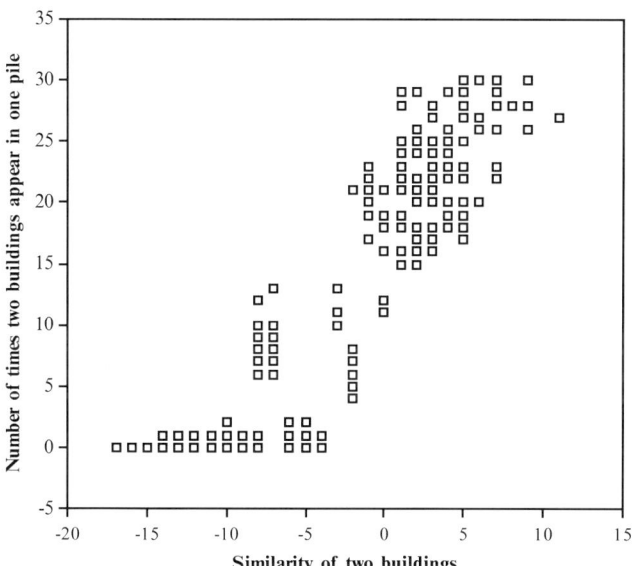

Fig. 4.14 Relationships between the similarity of two buildings and the number of times they appear in one pile (Source: Copyright Pion Limited, London. Figure 2, p. 228, reproduced with permission. Chan (1994), Websites: www.pion.co.uk, www.envplan.com)

of having x and y shown on the same pile is zero. Therefore, this model accounts for both similarity and dissimilarity. Results of the similarity value between objects and the number of times they would appear in one pile are graphically shown on Fig. 4.14. Statistically, the number of two buildings shown in one pile versus the value of similarity from $S(X, Y)$ is $y = 16.529 + 1.2828x$, $R^2 = 0.847$. These data suggest that style can be recognized by the common features between buildings. A larger set of common features appearing in objects will be easily identified as the same style.

4.3.1.4   Experiment Results D: Number of Piles

The "number of piles" corresponds to the total number of piles per style as sorted by the subjects. Its value symbolizes the strength of holding a style. Statistical data of the numbers of piles have been presented in Table 4.1. The numbers are very close in **Wright**'s (51 piles) and **Meier**'s (48 piles) styles, but the value of 75 piles increases by 56 % in **Moore**'s style, despite the minor difference of feature numbers (2–1) between **Meier** (5–6) and **Moore** (3–5).

Discussion In comparing the feature differences among three different styles, the 56 % increase in the number of piles from **Meier** to **Moore** is significant. This suggests that more common features in products would: (1) make a style coherent and less detachable (fewer piles); and (2) provide less possibility of splitting a style into multiple piles. Two other observations support this inference. First, 15 out of the 31 (48 %) subjects misplaced **Moore**'s pictures (which have the least number of features) into **Meier**'s, but only one subject misplaced **Wright**'s (with the largest number of features), and two subjects misplaced **Meier**'s (the medium number of features). This explains that fewer features in artifacts result in poor identification, so that errors occur more frequently. Second, an almost equal number of subjects (16 out of 31, 52 %) successfully identified **Wright**'s and **Meier**'s pictures as one style (putting the pictures into a single pile), but none successfully identified all Moore's pictures. This explains, again, that more common features tend to more cohesively define a style. Therefore, it can be concluded that the more common features a style has, the easier it will be for subjects to identify it.

4.3.1.5   Experiment Results E: Pattern Match by Key Features

As shown in Table 4.2, features in each picture can be coded by numbers, and a common set is categorized and labeled by a letter. For the Vernacular style represented by **Moore** and **Venturi**, there are five sets of features {A,B,C,D,E}, which have 31 possible ways of sorting them into one pile (set combination). If subjects are sorting randomly, the 31 ways are expected to appear with equal frequency. But data showed that only 20 differently combined patterns appeared within the total of 90 sorted piles (20 is 22 % of 90). Thus, it is assumed that there are certain rules governing the subjects' behavior. For example, there are ten possible ways of combining two sets from A, B, C, D, and E into one pile, (see Table 4.3). But results showed that only six different combinations appeared in 26 piles. To compare the probability of occurrence, a computer simulation was conducted to randomly generate pattern combinations 1,000 times (piles). Results showed that the probability ranged from 8.2 to 12.3 % for each pattern combination and that the number of piles ranged from 82 to 123. Another simulation generated 26 piles, in which each combination appeared at least once. And the probability ranged from 3.8 to 19.2 %, differing from the experiment result (0 to 42.3 %). These simulations suggest that the results

Table 4.3 Results from simulating a two-set combination

Pattern	AB	AC	AD	AE	BC	BD	BE	CD	CE	DE
Simulate 1,000 piles	98	101	99	123	87	103	101	118	88	82
Probability	9.8 %	10.1 %	9.9 %	12.3 %	8.7 %	10.3 %	10.1 %	11.8 %	8.8 %	8.2 %
Simulate 26 piles	3	1	2	5	5	4	1	1	3	1
Probability	11.5 %	3.8 %	7.6 %	19.2 %	19.2 %	15.4 %	3.8 %	3.8 %	11.5 %	3.8 %
Experiment result (26 piles)	3	0	0	0	11	1	0	2	4	3
Probability	11.5 %	0	0	0	42.3 %	3.8 %	0	7.7 %	15.4 %	11.5 %

Source: Copyright Pion Limited, London. Table 3, p. 230, reproduced with permission. Chan (1994), Websites: www.pion.co.uk, www.envplan.com

Table 4.4 Rules for sorting out pictures into piles

Rule 1: If key feature = {6,7,8},	Then piles = {A} and {B,C,D,E}
Rule 2: If key features = {6,7,8} and {2,3},	Then piles = {A} and {B} and {C,D,E}
Rule 3: If key features = {6,7,8} and {3,4},	Then piles = {A} and {C,B,D} and {E}
Rule 4: If key features = {2} and {5},	Then piles = {A,B} and {C,E} and {D}
Rule 5: If key features = {2,4} and {1,4},	Then piles = {A,B} and {C,D,E}
Rule 6: If key feature = {2,4},	Then piles = {A,B} and {C,D,E}
Rule 7: If key features = {2,4} and {4,5} and {1,3},	Then piles = {A,B} and {C,E} and {C,D}

Source: Copyright Pion Limited, London. Table 4, p. 230, reproduced with permission. Chan (1994), Websites: www.pion.co.uk, www.envplan.com

obtained from the experiment are not random creation, instead there are rules governing the sorting procedure.

It was also found from the data that if the number of common features between two sets of pictures is less than two, these two sets will not be merged into a pile. For instance, pictures of **Wright** and **Meier** share no common features in this experiment, so there is no chance that those pictures would be mixed in one pile. Because this rule accounts for 93 % of the data, it is inferred that the inter-set of two sets must have at least two common features. Furthermore, subjects assembled cards by using certain pivotal features, which became index keys for assembling cards. For example, if one focuses on features {6,7}, then all cards having features {6,7} will be put into the same pile and the rest will go to another pile. Table 4.4 lists some of these inferred rules that explain 65 % of the data.

The results show that the recognition process is dominated very much by matching key features. Even though the feature-grouping analyses only describe 65 % of the consistency in pattern-match behavior, it is assumed that the 35 % inconsistency is caused by the switch of attention to different pivot features, thus causing confusion. Or it may be that the limited span of short-term memory cannot hold the pivot features consistently throughout the sorting process. Since subjects had to work on three different styles for 10–15 min, each had to use at least three pivot features.

4.3.2 Summary of the Style Definition

Style is identified by grouping features for recognition, which proves that **a style is defined by the critical common features in art works**. Any set of features may represent a style by itself, or as inferred from the data analysis, it can combine with other sets to define other bigger styles. More features tend to make the style coherent and strongly hold the style together. In other words, the more features that make a style coherent, the greater its recognizability "between styles" and its strength of coherence "within style." On the other hand, some styles are visually more recognizable than others. They are more salient, providing a good approximation for the recognition of the style. For example, certain artists' works have a stronger stylistic tendency, whereas others are weak in expression. This also explains why some styles (**Meier's**) are easier to recognize than others (**Moore's**). However, other than the qualification of features, would the quantification of features affect the recognition of a style? This leads to the next study of **degree of style**, and it is assumed that degree of style relates to the number of common features present in objects. The degree of style means the level of perceiving the objects that have a style and the level of approximations of presenting that style in the objects.

4.4 Degree of Style: Experiment 2

The stimuli used in the last Experiment 1 consisted of different styles and different architects, thus the sets of features varied accordingly. It might be argued that certain features have a stronger visual impact than others, with distinct results. Experiment 2 explores whether a style has its own strength. Concentrating on the nature of the degree of style, the stimuli used were restricted to buildings designed by the same architect, and features were selected from a common set to balance out the differences between substantially different features from different architects. The hypothesis developed for this second experiment on studying the degree of style is that the degree of a style is in proportion to the number of common features presented in products.

4.4.1 Subjects and Materials

Subjects were the same group as in the last experiment. The experimental stimuli were 33 pictures of residential buildings designed by Frank Lloyd Wright. The pictures were selected to include the 11 features identified in Experiment 1. The 33 pictures were divided into 11 groups of different numbers of features, from 0 to 10 (see Figs. 4.15 and 4.16). Each group had three different pictures that had the same number but different features drawn from the common feature set. Pictures were mounted on 4 in. by 6 in. white index cards.

1-Millard House, 1923 (0-1). 2-Ennis House, 1924 (0-2). 3-Jones House, 1929 (0-3).

4-Jacobs House, 1937 (1-1). 5-Willey House, 1934 (1-2). 6-Winkler House, 1939 (1-3).

7-Winkler House, 1939 (2-1). 8-Hickox House, 1900 (2-2). 9-Coonley Playhouse, 1912 (2-3).

10-Bach House, 1915 (3-1). 11-Gale House, 1909 (3-2). 12-Winslow House, 1893 (3-3).

13-Hoyt House, 1907 (4-1). 14-Martin House, 1902 (4-2). 15-Husser House, 1899 (4-3).

16-Allen House, 1917 (5-1). 17-Hunt House, 1907 (5-2). 18-Dana House, 1903 (5-3).

Fig. 4.15 Building design by Frank Lloyd Wright with common features ranged from 0 to 5 (Photos adapted from Hitchcock (1942) for: **1**-**Millard House** (Figure 249); **2**-**Ennis House** (Figure 257); **3**-**Jones House** (Figure 297); **4**-**Jacobs House** (Figure 343); **5**-**Willey House** (Figure 316); **6**-**Winkler House** (Figure 181); **7**-**Winkler House** (Figure 377); **9**-**Coonley Playhouse** (Figure 185); **10**-**Bach House** (Figure 201); **11**-**Gale House** (Figure 160); **15**-**Husser House** (Figure 45); **16**-**Allen House** (Figure 217); **18**-**Dana House** (Figure 85). All these photos are ©2014 Frank Lloyd Wright Foundation, Scottsdale, AZ/Artists Rights Society (ARS), NY. Photos taken from Manson (1958), reproduced with permission of Wiley & Sons, Inc., include: **8**-**Hickox House** (p. 110, Figure 77B); **12**-**Winslow House** (p. 63, Figure 43); **17**-**Hunt House** (p. 174, Figure 117B). Photo of **13**-**Hoyt House**: Courtesy Oak Park Public Library, Oak Park, Illinois. Photo of **14**-**Martin House** taken from *Many masks, a life of Frank Lloyd Wright* © [1987] by Brendan Gill (1987), p. 144 (*top*). Used by Permission. All rights reserved. Photo of **15**-**Husser Residence**, S.046, from the *Frank Lloyd Wright Companion* by William Allin Storrer, Ph.D., ©1993 with permission)

Fig. 4.16 Building design by Frank Lloyd Wright with common features ranged from 6 to 10 (Photos adapted from Hitchcock (1942), for the following: **19-Evans House** (Figure 144); **21-Thomas House** (Figure 69); **22-Gridley House** (Figure 126); **23-May House** (Figure 162); **24-Boynton House** (Figure 142); **27-Barton House** (Figure 90); **29-Little House** (Figure 91); **30-Martin House** (Figure 102); **31-Fricke House** (Figure 64); **32-Willits House** (Figure 73); **33-Little House** (Figure 89); all these photos are ©2014 Frank Lloyd Wright Foundation, Scottsdale, AZ/Artists Rights Society (ARS), NY. Photo of **20-Adams House**: Courtesy Oak Park Public Library, Oak Park, Illinois. Photo of **25-Tomek Residence**, S.128, from the *Frank Lloyd Wright Companion* by William Allin Storrer, Ph.D., ©1993 with permission. Photo of **26-Martin House** adapted from Brooks (1984), Figure 17. Photo of **28-Robie House** taken from *Many masks, a life of Frank Lloyd Wright* © [1987] by Brendan Gill (1987), p. 196 (*top*). Used by Permission. All rights reserved)

4.4.2   Experiment Procedures

In the beginning, subjects were shown three buildings (the three pictures of 31, 32, and 33 on Fig. 4.16) that had ten features representing the typical style of the architect. After subjects looked at these pictures as long as they wanted, they were asked to use these pictures as referents to sort the other 30 pictures (with from 0 to 9 features) into 4 different piles that stood for 4 resemblance scales of "strongly like, like, unlike, or strongly unlike" the corresponding typical style shown earlier.

4.4.2.1   Result A: Distribution of Resemblance

Table 4.5 provides the name of stimuli (buildings), the number of responses on each resemblance scale, and the mean scores of resemblance for each feature group by assigning scores of 1 (strongly unlike), 2 (unlike), 3 (like), or 4 (strongly like). The distribution of the four resemblance scales obtained from plotting the number of features versus the total number of responses is shown in Fig. 4.17. The ranges of the scales in Fig. 4.17 were 0–4 for strongly unlike, 0–8 for unlike, 1–9 for like, and 3–9 for strongly like. Taking a 95 % interval of the distribution,[7] the four scales were ranged as 0–3 features (strongly unlike), 1–5 features (unlike), 2–9 features (like), and 5–9 features (strongly like). This indicates that the spreads of the scale fall into clusters that move upward along the scale as features increase. In other words, the scales of like, unlike, strongly like, or strongly unlike are associated with specific number ranges, and each has special relationships with particular number ranges.

To test whether the feature numbers are significant to the resulting resemblance scores, a general linear model was developed as:

$$Score = \beta_0 + \beta_1 F_1 + \beta_2 F_2 + .. + \beta_{10} F_{10} + \varepsilon_i,$$

where $\underline{F}$ represents features and ε_i represents random errors. Results showed that features scored in the model were significant, $[R^2 = 0.693,\ F(10,919) = 207.55,\ p < .00001]$, and each individual feature contributed a significant influence to the resulting score, (p values of features ranged from $p < .0185$ to $p < .0001$). This proves that features significantly correlate to scores. Figure 4.18 is a plot of the average scores per feature. When the number of features (X) and the resemblance scores (Y) were fit into a simple linear-regression model, the result of the equation was $Y = 1.246 + 0.309X$. The slope was significantly different from zero, with $[t(928) = 44.87,\ p < .00001]$, and the model explained 68.5 % of the variance in Y, the resemblance scores.

Discussion   The positive sign in the linear equation indicates the positive relationship between the number of features ($\underline{X}$) and the scale of resemblance ($\underline{Y}$). Because the scale of resemblance denotes the degree of similarity between the subject buildings and the referent buildings that stand for the prototypical style, and a higher score of resemblance represents a stronger similarity to the referent style.

---

[7] For two tails distribution (as Fig. 4.17), 95 % internal of the data is statistically recognized as significant.

Table 4.5   Data obtained from experiment 2

Card index	Name and date of building	Strongly unlike	Unlike	Like	Strongly like	Mean score
Number of features = 0						
1	Madison, 1923	31	0	0	0	1
2	Ennis, 1924	30	1	0	0	1.03
3	Jones, 1929	31	0	·0	0	1
Group mean score = 1.01						
Number of features = 1						
4	Jacobs, 1937	10	20	1	0	1.71
5	Willey, 1934	11	18	2	0	1.71
6	Winkler, 1939	19	11	1	0	1.42
Group mean score = 1.61						
Number of features = 2						
7	Winkler, 1939	11	18	2	0	1.71
8	Hickox, 1900	3	24	4	0	2.03
9	Coonley, 1912	7	18	6	0	1.97
Group mean score = 1.90						
Number of features = 3						
10	Bach, 1915	5	17	9	0	2.13
11	Gale, 1909	4	21	6	0	2.06
12	Winslow, 1893	8	10	12	1	2.19
Group mean score = 2.13						
Number of features = 4						
13	Hoyt, 1907	3	11	15	2	2.52
14	Martin, 1902	2	12	16	1	2.52
15	Husser, 1899	5	6	16	4	2.61
Group mean score = 2.55						
Number of features = 5						
16	Allen, 1917	0	6	16	9	3.10
17	Hunt, 1907	0	11	14	6	2.84
18	Dana, 1903	0	7	17	7	3.0
Group mean score = 2.98						
Number of features = 6						
19	Evans, 1908	0	5	16	10	3.16
20	Adams, 1913	0	7	16	8	3.03
21	Thomas, 1901	0	0	18	13	3.42
Group mean score = 3.20						
Number of features = 7						
22	Gridley, 1906	0	1	16	14	3.42
23	May, 1909	0	1	12	18	3.55
24	Boynton, 1908	0	0	11	20	3.65
Group mean score = 3.46						
Number of features = 8						
25	Tomek, 1907	0	1	10	20	3.61
26	Martin, 1902	0	0	7	24	3.77
27	Barton, 1903	0	0	14	17	3.55
Group mean score = 3.64						
Number of features = 9						
28	Robie, 1909	0	0	5	26	3.84
29	Little, 1903	0	0	10	21	3.68
30	Martin, 1902	0	0	6	25	3.81
Group mean score = 3.78						

Source: Copyright Pion Limited, London. Table 5, p. 232, reproduced with permission. Chan (1994), Websites: www.pion.co.uk, www.envplan.com

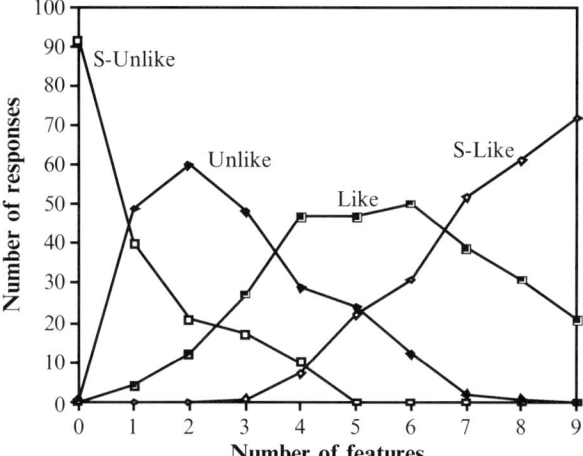

Fig. 4.17 Distribution of the four scales of resemblance (Source: Copyright Pion Limited, London. Figure 3, p. 233, reproduced with permission. Chan (1994), Websites: www.pion.co.uk, www.envplan.com)

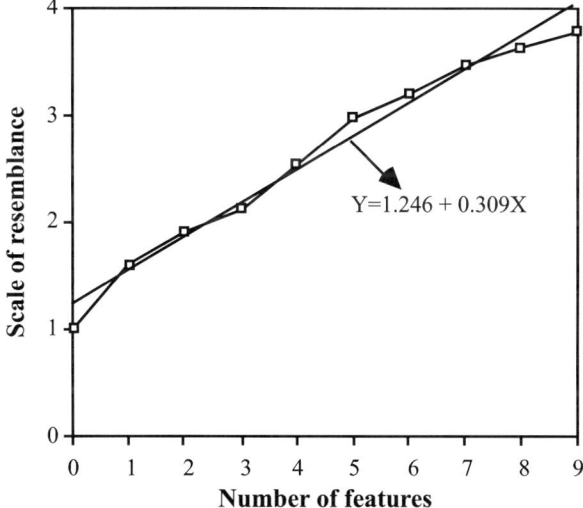

Fig. 4.18 The degree of style versus the number of features (Source: Copyright Pion Limited, London. Figure 4, p. 233, reproduced with permission. Chan (1994), Websites: www.pion.co.uk, www.envplan.com)

It also indicates that the degree of style is in proportion to the number of common features present. Another observation made from Fig. 4.17 is that four scales meet at the feature number of four, and pictures with fewer than four features tend to be unlike. Hence, four is proposed as the number of common features needed to represent a style.

4.4.2.2   Result B: Differences Between Subject Effect and Feature Effect

In Table 4.5, the mean score of resemblance varies from picture to picture in the same group, regardless of the magnitude of features. It is suspected that this could be caused by two factors: the subjects and the features. In other words, different responses between subjects and different features could cause the variations of score, and there are variabilities among subjects and features. Therefore, this section tests the effects caused by the two and a **general linear model** was designed in this analysis to test the significance of these two factors. In this new model, the resemblance score was modeled as the sum of student effect and ten feature effects.

$$Score = \beta_0 + \beta_1 SE + \beta_2 F_1 + \beta_3 F_2 + .. + \beta_{11} F_{10} + \varepsilon_i,$$

where $\underline{SE}$ denotes student effect, F is the number of different features, and ε_i is the random error. Statistical results showed that this model explained 73 % of the total variability, $[F(40,889) = 60.33, \ p < .00001, \ R^2 = 0.731]$. For the purpose of determining whether the features or the number of features determined the resemblance score, a reduced model was further developed. This model only fit the student effect and the number of features together to eliminate the difference of individual feature numbers, and had the following equation:

$$Score = \beta_0 + \beta_1 SE + \beta_2 NF + \varepsilon_i,$$

where SE denotes student effect, NF is the number of features, and ε_i is the random error. As a result, this model had a slightly reduced R^2 value, $[F(31,898) = 75.30, \ p < .00001, \ R^2 = 0.722]$.

Discussion   Comparison of these two models shows only an 0.86 % increase in explanation of total variability, which is not justified by 9 degrees of freedom. Thus, the second model is more suitable to explain the data. And statistically, the second model concludes that the differences between features (as in the full model) do not contribute to the resemblance scores as much as the number of features (as in the reduced model). Therefore a style is determined by the number of features more than by any particular feature. This result implies that in identifying a style, the low hip roof is no more important than the casement windows. It is possible, however, that the feature effect is balanced out by the individual student effect.

For the student effect, it is possible that some students may detect more similarities in pictures than others and give more responses toward the "like" or "strongly like" scale. Thus, different students give different scales of responses for the same pictures. This inference is supported by statistical results in both models. Both are significant at the 0.0001 level, $[\underline{F}(30,889) = 4.15, \ p < 0.0001]$ and $[F(30,898) = 4.06, \mathbf{p} < 0.0001]$. But the number of features is more significant than the student effect in the reduced model, $[\underline{F}(1,898) = 2212.60, \ p < .00001]$.

4.4.3   Summary of the Strength of Style

It can be concluded from the data analysis that the degree of a style is related proportionally to the number of common features that appear in the artifacts. More features in an artifact make a style more expressive. The metaphor is that the number of common features represents the strength of the glue—a larger number of common features will increase the strength of the glue to hold pictures together in a pile. Experimental results do not support the notion that certain features from the same style have a stronger impact than others. Rather, the number of features is more significant in identifying a style. The number of three is the lowest threshold. Thus, the degree of style is closely related to the number of features and varies among beholders.

4.5   Measurement of Style: Experiment 3

Reviewing the previous two completed experiments, the first experiment studied the nature of style and the strength "among styles" through recognizing features in products done by six different architects representing three major styles. Results show that if there are features repeatedly appearing in three products, they can be registered as a set of common features, for instance, the subsets of A, B, and C in **Moore**'s design (see Table 4.2 and Fig. 4.10). More features appearing in the common set will strongly hold styles together and could more easily be recognized. The second experiment concentrated on the recognition of the style "within style" through studying similarities between many product features created by the same architect. Experimental data indicated that four features repeatedly appearing in products created by the same designer would define the set of critical common features, which is the critical number for style recognition (see Fig. 4.17). As such, the complete results do suggest that if there is a set of four common features appearing in at least three objects designed by the same designer, it definitely signifies an *individual style*. Similarly, if the set appears in objects designed by a group of designers, it signifies a *group style*. Furthermore, a collection of the sets across time and geographical locations will symbolize a "regional style" and a *period style*.

Yet, in the process of identifying features, there are two key issues: first, at what point are two features claimed similar enough to be the same feature, and second, will the syntax among features affect the recognition of a style? Thus, the following two experiments were conducted to test this premise of whether changes in feature number and context would affect the *expressiveness* of a style, or its *recognizability* by viewers. Therefore, the experiment focuses on the style in one product to explore how its represented style can be "measured." Measurement is defined as the use of numbers to describe attributes of objects or events, and the described "common features" in this book can be used as a measurement unit with a number representing its magnitude. The magnitude of the number and its significance of representation set up certain restraints and expediencies to the sets. The restraint of the magnitude is the **measurement of a style**, or the threshold for recognizing a style, or to what degree a style could be recognized. Two concepts of feature frequency and recognizability are developed.

4.5.1 Feature Frequency

This experiment addresses style in one product to discover whether the number of features would determine the recognition of this style, and it uses feature frequency to do the measurement. The subject who participated in this experiment is a faculty member who taught architectural history in the Department of Architecture at Carnegie Mellon University when this experiment was conducted. He has a Ph.D. in Architectural History and is an expert on Frank Lloyd Wright's style.

Frank Lloyd Wright's side elevation of Little House, designed in 1903, with six recognizable features, was used for stimuli. These six selected features in the original—elevation of corner blocks, watertable, a row of horizontal casement windows, coping on the parapet, low hip roof, and symmetry—were taken away one at a time, two at a time, until all were gone. There were 64 permutations of taking these six features away, such as taking one to three features away in Fig. 4.19 and taking four to six features away in Fig. 4.20. These figures served the sources for the 64 pictures to be used as the experimental stimuli. Each picture was individually mounted on a 4 in. by 6 in. white index card.

Sixty-four pictures from Figs. 4.19 and 4.20 were shown one at a time to the subject, who was asked to make a judgment by answering yes or no about whether the picture could be considered Frank Lloyd Wright's style. The whole set was repeated six times, and each time cards were shuffled to avoid random error. In the interval between sessions, the subject would take a 5-min break to release his visual focus.

4.5.1.1   Result A: Measuring a Style from Features in an Object

In this experiment, there are seven categories (or groups) of numbers of features, ranging from zero to six. The number of stimuli per category is not the same across categories, which is determined by the permutation of the category number of features drawn from the set of six. The number of responses per category ranges from 6 (which are the 0 and 6 feature groups) to 120 (which is the 3 feature group), because each stimulus repeats six times. Results of the positive responses from six features present down to none are 6 (100 %), 14 (38.8 %), 16 (17.7 %), 6 (5 %), 4 (4.4 %), 0 (0 %), and 0 (0 %) respectively. When the number of features is three, the probability of positive responses is only 5 %, which totals 6 positive

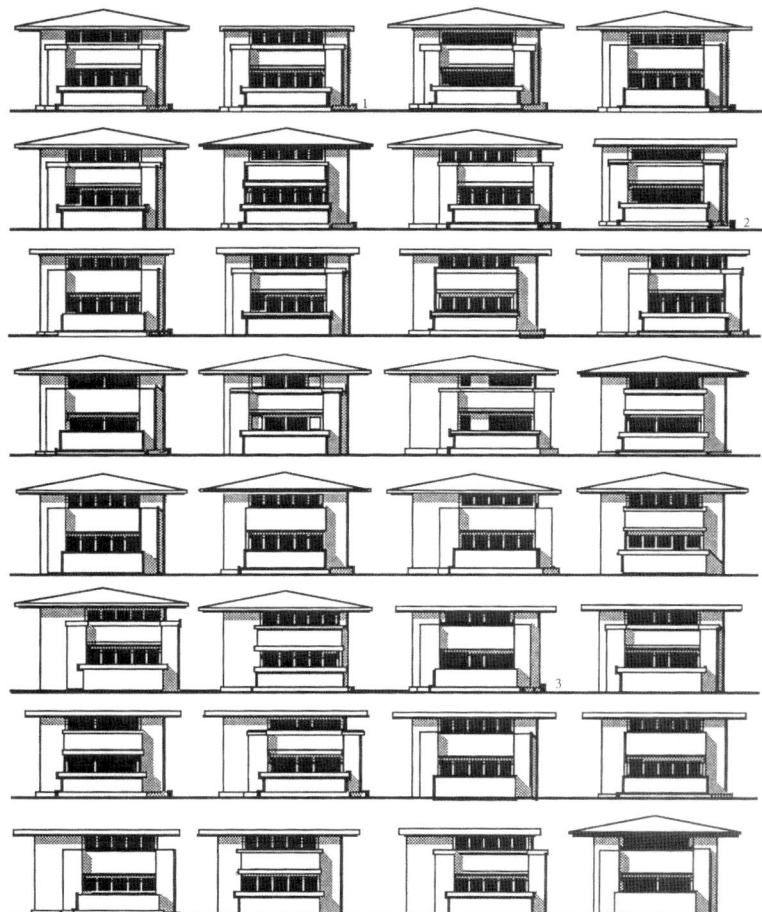

Fig. 4.19 The degree of style versus the number of features—changes of Wright's elevation in Little House design as stimuli. (The number represents the number of features being taken away from the elevation) (Source: Copyright Pion Limited, London. Figure 5, p. 236, reproduced with permission. Chan (1994), Websites: www.pion.co.uk, www.envplan.com)

responses out of 120 responses and is considered extremely poor (see Fig. 4.21). The probability remains constant at two features, and drops to 0 % with one feature is present.

Discussion Figure 4.21 shows that the positive responses decrease as the number of features drops. This suggests that the number of positive responses representing the probability of recognizability of a style is in proportion to the number of features present in a product. More features will make a style more visible. From the nature of the curve and the value of the probability shown in Fig. 4.21, it is inferred that

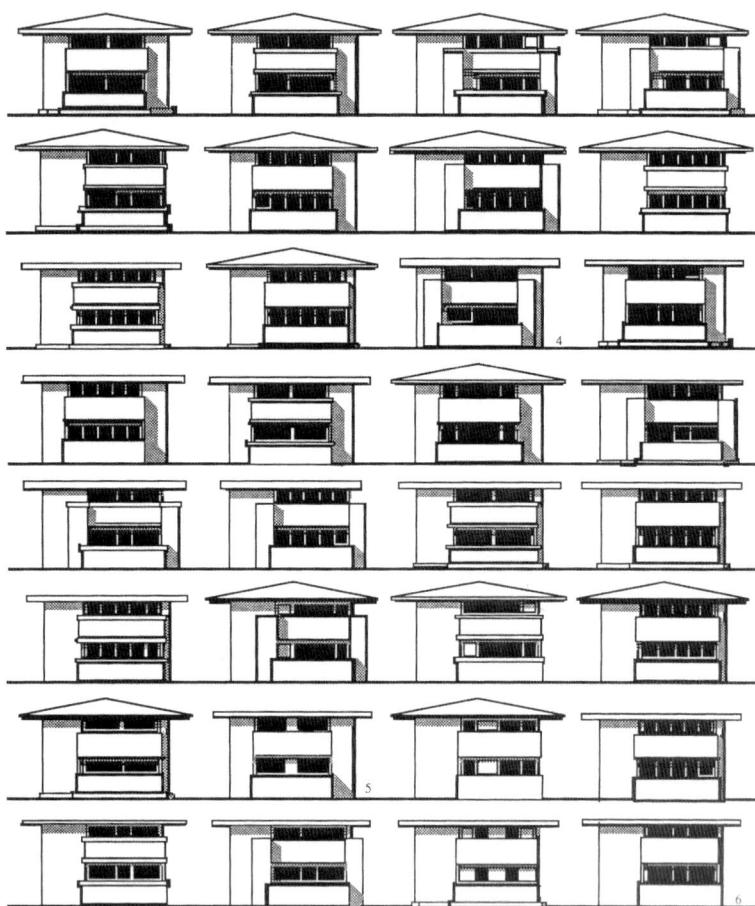

Fig. 4.20 The degree of style versus the number of features – changes of Wright's elevation in Little House design as stimuli. (The number represents the number of features being taken away from the elevation) (Source: Copyright Pion Limited, London. Figure 6, p. 237, reproduced with permission. Chan (1994), Websites: www.pion.co.uk, www.envplan.com)

three features could be the lower bound for style recognition. This suggests that a style is no longer recognizable when the number of features is three or fewer, despite the contents of features, and is measurable when there are more than three features. Therefore, as proven in experiment 2, four is the number of common features representing a style and when the number is down to 3, the represented style is not recognizable. However, if there are three features repetitiously appearing in a group of designs, as shown in experiment 1, then its style can barely be recognized with low probability. Yet, if the number is greater than three, then the style can positively be identified and measured.

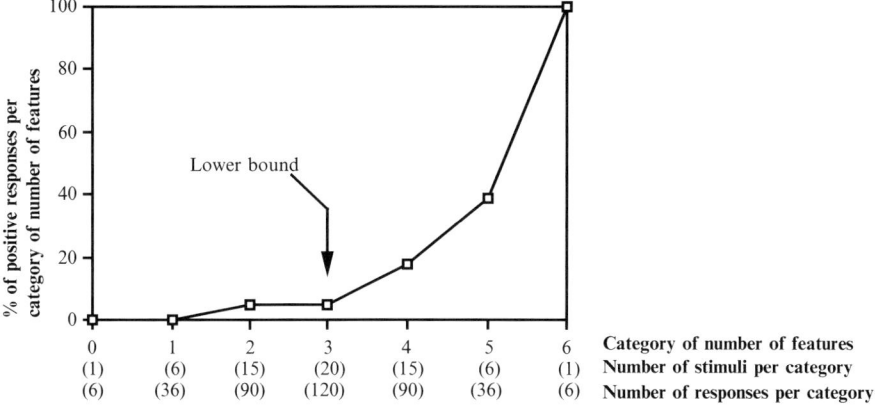

Fig. 4.21 Probability plot of responses per number-of-features category (Source: Copyright Pion Limited, London. Figure 7, p. 238, reproduced with permission. Chan (1994), Websites: www. pion.co.uk, www.envplan.com)

Table 4.6 Number of responses per cases and the grand mean of each feature

			CB				−CB				1. Mean Wd = 21.87 % (Casement window) Mean −Wd = 2.08 %
	Feature		Co		−Co		Co		−Co		
Feature			Rf	−Rf	Rf	−Rf	Rf	−Rf	Rf	−Rf	2. Mean Sy = 17.18 % (Symmetry) Mean −Sy = 6.77 %
Sy	WT	Wd	6	1	2	1	4	2	1	1	
		−Wd	0	0	0	0	0	0	0	0	3. Mean Co = 17.18 % (Coping) Mean −Co = 6.77 %
	−WT	Wd	5	1	3	0	2	1	2	1	
		−Wd	0	0	0	0	0	0	0	0	4. Mean Rf = 16.66 % (Low hip roof) Mean −Rf = 7.29 %
−Sy	WT	Wd	2	2	0	0	2	0	0	0	
		−Wd	0	0	1	1	0	0	0	0	5. Mean CB = 15.63 % (Corner blocks) Mean −CB = 8.33 %
	−WT	Wd	2	1	0	0	0	0	0	0	
		−Wd	0	2	0	0	0	0	0	0	6. Mean WT = 13.54 % (Watertable) Mean −WT =10.42 %

Source: Copyright Pion Limited, London. Table 6, p. 238, reproduced with permission. Chan (1994), Websites: www.pion.co.uk, www.envplan.com

4.5.1.2   Result B: Effectiveness of Features

Features shown in this experiment can also be treated as variables with the value of absence or presence. For example, a low hip roof is a variable representing its presence or absence in six repetitions of 64 stimuli. Table 4.6 arranges all features and their corresponding positive responses in a factorial design with a two-letter code standing for the presence and a prefix of negative sign for absence. Calculated from the data shown on the table, the grand means of each feature absence and presence corresponding to positive responses are rank ordered on the side of the table. Methods of calculation are based on the probabilities of a particular feature appearance within the same group. For instance, the subject gave 32 positive responses to the pictures that looked like Wright's design within 192 tests of having the roof present; whereas 14 positive responses were given among 192 tests of not having the roof shown in the stimulus. Results show that the probability of positive responses for the presence of a casement window is the greatest, about 21.87 % when it is present and 2.08 % when it is absent. The weight of six variables is, therefore, ranked by the value of probability of responses for feature presence in descendant order and absence in ascendant order. Thus, the order implies that some features are more attractive to the subject than others.

The grand mean of the absence and presence of each variable in Table 4.6 does not exclude the possibility of interactions among variables, which might occur when the setting for one variable in the test positively or negatively influences the setting of another variable. The interaction between variables might be caused visually by the syntax connections among variables, geometric comparison of each variable, or competition of attentions on the variables. To eliminate the effect of interaction, observation concentrates on the solo (or single) absence of a particular variable. This means studying the responses while a particular feature is absent. For example, when only the watertable is absent from the picture, the positive responses are five out of six trials, whereas when the casement window is absent, there is no positive response at all. A reasonable explanation is that the watertable is not an important feature to the subject for judging the style, but the casement window is. Thus, sorting out the responses by one absent feature and arranging variables by the number of "yes" responses, the degree of significance of the feature emerges as shown in Table 4.7. This result is very close to the last one in Table 4.6, thus, some features are more effective than others.[8]

Table 4.7 The number of responses per absent feature

Absent feature	Present features	Number of positive responses
WD (casement window)	(**SY, WT, RF, CO, CB**)	**0**
RF (roof)	(**SY, WT, WD, CO, CB**)	**1**
CO (coping)	(**SY, WT, WD, RF, CB**)	**2**
SY (symmetry)	(**WT, WD, RF, CO, CB**)	**2**
CB (corner blocks)	(**SY, WT, WD, RF, CO**)	**4**
WT (watertable)	(**SY, WD, RF, CO, CB**)	**5**

Source: Copyright Pion Limited, London. Table 7, p. 239, reproduced with permission. Chan (1994), Websites: www.pion.co.uk, www.envplan.com

---

[8] The contradiction between this result and the results on the Experiment 2 (that a style is determined by the number more than the identity of characteristic features present) may be explained by the fact that the feature effect is balanced by the variable of individual student effect which is not present in this experiment.

To further verify the significances of features to the subject, the analysis now focuses on a particular present feature and its relationship with other present features. This means that when a particular feature appears, the probabilities of responses counted by having 5, 4, 3, 2, 1, or 0 other features present together is recorded in Table 4.8 and plotted in Fig. 4.22. In the figure, each curve represents the correlation between responses and feature-numbers condition to a particular feature presented. In other words, this study concentrates on the significance of each individual feature appearance. The shape of the curves suggests a scale of WD>RF>CO=SY>CB>WT, which is similar to the result in Table 4.7. But interaction occurs when the number of features is three or less than three.

Table 4.8 The probability of responses for a particular present feature

Number of features	WD window	RF roof	CO coping	SY symmetry	CB corner blocks	WT watertable
0	0	0	0	0	0	0
1	0.033	0	0.067	0.033	0.100	0.033
2	0.083	0.050	0.033	0.067	0.033	0.033
3	0.267	0.167	0.183	0.167	0.150	0.133
4	0.467	0.433	0.400	0.400	0.333	0.300
5	1.000	1.000	1.000	1.000	1.000	1.000

Source: Copyright Pion Limited, London. Table 8, p. 239, reproduced with permission. Chan (1994), Websites: www.pion.co.uk, www.envplan.com

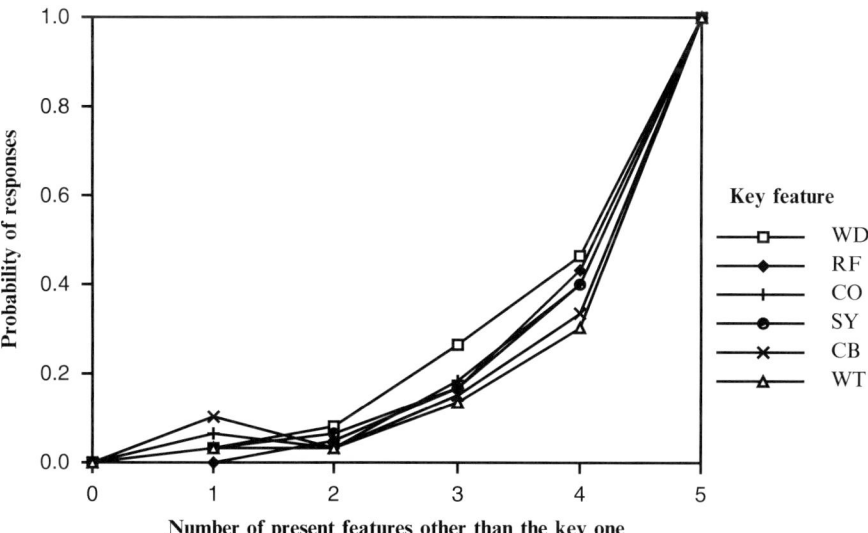

Fig. 4.22 Probability of subjects' responses to a particular feature being present (Source: Copyright Pion Limited, London. Figure 8, p. 240, reproduced with permission. Chan (1994), Websites: www.pion.co.uk, www.envplan.com)

4.5.1.3 Result C: Interaction Among Features

Figure 4.22 suggests that corner blocks and coping are two features generating interactions. This feature interaction result explains why, when the number of present features decreases to three or fewer, the perception of a style is less feasible in Fig. 4.21. Explained from another perspective, the feature interactions occur when the total number of features in an artifact is equal or less than three. Because the recognizability of a style in one object is based upon the perception of its consisted features, it is further inferred that three is the critical boundary for the measurement of a style in one object.

4.5.2   Summary on Measurement of Style

Data analyses suggest that there is a scale of importance among features in an artifact for a beholder to identify a style. This style judgment does not relate to the sizes or dimensions of features, but is determined by the number of features present.[9] For example, roofs have the largest volume among features, and they should be the most noticeable features. But roofs rank third and second in Tables 4.6 and 4.7, respectively. When features in an artifact are reduced to three or fewer, interaction occurs and the style in that object is no longer perceptible. This supports the findings obtained in experiment two that a set of four common features appearing in products would signify the existence of a style (see page 135). If the number of features presented is less than three, the style is not recognized.

4.6   Recognizability of Stylistic Features: Experiment 4

Another measurement of style is to measure if the style could be recognized after a feature has been modified. This depends on the extent to which a borrowed and modified feature remains recognizable as an element of a style. It is argued that whatever changes occur to a feature, as long as its shape is maintained within certain ranges of geometric proportion, it will retain its identity with a perceived style. Yet, topological character of the feature would make changes. In other words, **topological** characteristics of an artifact would provide a dominant key for identification and also provide a measurement for an individual style. The hypothesis is that a style can be identified within a certain range of geometrical distortion, yet, topological distortions disable the style identification. Experiment 4 tests this hypothesis.

---

[9] In this experiment, the size of a feature is not an issue as long as it is within the normal and expected range for that kind of feature. An unusually massive roof or a set of tiny windows will certainly be noticeable.

4.6.1 Subject and Experimental Materials

The subject was the same one who participated in Experiment 3. Wright's stylistic elements apparent in the Little House were used for stimuli (the original picture is shown in the uppermost left of Fig. 4.19). The dimension of each feature was either elongated or reduced 10 % at a time along either the $\underline{X}$ or $\underline{Y}$ axis. Thus, $\underline{Y} + 2\underline{a}$ meant that the vertical dimension of an object was increased by 20 %, and $\underline{X} - 2\underline{a}$ meant that its horizontal length was decreased by 20 %. The distorted features include horizontality, verticality, roof, coping, corner block, and watertable. The geometrical distortions, ranging from 10 to 50 %, are shown in Figs. 4.23, 4.24, and 4.25. Figure 4.23

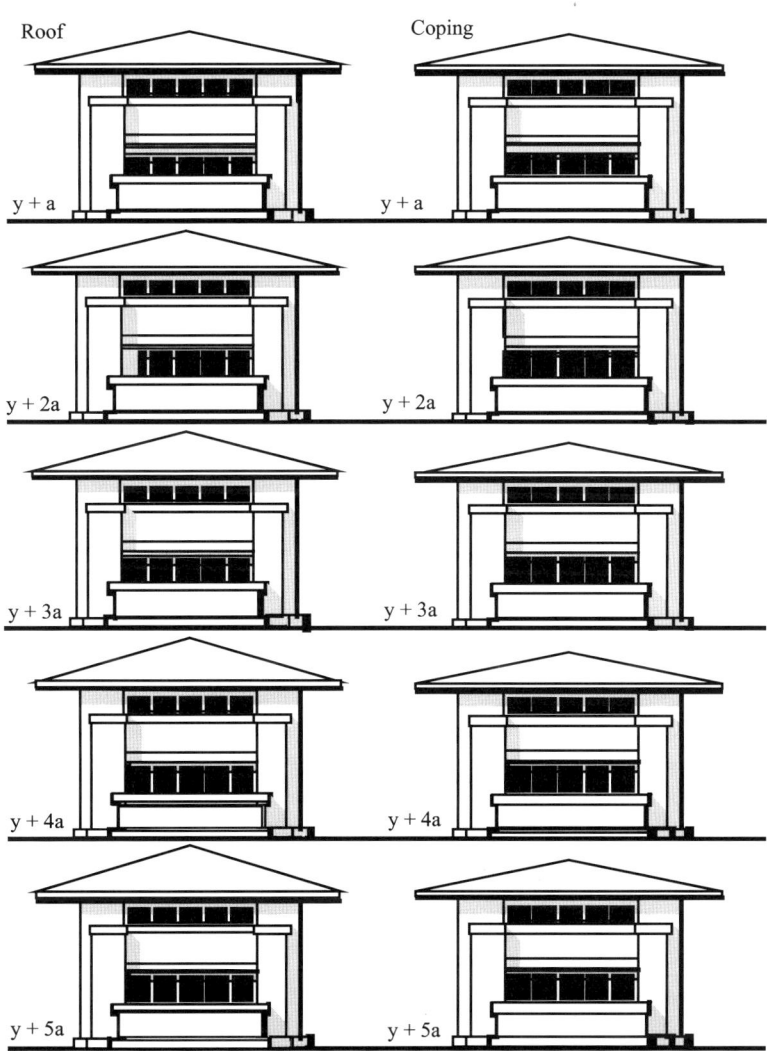

Fig. 4.23 Distorted features of roof and coping in the Little House

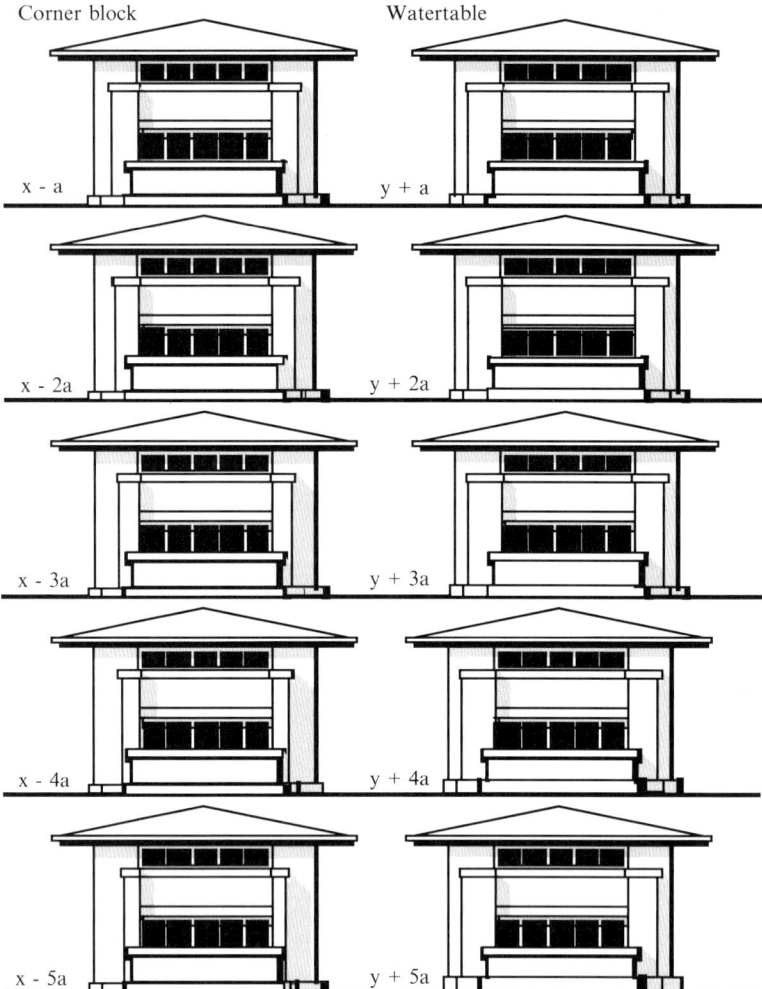

Fig. 4.24 The changed forms of Wright's corner block and watertable (Source: Copyright Pion Limited, London. Figure 9, p. 241, reproduced with permission. Chan (1994), Websites: www. pion.co.uk, www.envplan.com)

includes the roof and coping, the corner block and watertable are shown in Fig. 4.24, and verticality and horizontality are included in Fig. 4.25. There also were three pictures having a topological change, as shown in Fig. 4.26. Taking these 33 figures for stimuli, each picture was mounted on a white 4 in. by 6 in. index card.

4.6.2   Experimental Procedures

Thirty-three cards were shown to the subject, one card at a time, and the subject was asked to make a judgement about whether the picture could be regarded as Frank Lloyd Wright's style. This experiment was repeated six times, and cards were shuffled

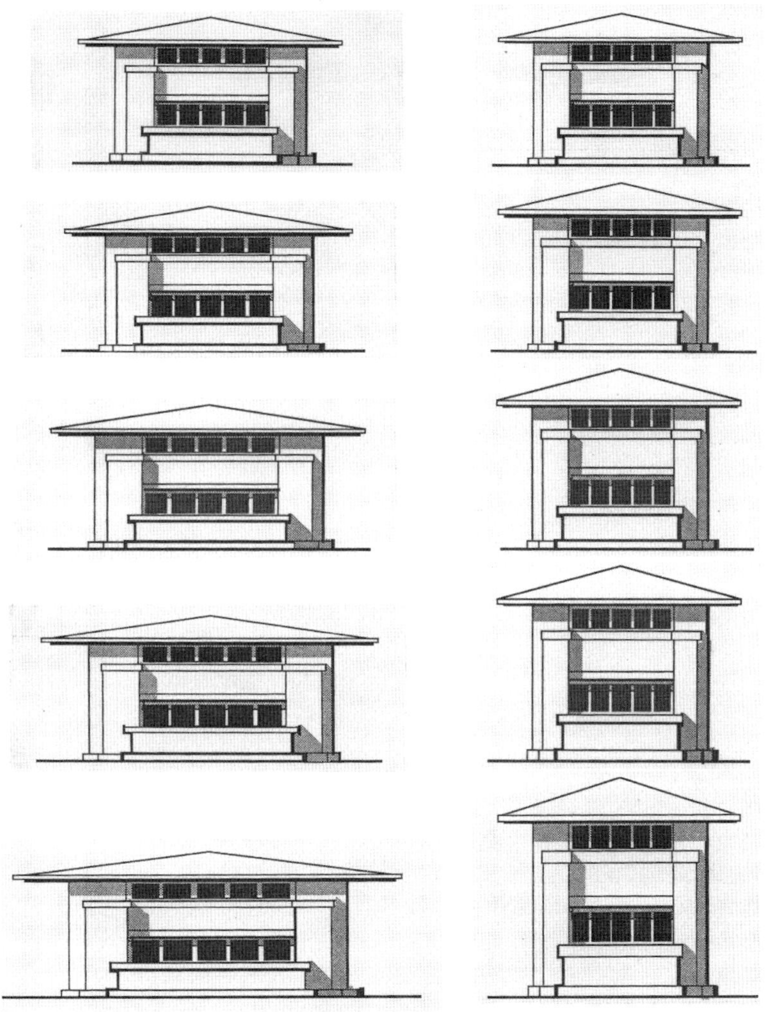

Fig. 4.25 Distorted horizontality and verticality of the mass of the Little House

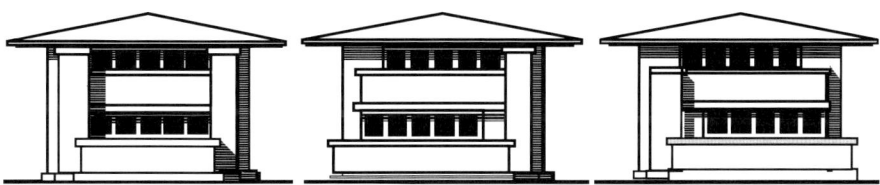

Fig. 4.26 Topologically distorted images of the Little House (Source: Copyright Pion Limited, London. Figure 10, p. 242, reproduced with permission. Chan (1994), Websites: www.pion.co.uk, www.envplan.com)

each time to avoid random sources of error. Between sessions, the subject took a 3-min break to release his visual focus.

4.6.2.1 Result A: Degree of Distortion

Plots of percentage of positive responses versus the degree of distortion are displayed in Fig. 4.27. Distortions of the massing, coping, and watertable shared a negative relationship. The curves in the plots of distortions of roof and corner blocks were not regular. For example, when the roof increased 40 % of its original height, the response increased. This phenomenon also occurred in the plot of corner blocks. To verify these results, the same set of these two stimuli was tested again on the same subject a month later. Results of the second experiment are displayed in the last two plots in Fig. 4.27. The curve of corner block exhibited a regular negative relationship, which means the positive answer to recognize the picture as Wright's style is linearly decreasing while the degree of distortion increases. But the result for roof remained the same. An interview with the subject afterward indicated that the picture with the 40 % increase in roof height resembles Wright's two early houses – the Heller House (1897) and the Husser House (1899). Thus, the subject gave positive responses. In response to the three topologically distorted pictures (total of 18 observations), the subject did not attribute any of these forms to Wright's style. Hence, it is reasonable to conclude that topological relationships are crucial in maintaining an individual style.

4.6.2.2 Result B: Threshold of Recognizability

All human sensory processes have a limited range. If the stimulus strength is sufficient to be detected, it has reached the threshold for detection. Obviously, recognition of an object has its own threshold that is commonly taken to be the stimulus intensity for which there is a 50–50 chance of detection, or for which the probability of detection is suggested as 0.5 (Kurtz 1966). But, the data shown in Fig. 4.27 has indicated thresholds for recognizability ranging from 10 to 30, and they vary from feature to feature. If geometric distortion exceeds 10, 20, or 30 %, it is possible to be judged as non-Wright's style. This range of recognizability is attributable to the fact that certain features stand out and thus would be identified immediately by the beholder.

Discussion The proportional distortion of a feature can be tolerated to a certain extent. But topological distortions change the relationships among features, which change the characteristics of the object and, consequently, alter the style. This suggests that there exists a topological structure (characteristic context) defined by the topological relationships among stylistic features. Any topological change would force the change of its representative role of a style, whereas the geometric distortion

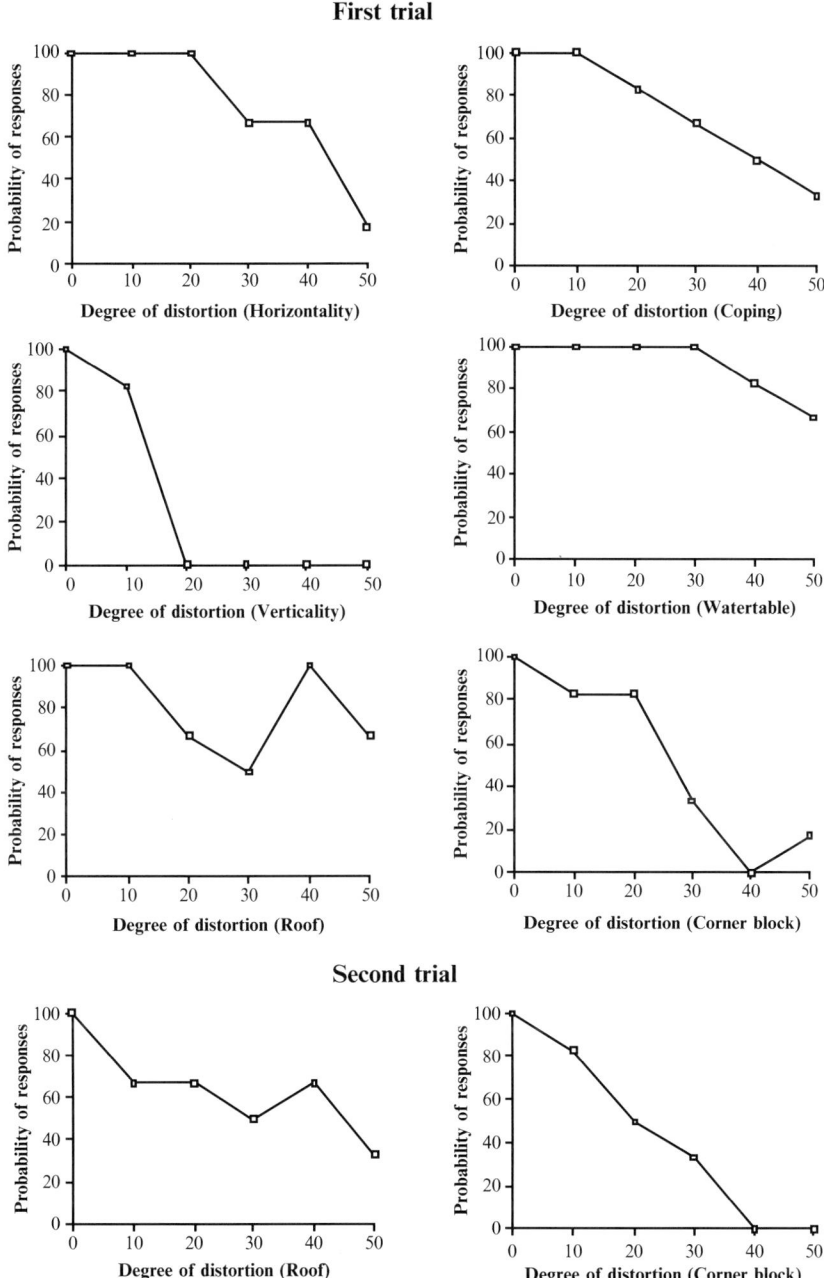

Fig. 4.27 Plot for results of each feature distortion and its probability (Source: Copyright Pion Limited, London. Figure 11, p. 243, reproduced with permission. Chan (1994), Websites: www. pion.co.uk, www.envplan.com)

of a feature has a more tolerable range of recognizability. Therefore, topological relationships among features are a crucial factor for sustaining a style.

The fluctuation of responses and curves among features implies that the recognizability of features used for identifying a style does not correlate to their dimensions but relates to the beholder's perception. Some features are more noticeable to the beholder than others. For example, a coping is a small element assumed to have the least conspicuousness among features. But it had a constant slope-of-response line (as shown in Fig. 4.27). In this experiment, the subject is more acutely aware of the changing properties of coping and corner blocks than of other features.

In this experiment, the geometric distortion has a function of recognizability of style, which can be explained by the *prototype model* derived from theories of *categorization*. In other words, the subject had mentally formed a concept (prototype) of the original picture presented in the beginning of the experiment and used it to categorize distorted pictures. The levels of distortion are analogous to the levels of *generality* suggested by **Rosch** et al. (1976) in the comparing of categories. Results obtained here were similar to their findings that it was the most general level at which a prototype could be formed to represent the category as a whole. Thus, general level is the representative of a category (style), whereas a less general (more distorted) level will be less representative.

4.7   The Degree of Style Within Style and Across Styles

The theory developed in this chapter proves the set of common features in objects represents a style. A large number of common features appearing in an object would more strongly represent its style than the object having fewer common features. This is the factor of the quantity of representation determining image recognition "within" a style. If the set of common features exists in many objects in large quantity, then these objects will more easily be recognized as having the same style, and the style is strongly manifested by them. This factor of the quantity of representation also determines the pattern recognition "across" styles, which means that objects equipped with a large number of common features not only are more easily determined to have the same style, but also are strongly recognized. For instance, the study on comparing the styles of **Wright**, **Meier**, and **Moore** from three sets of 10 buildings showed that **Wright**'s Prairie Style, with 8–11 common features, is stronger than **Meier**'s style with 5–6 features and surpasses **Moore**'s style with 3–5 common features. Results of the experiments demonstrated that the styles of **Wright** and **Meier** were identified without any mistakes, whereas **Moore**'s style elicited some confusion. Metaphorically, the number of common features represents the strength of the glue that holds a style together. A larger set of features will more strongly hold the style together. Thus, the styles of **Wright**, **Meier**, and **Moore** have different strengths, signifying the phenomenon of the degree of style.

The degree of style is not only influenced by the quantity of the critical common features, but also by their quality. A style is judged by the features perceived, and two factors determine how the critical common features can be perceived: (1) the size of the features in an object, and (2) the significance of perceptibility. The size of a feature correlates to its dimensions in proportion to the entire object. Large-sized features will attract more attention than small-sized features. On the other hand, the visual significance of perceptibility relates to the complexity and the visual impact of the feature's shape. Some features are more appealing and attractive than others. Interesting features are more easily visualized, and a style that has such features is easier to memorize and recognize. For instance, **Picasso**'s cubic style with strong, bold, and heavy color and the Rococo building style with curvilinear shapes are easier to identify than **Pierre-Auguste Renoir**'s style of Impressionism (1841–1919) and the Modern International building style. Therefore, critical common features should have evaluation weight and can be used to rank how critical they are for recognizing a style.

The more perceptible a feature is, the easier it is to be located and the more easily its represented style can be identified. This suggests variations in perceiving and expressing styles. Such variations may occur because the critical common features of a style appear in objects in different quantities and combinations: one object consists of certain features drawn from the set, while different objects of the same style have different combinations of features from the same set. For example, incorporating four features from Wright's common set into a residence design would characterize the Prairie Style. However an observer's ability to perceive the style depends upon the specific features applied. Based on the notion of strength or degree of style, the combination of a low-hip roof, a band of casement windows, a continuous band of sill, and an extended terrace with low parapet would be more perceptible as Prairie Style than the combination of planting urn, massive brick chimney, corner blocks, and watertable. In sum, different combinations of features yield different expressions and perceptibility: if the stylistic features in A are stronger than the same number of features in B, its style is stronger than its counterpart. The same is true for objects in the same style as well as the objects across styles. For instance, four strong features from Wright's set could be more easily identified than four weak features from Meier's set.

Taking the number of times a style can be recognized to represent perceptibility, the degree of style in a style and between styles is diagrammed in Figs. 4.28 and 4.29. Figure 4.28 is set up under the hypothesis that while objects A and B both represent the same style, A with n number of strong features would be recognized more frequently than B with the same number of weak features. For example, three features having strong texture, color, and curvilinear shape will be more easily identified than three flat and monotonous features. Figure 4.29 shows the degree of style between styles. If the object A representing style A′ has less significant critical common features than the object B of style B′, A might be less perceptible and A′ is a weaker style. On the other hand, if object C of style C′ has a greater number of features than the object D of style D′, but they are weaker features, object C may be less perceptible and its style C′ is probably weaker than style D′. Therefore, the degree of style is determined not only by the quantity but also by the quality of features.

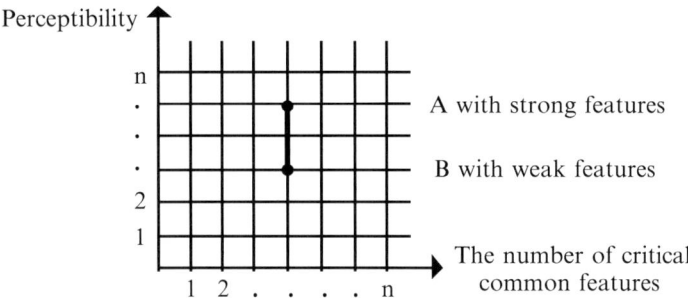

Fig. 4.28 The degree of style between objects with strong and weak features (Source: Reprinted from Chan (2000) Figure 8, p. 289, Copyright (2000), with permission from Elsevier)

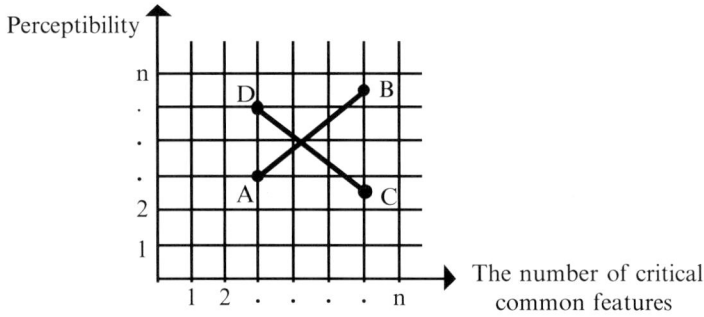

Fig. 4.29 The degree of style between styles (Source: Reprinted from Chan (2000) Figure 9, p. 289, Copyright (2000), with permission from Elsevier)

4.8   Conclusions

The series of experiments provide some understanding of the essential nature of style. Similar analyses of architectural style as determined by sets of features have been done by **Atkin** (1974, 1975) using Q-analysis to sort out the clusters of features that characterize Tudor-style architecture in Lavenham (England). One of the criticisms of that work (Pinkava 1981; Couclelis 1983) is that mere sets of features do not make a style, because features must also exist within a characteristic context (topological structure). Experiment 4 shows that any topological change will distort a style, and the style will not be recognizable anymore. Therefore, the characteristic context also plays an important role for the identification of a style. This is especially important for the study of historical, regional, and group style. Quite often, features from many different styles and periods may co-exist in an untypical spatial relationship to each other, as in these days of eclectic and post-modern architecture.

Concepts developed in this chapter focus on the basic notions of style in fine arts, with the exception of discussing style in performing arts, which relates to the issues of patterns of behavior. Similar methods of identifying common features across objects to label a style can be used to study painting, sculpture, furniture, interior design, and architectural design styles. In painting, for example, **Johannes Vermeer** (1632–1675), a famous Dutch painter, applied figures in seventeenth-century rooms with combinations of light, color, and proportion of objects in the room. Features appearing repeatedly in his 35 certified paintings[10] (12 pictures are shown in Fig. 4.30) include: light sources come from the upper left corner; black-and-white chessboard

---

[10] Names of the 35 certified paintings of Johannes Vermeer and the resources holding the collections are itemized chronologically on the following list. Names of paintings in bold font are included in Fig. 4.30 for reference. Most of them had been displayed and could be found on the Internet Wikimedia public domain Web pages. They are all accessed on 10 Oct 2013.

1. *Christ in the House of Martha and Mary* – Edinburgh, National Gallery of Scotland – 1654/55
2. *Saint Praxidis* – Private Collection – 1655
3. *Diana and her Companions* – The Hague, Mauritshuis – 1655–56
4. *The Procuress* – Dresden, GemÄ¤ldegalerie – 1656
5. ***Girl reading a Letter at an Open Window*** – **Dresden**, **Gemäldegalerie Alte Meister** – **1657**, http://en.wikipedia.org/wiki/File:Jan_Vermeer_-_Girl_Reading_a_Letter_at_an_Open_Window.JPG.
6. *A Girl Asleep* – New York, Metropolitan Museum – 1657
7. *The Little Street* – Amsterdam, Rijksmuseum – 1657/58
8. ***Officer and a Laughing Girl*** – **New York**, **Frick Collection** – **1658**, http://en.wikipedia.org/wiki/File:Jan_Vermeer_van_Delft_023.jpg.
9. ***The Milkmaid*** – **Amsterdam**, **Rijksmuseum** – **1658/60**, http://en.wikipedia.org/wiki/File:Vermeer_-_The_Milkmaid.jpg.
10. *The Glass of Wine* – Berlin, Gemäldegalerie Alte Meister – 1658/60
11. *The Girl with the Wineglass* – Braunschweig, Herzog Anton Ulrich Museum – 1659/60
12. *Girl Interrupted at her Music* – New York, Frick Collection – 1660/61
13. *View of Delft* – The Hague, Mauritshuis – 1660/61
14. *Woman in Blue reading a Letter* – Amsterdam, Rijksmuseum – 1662/64
15. *A Lady writing a Letter* – **Washington DC**, **National Gallery of Art** – **1662/64**, http://en.wikipedia.org/wiki/File:DublinVermeer.jpg.
16. *The Music Lesson* – **London**, **Buckingham Palace** – **1662/65**, http://en.wikipedia.org/wiki/File:Jan_Vermeer_van_Delft_014.jpg.
17. *Woman with a Lute* – New York, Metropolitan Museum – 1663
18. *Woman with a Pearl Necklace* – Berlin, Gemaldegalerie – 1664
19. ***Woman with a Water Jug*** – **New York**, **Metropolitan Museum** – **1664–65**, http://en.wikipedia.org/wiki/File:Jan_Vermeer_van_Delft_019.jpg.
20. *The Girl with a Pearl Earring* – The Hague, Mauritshuis – 1665
21. *The Concert* – **Boston**, **Isabella Stewart Gardner Museum** – **1665/66**, http://en.wikipedia.org/wiki/File:Vermeer_The_concert.JPG.
22. *A Woman Holding a Balance* – **Washington DC**, **National Gallery** – **1665/66**, http://commons.wikimedia.org/wiki/File:Johannes_Vermeer_-_Woman_Holding_a_Balance_-_Google_Art_Project.jpg
23. *Portrait of a Young Woman* – New York, Metropolitan Museum – 1666/67
24. *The Allegory of Painting* – **Vienna**, **Kunsthistorisches Museum** – **1666/67**, http://en.wikipedia.org/wiki/File:Jan_Vermeer_van_Delft_011.jpg.
25. *Mistress and Maid* – New York, Frick Collection – 1667/68
26. *The Astronomer* – **Paris**, **Louvre** – **1668**, http://en.wikipedia.org/wiki/File:JohannesVermeer-TheAstronomer(1668).jpg.
27. *Girl with a Red Hat* – Washington, National Gallery – 1668

1.Girl at Window, 1657.

2.Officer and a Laughing Girl, 1658.

3. The Milkmaid, 1658/60

4. Lady Writing a Letter, 1662/64.

5. Music Lesson, 1662/65.

6.Woman with a Water Jug, 1664/65.

7.The Concert, 1665/66.

8. Woman Holding a Balance, 1665/66.

9.Allegory of Painting, 1666/67.

10.Geographer, 1668/69.

11. The Astronomer, 1668.

12.The Love Letter, 1669.

Fig. 4.30 Examples of paintings by Johannes Vermeer

patterned marble floor tile, figures have their own visual focus point in the painting, windows in the painting are located on the left side of the wall, nine out of ten have pictures on the back wall. These features can be used to define the set of critical common features signifying **Vermeer**'s style.[11] Similar methods of identifying constant features in objects have been used to study the product style in industrial designs (Chuang and Chen 1994).

Of course, features can change over time due to changes in social context, convention, custom, knowledge, mental image, and personal preference. For instance, **Wright**'s style changed from his earliest residential design (Bootlegged Houses), which he developed while employed by **Louis Sullivan** from 1889 to 1894 (Manson 1958), to the Prairie houses, to the Usonian houses (Storrer 1978) throughout his career. Each stylistic period shared some different sets of features. Changes do occur with special intentions. Thus, the change of the set of common features over time marks the change of an artist's individual style. The more new features that can be created by a designer who puts them into different organizational syntax will signify more styles generated by the same designer over time. These changes of style and emergence of new common features can be used as an index to mark creativity and as a scale to measure the degree of creativity of an individual designer.

Research conclusions on style in this chapter drawn from experiments could be synthesized in short by experiments into the following four major points: Experiment 1 demonstrates that a style is represented by a common set of features that appear in objects. This set of common features defines a style; Experiment 2 explains the notion of the degree of style. The more features appearing in objects, the higher degree their style, which can be recognized easily; Experiment 3 shows that style is measured by the number of features present in a given set of artifacts. If the number of features drawn from a common set is fewer than three, then beholders will not be able to recognize the style of the artifact, and the recognizability of style is not measurable; Experiment 4 shows the tolerance of geometric change of a feature design. If a feature drawn from a common set has been geometrically distorted horizontally or vertically up to approximately 40 %, it is still recognizable as the representative stylistic feature. Distorted beyond this level, a feature should not be considered as representative of a style.

Finally, conclusions can be condensed into two sentences to explain the operational concepts of style: (1) if an artifact consists of at least three recognizable

---

28. *The Geographer* – **Frankfurt am Main, Steadelsches Kunstinstitut** – **1668/69**, http://en.wikipedia.org/wiki/File:Jan_Vermeer_van_Delft_009.jpg.
29. *The Lacemaker* – Paris, Louvre – 1669/70
30. *The Love Letter* – **Amsterdam, Rijksmuseum** – **1669/70**, http://en.wikipedia.org/wiki/File:The_Love_Letter_Vermeer.jpg.
31. *Lady writing a Letter with her Maid* – Blessington, Beit Collection – 1670
32. *The Allegory of Faith* – New York, Metropolitan Museum – 1671/74
33. *The Guitar Player* – London, Iveagh Bequest – 1672
34. *Lady Standing at the Virginals* – London, National Gallery – 1673/75
35. *Lady Seated at the Virginals* – London, National Gallery – 1673/75

[11] Of course, there certainly are other features existing in the pictures relating to the use of color and techniques of stroke.

features (result from Experiment 3) within at least one topological structure (result from Experiment 4), then a style exists in this artifact; (2) if there are four features (result from Experiment 2) with the same topological structure(s) replicated in at least three different artifacts (results from Experiments 1 and 4), then these four features and the topological structure(s) represent an individual style. In some cases, some objects have more members from the set of common features than others. For instance, 10 features appearing in a design will more strongly suggest a style than having five features present. Thus, the number of features appearing in an object affects how easily a style can be recognized, which suggests the notion of *perceptibility*. In other words, greater or fewer numbers of features existing in an object will alter the perception of a particular style, which signifies the degree of *style within a style* class. In the same way, ten common features appearing in design objects will more strongly suggest the style than five common features. The number of the set of critical common features indicates the notion of *expressiveness*. Greater or fewer members in the set appearing in a group of objects will affect the expression of styles, which denotes the degree *between style* classes. All in all, style is represented and measured by features to show the degree of style, which signifies the strength or weakness of a style.

References

Ackerman JS (1963) Style. In: Ackerman JS, Carpenter R (eds) Art and archaeology. Prentice-Hall, Englewood Cliffs, pp 174–186

Alexander C (1963) The determination of components for an Indian village. In: Jones JC, Thornley DG (eds) Conference on design methods. Pergamon Press, Oxford, pp 83–114

Allen G (1980) Charles Moore: monographs on contemporary architecture. Whitney Library of Design, New York

Atkin RH (1974) Mathematical structure in human affairs. Heinemann Educational Books, London

Atkin RH (1975) An approach to structure in architectural and urban design. Environ Plann B 2:21–57

Beazley M (1988) The world atlas of architecture. Portland House, New York

Bloomer KC, Moore CW (1977) Body, memory, and architecture. Yale University Press, New Haven

Brooks HA (1984) Frank Lloyd Wright and the prairie school. George Braziller, New York, p 54

Brunner JS, Goodnow JJ, Austin GA (1956) A study of thinking. Wiley, New York

Chan CS (1992) Exploring individual style through Wright's design. J Archit Plann Res 9(3):207–238

Chan CS (1993) How an individual style is generated? Environ Plann B 20(4):391–423

Chan CS (1994) Operational definition of style. Environ Plann B 21(2):223–246

Chan CS (2000) Can style be measured? Des Stud 21(3):277–291

Chan CS (2001) An examination of the forces that generate a style. Des Stud 22(4):319–346

Chen K, Owen C (1997) Form language and style description. Des Stud 18:249–274

Chuang MC, Chen CC (1994) Exploring the perception and recognition of the eastern and western style: Using chairs as examples. In: Proceedings of the 1994 conference on technology & teaching – industrial design section, Taiwan (in Chinese)

Couclelis H (1983) On some problems in defining sets for Q-analysis. Environ Plann B 10(4):423–438

Dunster D (1979) Michael Graves, architectural monographs 5. Academy Editions, London, p 38

Gill B (1987) Many masks, a life of Frank Lloyd Wright. Putnam, New York

Gombrich EH (1960) Art and illusion: a study in the psychology of pictorial representation. Pantheon, New York

Gombrich EH (1968) Style. In: Sills DL (ed) International encyclopedia of the social sciences. Macmillan, New York, pp 352–361

Greene TM (1940) The arts and the art of criticism. Princeton University Press, Princeton

Hampton J, Dubois D (1993) Psychological models of concepts: introduction. In: Mechelen IV, Hampton J, Michalski RS, Theuns P (eds) Categories and concepts: theoretical views and inductive data analysis. Academic, London, pp 11–33

Hayes-Roth B, Hayes-Roth F (1977) Concept learning and the recognition and classification of examples. J Verbal Learn Verbal Behav 16:321–338

Hitchcock H (1942) In the nature of materials: 1887–1941; the buildings of Frank Lloyd Wright. Duell, Sloan and Pearce, New York

Jencks C (1977) The language of post-modern architecture. Rizzoli, New York, p 80

Johnson SC (1967) Hierarchical clustering schemes. Psychometrika 32:241–254

Johnson E (1986) Charles Moore: buildings and projects. 1949–1986. Rizzoli, New York

Jordan RF (1969) A concise history of western architecture. Harcourt Brace Jovanovich Inc., London

Kroeber A (1957) Style and civilizations. Cornell University Press, Ithaca

Kroeber A (1963) An anthropologist looks at history. University of California Press, Berkeley

Kurtz KH (1966) Foundations of psychological research: statistics, methodology, and measurement. Allyn and Bacon, Boston

Lakoff G, Johnson M (1980) The metaphorical structure of the human conceptual system. Cognit Sci 4(2):195–208

Littlejohn D (1984) Architect, the life and work of Charles W Moore. Holt, Rinehart and Winston, New York

Manson GC (1958) Frank Lloyd Wright to 1910: the first golden age. Van Nostrand Reinhold, New York, p 122

Meier R (1976) Richard Meier, architect, buildings and projects, 1966–1976. Oxford University Press, New York, p 21

Meier R (1984) Richard Meier, architect, 1964/1984. Rizzoli, New York

Moos S (1987) Venturi, Rauch, & Scott Brown building and project. Rizzoli, New York, p 257

Museum of Modern Art (1975) Five architects. Oxford University Press, New York

Newton E (1957) Style and vision in art. The Listener 57:467–469

Pinkava V (1981) Classification in medical diagnostics: on some limitations of Q-analysis. Int J Man-Mach Stud 15:221–237

Posner MI, Keele SW (1968) On the genesis of abstract ideas. J Exp Psychol 77:353–363

Pothorn H (1982) Architectural styles: an historical guide to world design. Facts On File, New York

Reed SK, Friedman MP (1973) Perceptual vs. conceptual categorization. Mem Cognit 1:157–163

Rosch E (1973) Natural categories. Cogn Psychol 4:328–350

Rosch E, Mervis CB (1975) Family resemblances: studies in the internal structure of categories. Cogn Psychol 7:573–605

Rosch E, Mervis CB, Gray WD, Johnson DM, Boyes-Braem P (1976) Basic objects in natural categories. Cogn Psychol 8:382–440

Schapiro M (1962) Style. In: Tax S (ed) Anthropology today: selections. University of Chicago Press, Chicago, pp 278–303

Schwarting JM (1984) Teaching style: the first term at Columbia. J Grad School Archit Plann 5:7–24

Shepard RN (1974) Representation of structure in similarity data: problems and prospects. Psychometrika 39:373–421

Simon HA (1975) Style in design. In: Archea J, Eastman C (eds) Proceedings of the 2nd annual environmental design research association conference. Dowden, Hutchinson & Ross, Stroudsburg, pp 1–10

Smith EE (1989) Concepts and induction. In: Posner MI (ed) Foundations of cognitive science. MIT Press, Cambridge, pp 501–526

Smith EE, Medin DL (1981) Categories and concept. Harvard University Press, Cambridge

Smithies KM (1981) Principles of design in architecture. Van Nostrand Reinhold, New York

Sparshott F (1965) The structure of aesthetics. University of Toronto Press, Toronto

Sprague P (1976) Guide to Frank Lloyd Wright and Prairie School architecture in Oak Park. Oak Park Bicentennial Commission of the American Revolution, Oak Park, p 87

Storrer WA (1978) The architecture of Frank Lloyd Wright. MIT Press, Cambridge

Tovey M (1997) Styling and design: intuition and analysis in industrial design. Des Stud 18:5–31

Tversky A (1977) Features of similarity. Psychol Rev 84:327–352

Chapter 5
Style Approached from the Design Process

Studies of style can be approached from two directions: the end and the means. From the end point of view, a style is a cluster of features present in artifacts; scholars usually classify the features in products to differentiate styles (Newton 1957; Finch 1974; Scott 1980; Smithies 1981; Chan 1994, 2000). Similar approaches used to examine features for further exploring the nature of style, the degree between styles, and the systematic measurement within style were extensively covered in Chap. 4. From the means point of view, a style is a mode by which designers' personal and professional preferences are expressed, and studies attempt to deliberate the mode of expression to mark styles (Torossian 1937; Evans 1982; Cleaver 1985). Although most style researchers have studied both directions, their efforts cannot provide clear explanations of how a style is generated. That is because not enough research has been devoted to the study of the means used that create a style. This chapter begins to explore, through a case study, the aspects of style creation and the forces that generate a style (Chan 1995, 2001). Studies of style approached from the means point of view conducted in various fields and the factors determining the generation of style are reviewed first.

5.1 Studies Approached from the Means Point of View

As explained in the beginning of Chap. 3, art is highly abstract, conceptual, and metaphysical, and it is examined by philosophers and scholars as a means to differentiate between artists, schools and periods. Thus, style is a norm developed since Plato for studying art. Since then, the concept of style has exerted an influence on the discipline of art history. The field of art history involves the study of art in its historical development and stylistic context, and was developed in the West to study European art. Studies in this field, in short, attempt to explain artifacts using the cultural values in which the artifacts were created. Thus, style has been used in art

© Springer International Publishing Switzerland 2015   157
C.-S. Chan, *Style and Creativity in Design*, Studies in Applied Philosophy,
Epistemology and Rational Ethics 17, DOI 10.1007/978-3-319-14017-9_5

history to identify the cultural connections in artifacts, and the fundamental nature of art revealed in artifacts.

After a long period of theoretical development in fine arts research, style has been utilized in the fields of painting, literature, architecture, sculpture, music and cinema to study form and expression extensively. Form relates to the final products, whereas expression relates to the mode of expression delivered and manifested through form. Only after the twentieth century did theorists begin to advocate that the study of art should not merely concentrate on either the forms in products or meanings expressed, but should also study the mode of expression executed through the creative process. The mode of expression is defined as the way that an inner peculiarity shows itself in the outward representation through characteristic marks (Wackernagel 1923). The creative process is any process issuing in the production of a work of art (Sparshott 1965). Since then, art critics, who criticize art in the context of aesthetics or the theory of beauty, together with art historians have attempted to answer the questions: (1) how did the artist come to create the artifact? (2) what are key features of the style associated with the artifact? (3) what meaning did the artifact communicate? (4) how does it function visually? (5) what symbols are involved? (6) did the artist meet the goals well? and, (7) does it function correctly? These questions all relate to the theme of the mode of expression utilized in the design process.

Since studies began exploring the relationships between the creative process and the production of art, or how a style is generated during the creative process (Wollheim 1979), there have been several discoveries (Whyte 1961; Sparshott 1965; Weitz 1970; Kubler 1979; Chan 1995). For example, Newton (1957, p. 467) indicated that "style is the outward manifestation of the artist's temperament." Schapiro (1962, p. 278) argued that by style is meant the constant form—and sometimes the constant elements, qualities, and expression—in the art of an individual or a group. He proposed that the description of a style should refer to three aspects: form elements or motives, form relationships, and qualities. Form elements or motives were essential for expression, yet alone they were not sufficient for characterizing a style. To distinguish styles, it is necessary to look for the different ways of combining elements, which are form relationships. From the aspect of quality, form elements and form relationships are intrinsically expressive, and they tend to constitute a coherent whole. Thus, a style is like a language, with an internal order and expressiveness, admitting a varied intensity or delicacy of statement. Furthermore, Ackerman (1963, 1967) explained that a distinguishable ensemble of the characteristics of: (1) conventions of form and of symbolism; (2) material; and (3) technique can be called a style. These concepts share the same territory in the study of forms that characterize an artistic style, and in the study of aesthetic principles or some internal orders that make an artifact beautiful. Even though these theories explain the fundamental concept of style, in order to understand the origin of common forms and how these forms are created, it is essential to look at the methods of creation in the creative process, defined by Sparshott (1965) that "style is a way of doing things" (1965, p. 99).

Gombrich, on the other hand, set up the further notion of style by means of **choices**. In his interpretation of arts approached from Gestalt psychology, he proposed a thorough theory to explain the phenomena occurring in an artist's creative processes as well as in a beholder's appreciation of arts. His concept can be roughly divided into two parts: style, and the explanations of an artist's creative processes (Gombrich 1960, 1968, 1971, 1982). He considered style to be "any distinctive, and therefore recognizable, way in which an act is performed or an artifact made or ought to be performed and made" (1968, p. 352), whereas the explanation of an artist's creative processes is the beholder's interpretations of the creative processes. His concept on style proposes that style arises from the act of choice, which is further defined by **Simon**: "A style is some one way of doing things, chosen from a number of alternative ways" (Simon 1975, p. 1). **Gombrich** and **Simon** both held the same belief that style should be confined to instances where there is a choice between ways of performance or procedures (Simon 1975), and only against the background of alternative choices can the distinctive way also be seen as expressive (Gombrich 1968). The central notion is that there are alternative choices, and the particular selection made from them can be seen as the artist's personal expression. But for **Gombrich**, the choices refer to (outer) choices of perceptual cues available for representing the external reality, whereas the choices for **Simon** are (inner) choices imposed in the design process.

Meyer (1979) has taken one more step to explain explicitly the nature of **choice**. His definition of style is "a replication of patterning, whether in human behavior or in the artifacts produced by human behavior, that results from a series of choices made within some set of constraints" (1979, p. 3). He indicated that the presence of choice is stipulated because there is replicated patterning in the world that would not normally be thought of as being stylistic. The word choice tends to suggest "conscious awareness and deliberate intent (1979, p. 4).

In anthropology, style has been defined as "a system of coherent ways of patterns of doing certain things" (Kroeber 1963, p. 66). **Kroeber** viewed civilizations and cultures as stylistic expressions. In music, **LaRue** indicated that "the style of a piece consists of the predominant choices of elements and procedures a composer makes in developing movement and shape" (1970, p. IX). **LaRue** argued that one can perceive or distinguish style in a group of pieces from the recurrent use of similar choices, and that a composer's style as a whole can be described in terms of consistent and changing preferences in his or her use of musical elements and procedures.

In the field of architectural design, **Simon** has suggested that design constraints are one of the influential factors on style. He indicates that constraints set up limitations on the degree of freedom of selections, and do most of the work of determining the solution in design (1975). **Akin** has further elaborated that "style is used to reduce the number of alternatives to be considered" (1986, p. 94) and "hence the method for determining the acceptability of a solution is governed by stylistic choices" (p. 96). All these notions in various fields emphasize that **choices** are a major factor determining a style in the process of creation.

The other factor determining a style that happens in the creative processes is the utilization of constraints for making choices. This refers to the concept of criticality used by **Rapoport** to represent relations between the degree of choice and freedom and the degree of constraints imposed into design. Rapoport's uses a house form to show the result of choice among existing possibilities. If the number of possibilities is greater, the choice is also greater, and the criticality is lower because of fewer constraints imposed. On the other hand, if the constraint is greater, the criticality is also greater and the degree of choice is less. Similarly, a rocket has greater criticality than an airplane because it is more severely constrained by technical requirements; slow-speed airplanes have a greater degree of freedom (i.e., lower criticality) than rapid ones (Rapoport 1969). As new building technologies become available, the criticality of design constraints diminishes and the extent of freedom in choices of alternative solutions increases.

The third factor determining the generation of style in the creative processes is the search order indicated by **Simon** (1975), who proposes that human designers have certain sets of procedures to determine which design unit, or design constraint, or goal to start with. Because design is a satisfying process, the first object that is taken into consideration will satisfy a specified set of constraints and yield a first satisfying solution. Any successive object taken into account next will be based on the first solution to generate a further solution that also is satisfying. Thus, the order in which possibilities are examined will have a major influence on the solution and ultimately casts an influence on forms.

The concept of search order can be interpreted as an analogy for the order of thinking processes. In writing, some scholars (Buffon 1923) believed that style is simply the order and movement one gives to one's thoughts. Thus, the style, or the order in which a man arranges his thoughts on the subject-matter springs from the man himself; the style is so much of the man as exists in the ordering of his thoughts. To present a natural style in writing, it is necessary for a writer to make a plan to collect and to put in order all the essential thoughts on his or her subject (Buffon 1923). The movement (order) of thoughts yields a source for a style in literature. However, these concepts are not able to explain the detailed mental processes creating a style. For an explicit elaboration of the phenomenon of style, including the causes and changes of style, one must focus on how designers process information.

5.2   Style Created in the Design Processes

When exploring style from the information processing point of view, it is necessary to discover how an individual style is generated from a series of design processes. A *design process* is defined as a sequence of states for achieving a goal. A process may create new processes, it may share some information with others, and it may allocate some of its information to others for operation as well. An architectural design consists of a number of such design processes, which are the procedures for handling information in the designer's mind.

Apart from any doctrine, style is operationally defined here as "**any distinctive and recognizable way of design that is repeatedly manipulated in design processes and, thus, generates certain common features across design products**." In this regard, a style is argued to be represented by some features in a product that result from certain mental activities. A work of art is not a natural product, but a thing brought into being by human activity, created by the mind. These mental activities can be an individual or group effort. Coming from an individual effort, it is termed individual style. For an individual style, distinctive features in a design work enable observers to visually link a design work with other works by the same architect or with other works by a different architect.

Regardless of whether style is generated from single or group efforts, mental activities and design decisions are practically operated in the design process. Thus, the study of style should ultimately concentrate on the investigations of the cognitive activities utilized in the design process. This content, however, differs from the cognitive style that refers to the characteristic ways in which individuals conceptually organize the environment (Goldstein and Blackman 1978) or the consistent patterns of organizing and processing information (Messick 1976). The differences are that the cognitive style emphasizes the cognitive structure—how cognition is organized—rather than the content of thought—which knowledge is available (Suedfeld 1971); whereas individual style in this chapter refers to the style expressed in art works, including architectural design.

Therefore, there are two groups of components involved in the overall definition of style defined in this chapter—the common features present in the design products, and the procedures and factors that repeatedly operate in the design processes to generate common features. The first component has been explained previously, whereas the second component of the definition refers to certain elements that can be observed and recognized in a number of design processes executed by the same designer. These elements include executing some fixed sequences of procedures in solving a design problem, using the same set of design constraints, and using certain geometric forms. Two examples are used to verify these notions. **Frank Lloyd Wright**'s Prairie House Style is the first one representing a strong style covered in this chapter. The second example is a weak style of a practicing architect studied from the perspective of design cognition analyzed in Chap. 6. The term *weak style* relates to the quantification of style but not qualification. Eventually, these examples should provide an overall understanding of style, approached from the design processing point of view.

5.3   Case Study: Prairie Houses Style

The fundamental notion starts from a premise that a design product is a function of a design process. Then following this notion and explaining from the computational aspect, executions of design processes would execute series of functions that would generate a number of design products. Thus, if a design process is repeated in different designs under similar contexts, then similar products will be generated. Here, the design process is not necessarily the entire design process, but a sub-process within any process. As long as the factors implemented in the design processes that create the features signify a style, the phenomenon of style creation can be explored. Frank Lloyd Wright is selected as a subject of study because: (1) Wright's designs possess abundant common features, which reflect a stronger degree of style (Chan 1990a, 1992); (2) Wright had by far the most individualistic style of his generation; and (3) Wright developed a complete vocabulary of his own. The study of his works helps demonstrate the concepts developed in this book. Also, the Prairie Style was Wright's first style, and the Prairie Houses were designed consecutively and intensively within a 10-year period. This period provides a relatively pure context for observations. The task is further aided by the fact that so much has been written by and about him, and thus far, more documentation exists for Wright than for most architects in the United States.

5.3.1 Background of Frank Lloyd Wright

Frank Lloyd Wright (1867–1959) can be considered as primarily a residential architect. During his 70 years of architectural production (1888–1959), from the earliest houses Wright was responsible for designing in the employ of **Louis Sullivan** to the buildings carried to completion by the **Taliesin** Associated Architects from Wright's extant designs after his death in 1959 (Storrer 1978), Wright designed 929 buildings, including designs actually built as well as unexecuted projects (Streich 1972). Among them, 653 are residential and 276 are non-residential (Streich 1972). Approximately 70 % of his designs are single-family residential. His residential designs during the years 1901–1910 have been called "Prairie Houses." The Prairie Houses shared some common prominent characteristics dissimilar to those in his late residential designs, the so-called Usonian houses. From 1902 to 1906, he also did three non-residential designs (Yahara Boat Club, Larkin Building, and Unity Temple) that have been discussed by many critics. These three non-residential designs also share certain common characteristics. In this case study, examples of Prairie Houses and these three non-residential designs have been selected as representative subjects of Wright's early design works.

5.3.2 Method of Data Collection

Methods on exploring an individual style can be approached by using three procedures. The first step is to collect many designs done by the same architect and to identify the replicated features in these designs. The second task is to trace constant design concepts underlying these constant features. This is accomplished by collecting data about design concepts explained by the architect, and observing how these concepts are implemented in his or her designs. By doing so, it is possible to get insight into the causes of form repetitions. The last step is to study whether some constant design methods or certain sequences of design processes are used by this architect, and to analyze whether these particular methods or processes yield particular characteristics in forms that mirror the features found in step one.

Data could be collected mainly from publications instead of psychological experiments. In most architectural periodicals, architectural designs are presented with photographs and illustrations of the project, together with the author's interpretation. Under these circumstances, it is not difficult to obtain pictures or drawings for a particular design. It is, however, a bit difficult to get data about the design processes, except through protocol analysis, because architects seldom explain their design processes or design methods. If they do so, their explanations given after design are retrospections rather than reflections of what may have actually happened while they were designing. Thus, their given explanations of design processes or methods should be seen as descriptions of their "general" ways of doing things. Therefore, after collecting information available from critics and from the designer's writings and speeches, a picture of an architect's general design processes, methods, and the corresponding features generated will appear.

It is debatable to say that putting all the pieces of second-hand information from scholarly reports together would generate a perfect picture to portray a person's design creation, and such a picture could really represent a true story of a design style. Especially, each design is unique in its own way and a designer's design thinking does change from case to case and from time to time. Therefore, there are variations of design generation from project to project, and pictures of design principles behind each project are also difficult to be identified by anyone other than the creator. Yet, the purpose of such an approach is to use a story to explain a cognitive phenomenon of style. As long as the collected information, evaluated by experts who are specialized in **Wright**'s design and are verified as true, then it is justifiable and legitimate to say that the stories told are reliable to certain degree in representing a cognitive phenomenon.

5.3.3   Replicated Features Appeared in Design

The replicated features appearing in Wright's designs are first classified into residential and non-residential groups. The residential group refers to Prairie Houses designed from 1901 to 1910, whereas the non-residential group includes the Yahara Boat Club, Unity Temple, and the Larkin Building designed during the same period, from 1902 to 1906. Features commonly appearing in the Prairie Houses can be categorized by groups of floor plan, elevation, wall, materials, and structures. These features relate to the context of layout, spatial configurations, particular architectonic elements utilized, and special methods used for form generation. Six examples of floor plan and elevation drawings are reproduced in Figs. 5.1 and 5.2 for reference.

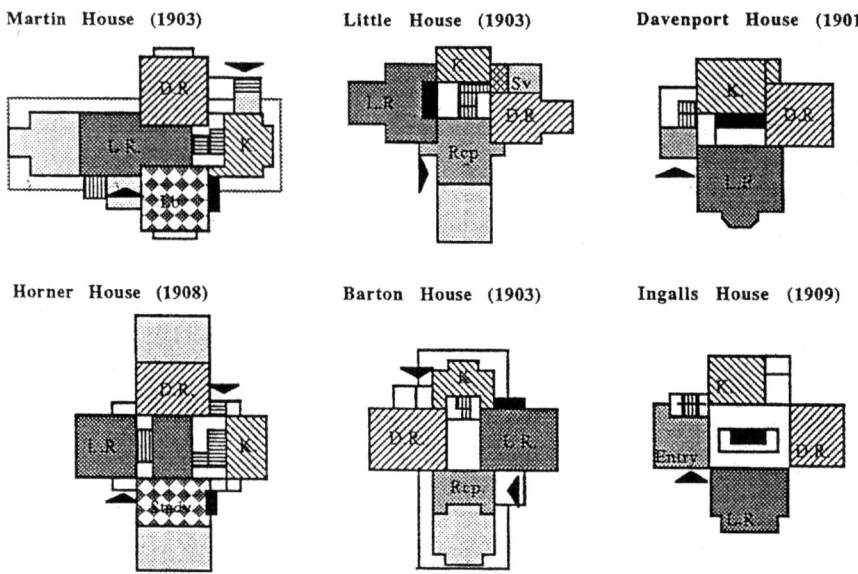

Fig. 5.1 Examples of Prairie Houses design with similar patterns (Source: Copyright Locke Science Publishing Company, Inc. Figure 13, p. 229, reproduced with permission. Chan (1992))

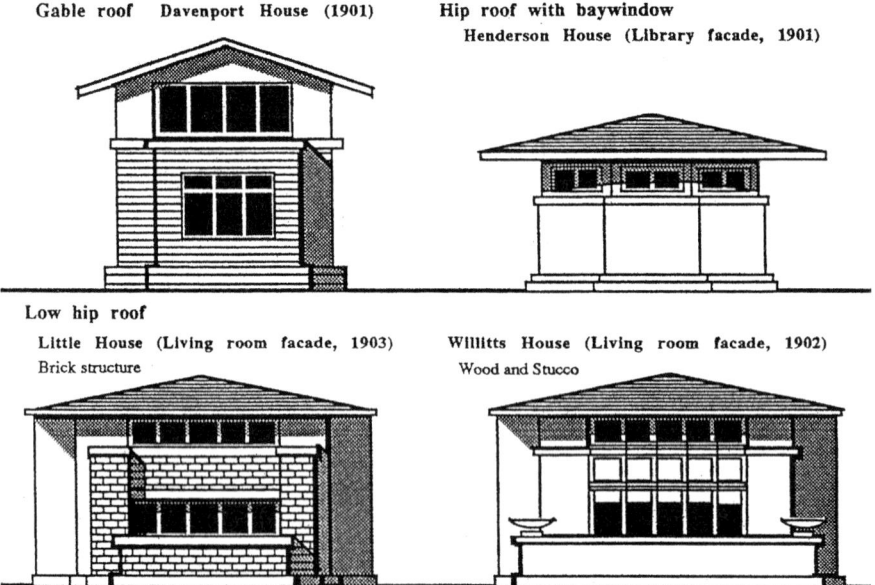

Fig. 5.2 Examples of Wright's elevations (Source: Copyright Locke Science Publishing Company, Inc. Figure 1, p. 211, reproduced with permission. Chan (1992))

5.3.3.1   Features of Residential Design in Prairie Houses (1901–1910)

Floor Plan

(a) At the heart of every Prairie House there was a fireplace, often of brick, always broad and firmly anchored at the center of the composition. From its hearth all spaces would extend, radiating throughout the site.
(b) The principal spaces—living room, dining room, kitchen and entrance hall—each occupied one axis and usually comprised a cross plan (see Fig. 5.1). The servants' quarters were located next to the kitchen, which often was sealed off from the living area.
(c) No underground basement (unless a client asked for it), no attic.
(d) Interior walls were extended into the landscape to form terraces and courts outside the line of the sill. These terraces and courts were surrounded by low parapets to prevent direct access from outdoors.
(e) There was a major shape in the plan that was long and narrow. This elongated shape led to stringing out rooms in their longest possible form, creating an extended horizontal line. Therefore, much of a Wright house was only one room in depth. Rooms, hence, have light from both sides, cross ventilation, large exposure to the outside, and access to different parts of the site.

Elevations

(a) The roof overhangs were deep and thin-edged.
(b) The windows, often casement windows opening out, were shaped as continuous ribbons of glass, starting directly at the underside of the roof and continuing down to a common, horizontal sill line.
(c) The houses had a base, called a water table by Wright. They were surrounded with gradually descending terraces that finally brought the floor line down to natural grade.
(d) The parapets of the terraces had continuous horizontal copings of concrete or limestone that often would become planting boxes or planting urns.
(e) There would be a central element somewhat taller than the rest—a bedroom area or sometimes a two-story living room. From this tall, central mass, wings would extend in all directions—first roof planes, then the wall, the planes of parapets or porches, and finally the low slabs of terraces—so that the entire house had several horizontal elements or planes.

Walls

(a) Walls were screens or vertical planes, never the sides of a box. Often there were corner windows.

(b) There was a horizontal band inside at the height of a door opening. This band continued all the way around the walls and finally emerged on the exterior as the roof fascia. All added heights were developed above this low plane, so that many rooms would have low ceilinged areas (usually around the fireplace).

Materials and Structures

(a) Most houses either were built of brick or were plaster-surfaced, wood-trimmed structures.
(b) Oak was used in the floors, doors, and wooden trim, as well as in the furniture, window frames, and lighting fixtures.

5.3.3.2   Features of Non-residential Designs (1902–1906)

The Yahara Boat Club project was never erected. The Larkin Building and Unity Temple are two designs of outstanding importance done during the period of the Prairie Houses design stage.

1. Yahara Boat Club (Madison, Wisconsin, 1902): It consisted of a simple, rectangular block, topped by a long band of glass, and finished off with a deeply cantilevered, flat, slablike roof. The entire composition sat on an extended base of retaining walls (Blake 1960, pp. 325–326). The horizontal character that dominated Prairie Houses was found in the horizontal roof plane and the long side walls (see Fig. 5.3).
2. Larkin Building (Buffalo, New York, 1904): It was as vertical as the Boat Club was horizontal. The exterior and interior materials were both of brick, with floors and ceilings of concrete. The inner court was lit by a skylight

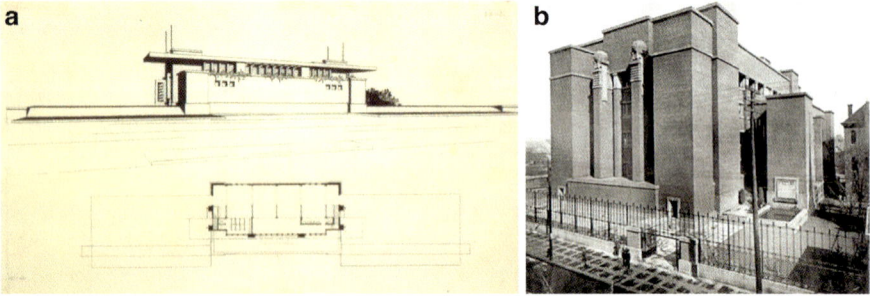

Fig. 5.3 Perspectives of Yahara Boat Club and Larkin Building. Perspective drawing of Yahara Boat Club in (**a**) Source: © The Frank Lloyd Wright Foundation, AZ/Art Resource, NY. Photo of (**b**) Source: http://en.wikipedia.org/wiki/Larkin_Administration_Building. Accessed 10 Oct 2013

Fig. 5.4 Perspective view of Unity Temple (Source: Aude, http://commons.wikimedia.org/wiki/ File:Oak_park_unity_temple.jpg. Accessed 10 Oct 2013)

above. The principles of the Prairie House, in which a solid core is surrounded by spaces, were dismissed here. Stairs looking like towers were located in the corners (see Fig. 5.3).
3. Unity Temple (Oak Park, Illinois, 1906): The entire building was of poured concrete. It was not horizontal in emphasis. It had a solid base, topped by a band of windows that in turn, were held down by the projecting lid of a flat roof slab. The volumes were vertical. Skylights were used to light the central space. Stairs, looking like massive blocks, were in the four corners of the building (see Fig. 5.4).

Common features found in these three projects included a symmetric floor plan, flat roof, and massive corner blocks. Factors that generated these features in **Wright**'s designs during this period are discussed in the following sections.

5.4   Constant Concepts and Principles Used in Design

With common features identified, the methodologies and thinking patterns applied in the design projects completed during this Prairie Houses period are the next task of investigation. These methodologies include design concepts, principles, constraints, or design elements applied for the generation of the identified representative common features. The constant concepts and principles of methodologies used by **Wright** are listed below.

Design Concepts: Design concepts are general and abstract ideas about how to design a part of a building or the building as a whole. Wright wrote in his autobiography: "I had an idea that the horizontal planes in buildings, those planes parallel to earth, identify themselves with the ground—make the building belong to the ground. I began putting this idea to work" (Wright 1943, p. 140). This was his retrospective view of one of the design concepts developed originally for his Prairie Houses. The concept, in short, is the expression of horizontality. **Wright** called it the Earth Line and associated it with emotional qualities, identification with the earth, stability, and shelter (Blake 1960). Therefore, a design concept, explained from a design cognition (see Chap. 2) point of view, is a piece of knowledge that is abstract and sometimes metaphysical. This example also explains that Wright had associated the notion of the horizon to the earth for creating the character of horizontality.

Design Principles: A design concept could become a *design principle*, defined as the consistent application of certain design concepts in a designer's works. For instance, the concept of horizontality is categorized as one of **Wright**'s design principles because it has been used in all Prairie Houses. Design principles also can be seen as several design concepts combined together and used in many designs. They are fundamental guidelines and are philosophical in nature. For example, **Wright** formulated six design concepts, which he called six propositions, early in 1894 and first published them in 1908 (Wright 1908). These propositions were to: (1) show simplicity; (2) achieve individuality of houses; (3) have a building grown from its site; (4) express color schemes from nature; (5) show the nature of materials; and (6) have the character of a house. In 1931, he mentioned the concepts of Prairie Houses again and listed nine items, which he called motives and indications (Wright 1931). These motives were to: (1) achieve simplicity; (2) associate the building with its site; (3) eliminate the room as a box; (4) set houses on a platform; (5) achieve plasticity; (6) use a mono-material; (7) incorporate heating, lighting, and plumbing as a whole system; (8) incorporate furnishings with the building; and (9) eliminate the decorator. Later, in his 1932 autobiography (first edition), he once again mentioned three concepts of Prairie Houses—simplicity, plasticity, and the nature of materials. Among his publications of 1908, 1931, and 1932, the concepts of simplicity, plasticity, and the nature of materials are repeatedly emphasized and hence, illustrate **Wright**'s design principles for Prairie Houses. As expressed in his autobiography, his explanations on these three principles are as follows (Wright 1943, pp. 144–149).

1. **Simplicity**. Wright said, "One must achieve simplicity as a perfectly realized part of some organic whole. Only as a feature or any part becomes a harmonious element in the harmonious whole does it arrive at the state of simplicity" (Wright 1943). One of his apprentices in the Oak Park office, Charles E. White, Jr., wrote in 1904 that "[Wright's] tendency of the last 2 years (1902–1904) has been to simplify and reduce to the 'lowest elements' in his

design" (White 1971, p. 105). This shows that Wright had tried to achieve this design principle in his early works.

2. **Plasticity and continuity**. On plasticity and continuity, Wright said, "plasticity may be seen in the expressive flesh-covering of the skeleton as contrasted with the articulation of the skeleton itself… In my work the idea of plasticity may now be seen as the element of continuity… Let walls, ceilings, floors become seen as component parts of each other, their surfaces flowing into each other… I have since concentrated on plasticity as physical continuity, using it as a practical working principle within the very nature of the building itself in the effort to accomplish this great thing called architecture"(Wright 1943, pp. 146–147).

3. **The nature of materials**. Prairie Houses might be grouped in various ways— by plan, by size and cost, by location, by materials, or even by roof types. But basic to Wright's architecture was his feeling for the nature of materials. He said, "bring out the nature of the materials, let their nature intimately into your scheme. Strip the wood of varnish and let it alone; stain it. Develop the natural texture of the plastering and stain it. Reveal the nature of the wood, plaster, brick or stone in your designs; they are all by nature friendly and beautiful" (Wright 1908, p. 55). He also said that "the materials of which the building is built will go far to determine its appropriate mass, its outline and, especially, proportion. The expression of structure as a pattern must be true to the nature of the materials out of which it was made" (Wright 1943, p. 345). Hence, the innate characteristics of the materials became a principle medium of architectural expression.

5.5   Constant Constraints Used in Design

Design constraints are more specific information used in design. They are defined as certain functional requirements that must be fulfilled in designing a design unit or a group of design units. They also can be defined in a negative sense as certain requirements that must be avoided. Thus, design constraints represent design rules, relations, conventions, structural properties, natural laws, and building codes to be considered in a design.

On the other hand, design concepts, principles, and constraints also can be explained by the level of abstraction. For example, concepts and principles are more abstract than constraints. But concepts could be ideas developed for a particular design, whereas principles are groups of concepts that are general in character and are used in different types of designs. Design constraints are more specific, and have particular attributes and values. The following constraints that **Wright** used in designing the floor plans of Prairie Houses are identified from his writings.

5.5.1   Constraints Used in Making Floor Plans

Wright (1928) mentioned that the several factors most important in making the plans, after general purpose or scheme or "project" was decided, were materials, building methods, scale, articulation,[1] and expression or style. These six factors can be seen as six design constraints he used while he was working on floor plans.

According to **Wright**, the general purpose of the building came before anything. Then, among the other five factors, scale was the most important one, which dominated the making of plans. Then, materials determined the structural method and consequently resolved and affected scale. Different building methods yielded different forms and shapes, and ultimately shaped the plans. For articulation, he argued that each separate portion of the building devoted to a special purpose should assert itself as an individual factor (unit) in the whole (building). Finally, after all was set, the architect emphasized what he loved, and that was expression.

To **Wright**, scale was tightly related to three other factors: human proportions, the nature of materials, and the method of building. All together, they determined the overall shape of plan. The articulation and expression were subordinate to these three factors. Hence, a certain priority existed while **Wright** considered these design factors.

According to **Manson**, "whenever the client could afford to give him carte blanche, he thought in terms of brick and stone, but he had, more often, to work out his ideas in wood and stucco… These things are all, however, variations on a single theme (which is symmetry)" (Manson 1958, p. 111). Here, **Manson** reported that **Wright** often would use brick and stone as materials, and another design constraint, symmetry. The example given by **Manson** is **Willitts** House (1902). In Willitts House, **Manson** indicated that **Wright** had established the precedent for Prairie Houses with symmetrical wings.

Although these six factors (constraints) are interrelated and complicated in nature, **Wright** had a method for handling them. His method was to use a **unit-system** by which all the factors (constraints) are addressed and tied together. On the basis of the unit system, he would then develop a geometric pattern for the plan. Thus an integrated floor plan was anticipated.

5.5.2   Constraints Used to Determine the Wall Location

Several scholars indicate that Wright's major innovation was to design interior spaces that were not enclosed in the traditional sense. This is the so-called "destruction of the box" (Scully 1960; Brooks 1984). For example, in the **Ross House** (see

---

[1] Articulation can be seen as certain specific and unique expressions shown in form, which could be specially arranged patterns or structural features.

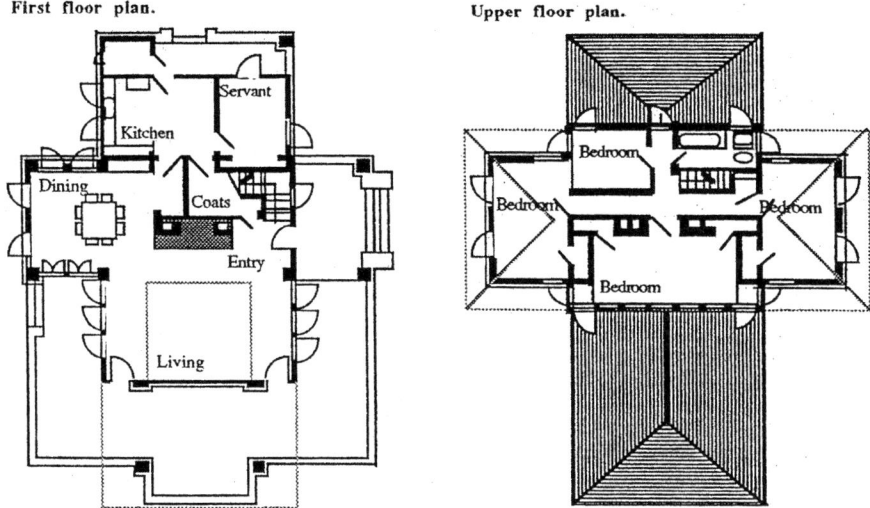

Fig. 5.5 Floor plans of Charles S. Ross House, Wisconsin, 1902 (Source: Copyright Locke Science Publishing Company, Inc. Figure 2, p. 215, reproduced with permission. Chan (1992))

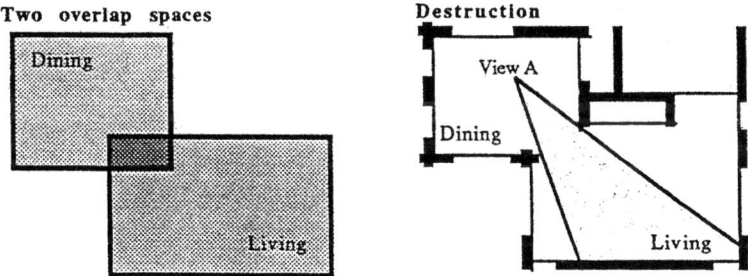

Fig. 5.6 Destruction of the box (Source: Copyright Locke Science Publishing Company, Inc. Figure 3, p. 216, reproduced with permission. Chan (1992))

Fig. 5.5) and the **Willitts** House, the living and dining rooms overlapped at their corners. Wright created a diagonal view between rooms and then obliterated all corners. The dissolved corners became areas for circulation or for corner windows. The wall, which no longer served to connect the corners, became more like a screen that delimits or defines a space rather than enclosing it (see Fig. 5.6). This can be seen as a constraint that Wright used in handling the detail plan, and is interpreted as a local constraint applied at local level or is used in determining the room shapes. This concept, according to Wright, was first implemented in the plan of Unity Temple (Wright 1953).

5.6   Constant Design Methods and Processes

A design method refers to a manner of design procedure. It can be seen as a systematic means used in a particular stage to process a task. The design methods used by **Wright**, according to explanations given by him, his apprentices, and critics, can be categorized into the methods for making floor plans, elevations, and forms, and for form determinations.

5.6.1   Design Methods

5.6.1.1   In Making Floor Plans: Unit System and Grid System

Unit System of Design

Charles E. White, Jr. pointed out in a letter to his friend in 1904 that "all [**Wright**'s] plans are composed of units grouped in a symmetrical and systematic way. The unit usually employed is the casement window unit of about the following proportions. These units are varied in size and number to suit each particular case, and the unit decided upon, is consistently carried through every portion of the plan" (White 1971, p. 105) (see the right image of Fig. 5.7).

In 1908, **Wright** wrote that "in laying out the ground plans for even the more insignificant of these buildings a simple axial law and order and the ordered spacing upon a system of certain structural units definitely established for each structure in accord with its scheme of practical construction and aesthetic proportion, is practiced as an expedient to simplify the technical difficulties of execution, and although the symmetry may not be obvious always the balance is usually maintained" (Wright 1908, p. 160).

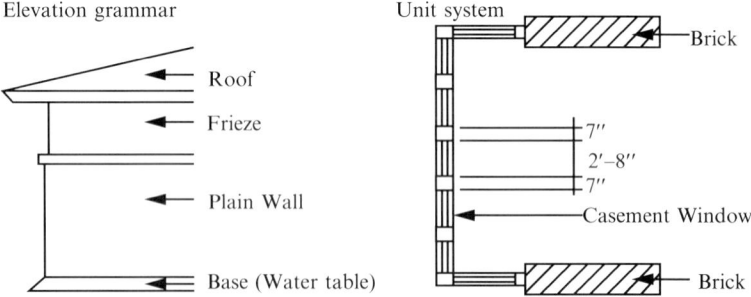

Fig. 5.7 Unit system and module used by Wright (Source: Copyright Locke Science Publishing Company, Inc. Figure 4, p. 217, reproduced with permission. Chan (1992))

In 1925, **Wright** explained, "all the buildings I have ever built, large and small, are fabricated upon a unit system—as the pile of a rug is stitched into the warp. Thus each structure is an ordered fabric. Rhythm, consistent scale of parts, and economy of construction are greatly facilitated by this simple expedient—a mechanical one absorbed in a final result to which it has given a more consistent texture, a more tenuous quality as a whole" (Wright 1925, p. 57).

Grid System

Wright consistently used a geometric grid (rectangles, triangles, diamonds, hexagons, etc....) as a basis for developing his floor plans (Streich 1972). The **grid system** and the unit system of design were physically correlated with each other in making plans. They were used by **Wright** to handle design factors being considered in his design of plans. **Wright** indicated that "the scale or unit-of-size of the various parts varies with the specific purpose of the building and the materials used to build it. The only sure way to hold all to scale is to adopt a unit-system, unit-lines crossing the paper both ways, spaced as pre-determined, say 4′-0″ on centers (grid system)— or 2′-8″ (casement window size in unit system) or whatever seems to yield the proper scale for the proposed purpose" (Wright 1928, p. 50).

It seemed to **Wright** that the development of a grid system was based on the specific purpose of the building, its proportions, materials, and the structural methods used to build it. Examples of **Wright**'s various grid systems for different materials and building methods are collected in Table 5.1.

5.6.1.2 In Making Forms

Froebel Kindergarten System

Rectangular spaces, along with geometric blocks, were used as space-planning units and means for form determination. In **Wright**'s autobiography (1943), he praised the positive aspects of his **Froebel Kindergarten education**, to which he attributed much that governed his own method of design. The **Froebel system** consisted of geometric blocks and colored cardboard shapes with which the child made patterns and constructions upon a squared unit grid. The first system consisted of a single cube, sphere, and cylinder and only when the potential of this system was completely mastered would an additional system be meted out. The emphasis was upon pure geometric forms and the abstract, symmetrical patterns they produce upon the grid (Manson 1958; Scully 1960; MacCormac 1974). Because of the manipulation of blocks upon a grid system, forms were proportionally and symmetrically interlocked and expressed massively. This geometric proclivity, relying on a grid system, is one of the characteristics that, according to **Manson**, has clearly been reflected in the Unity Temple and Larkin Building.

Table 5.1 Wright's various grid systems

Material	Building method	Grid system	Example
Wood	Plaster surfaced, wood trimmed (Lath and plaster)	4′-0″ (grid) unit works with 16″ centers for the length of lath	Coonley House (1907)
Concrete	Cast block and slab	7′-0″ (grid) horizontal divisions. Multiple of 16″ unit for lumber work	Unity Temple (1906)
Brick	Brick-pier	4′-6″ grid	Martin House (1904)
			Ullman House (1910)
Concrete support and floor slabs	Steel-and-glass	4′-0″ grid	
Pre-cast blocks	Double-wall construction	Multiple of 16″ square both horizontal and vertical	Ennis House (1923)
Concrete	Concrete slab	Multiple of 7′-0″	Merchandise Building
Concrete and brick	Concrete mushroom column and brick extension curtain wall	20′-0″ grid system both ways and a vertical unit of a brick course of 3.5″	Capital Journal Project (1931)
			Johnson Office Building (1936)

Source: Copyright Locke Science Publishing Company, Inc. Table 1, p. 218, reproduced with permission. Chan (1992)

Roof as a Determining Factor

Rorick (1975) proposed that the forms of Prairie Houses were defined by roofs. He analyzed the rules used by **Wright** in generating forms and applied them in a computer program that successfully imitated **Wright**'s style. He reported that rules used in generating forms include: (1) the major forms of the building are defined by the roofs (simple roofs over more complex plans); (2) there is a major form (roof) that is long and narrow—it is the lowest roof (Rorick 1975, p. 57).

5.6.1.3   In Making Elevations: Elevation Grammar

After floor plans had been developed and forms generated, **Wright** would use a technique to finish the elevations. This technique was an application of an elevation grammar. **Charles White Jr**. pointed out in a letter to a friend in 1904 that "[**Wright**'s] grammar, which he may be said to have invented, is such as he used in the **Winslow house**, consisting of a base, a straight piece of wall up to the second story window sills, a frieze from this front to the roof, and a cornice with a wide overhang. He never cuts anything above the cornice line, like dormers. Here is his grammar (see the left part of Fig. 5.5), roughly sketched, and all his buildings today are built along these lines" (White 1971, p. 105).

Wright also mentioned his own grammar in 1908: "There is good, substantial preparation at the ground for all the buildings and it is the first grammatical expression of all the types. This preparation, or water table, is to these buildings what the stylobate was to the ancient Greek temple … with this innovation established, one horizontal stripe of raw material, the foundation wall above ground, was eliminated and the complete grammar of type one made possible. A simple, unbroken wall surface from foot to level of second story sill was thus secured, a change of material occurring at that point to form the simple frieze that characterizes the earlier buildings … the matter of fenestration… as elementary constituents of the structure grouped in rhythmical fashion… The groups are managed, whenever required, so that overhanging eaves do not shade them… soon the guillotine window became apparent, and I waged a determined battle for casements swinging out… with the grammar so far established came an expression pure and simple" (Wright 1908, pp. 159–160).

Again in 1954, he said that "every house worth considering as a work of art must have a grammar of its own. 'Grammar' in this sense, means the same thing in any construction—whether it be of words or of stone or wood. It is the shape relationship between the various elements that enter into the constitution of the thing. The grammar of the house is its manifest articulation of all its parts" (Wright 1954, p. 181).

5.6.1.4 In Determining Form: Perspective Proof

After elevations were determined, **Wright** would check the whole form by perspective renderings. Hence, perspective drawings appeared late in **Wright**'s process of design. **Wright** said: "No man ever built a building worthy of the name of architecture, who fashioned it in perspective sketch to his taste and then fudged the plan to suit." **Wright** indicated, "such methods produce mere scene-painting. A perspective may be a proof but it is no nurture" (Wright 1908, p. 161). Thus, to **Wright**, perspectives were used to present ideas rather than to generate ideas (Connors 1984).

Wright also was always faithful to his belief that when a building was organized in an organic way with right proportions, then the picturesqueness of the building would take care of itself. Buildings are seen from ground level, but their view is determined by the conditions of approach. So, sometimes after the plan with section or elevation was completed, he would construct a little perspective proof of the building to see what the building would look like, and then returned to plans for corrections and revisions. For example, in planning the **Ullman** House in Oak Park in 1904, two study drawings had shown messy plans. In one drawing, he covered the plan with a series of radiant lines that stemmed from a vantage point. These lines were used to construct a perspective proof shown in another drawing. It seems that **Wright** was discontented with the result, so the plan was pulled and stretched on the living room wing (Connors 1984). It was exactly the final drawing shown in the **Architectural Record** (Wright 1928). It is not clear whether in this particular instance, the resulting form was determined by the perspective rendering, but it is possible that perspectives were used to perceive the generated form and possibly for remedying it.

In the design of Unity Temple, the perspective drawings were done at the very end of the design. In this instance **Wright** used perspectives to present and convey his design to clients. For instance, the drawing later printed in the Wasmuth portfolio of 1910 came after the design was completed, and **Wright** was concerned with presenting the building rather than shaping it (Connors 1984). **Wright** mentioned in his autobiography that after the plans, sections, and elevations were completed, "we have enough now on paper to make a perspective drawing to go with the plan for the committee of 'good men and true' to see. Usually a committee has only the sketch to consider. But it is impossible to present a 'sketch' when working in this method. The building as a whole must be all in order before the 'sketch' not after it" (Wright 1943, p. 51). It is clear that only after the form or the mass had been generated would **Wright** develop a perspective to confirm the visual condition of the building or to show his concept.

5.6.2 Design Processes

Design process refers to the sequential states of accomplishing a design work. A design process is the concatenation of states in which design methods are applied to complete certain tasks. A designer may have his or her own method of approaching a design. Such methods are called *methods of design processes*. Sometimes, a method of design process is a particular sequence of designing. In this section, **Wright**'s design process is described by analyzing documentation provided by his apprentices, clients, or **Wright** himself.[2] It is hoped that this section will present the necessary information to give a clear picture about **Wright**'s particular method of design.

5.6.2.1 White's Interpretation

Charles E. White Jr. wrote in 1904: "His [**Wright**'s] process in getting up a new design is the reverse of that usually employed. Most men outline the strictly utilitarian requirements, choose their style, and then mold the design along those lines, whereas **Wright** develops his unit (unit system) first, then fits his design to the requirements as much as possible, or rather, fits the requirements to the design. I do not mean by this that he ignores the requirements, but rather that he approaches his work in a broad minded architectural way, and never allows any of the petty wants of his client to interfere with the architectural expression of his design" (White 1971, p. 106). **White**'s message about **Wright**'s design process is diagrammed in Fig. 5.8.

---

[2] **Wright** once reported his design processes for Unity Temple (1904) in the first edition of his autobiography of 1932. But this description was written some 18 years later, and cannot be regarded as a reflection of the real thinking processes that occurred while he was designing the Temple. Instead, it is a retrospection of what may be logically related events in design. Thus, this information is used as data for reference rather than for analysis.

Fig. 5.8 White's description of Wright's design process (Source: Copyright Locke Science Publishing Company, Inc. Figure 5, p. 221, reproduced with permission. Chan (1992))

White's interpretation refers to designs for the early Prairie Houses. But he only referred to **Wright**'s process of designing plans, which is to have a unit system first and then fit in functions. White did not explain how **Wright** would generate forms, nor how Wright would use grid system and elevation grammar. But it is obvious that unless a floor plan has been developed to a certain extent, the elevation cannot be worked out. Therefore, it is inferred that Wright would develop a grid system first for a floor plan, then apply the elevation grammar to handle elevation. This is confirmed by **Wright** in 1908 and by a note by **John Howe**, **Wright**'s former apprentice in **Taliesin West**. **Wright** said: "I have endeavored … to establish a harmonious relationship between ground plan and elevation of these buildings, considering the one as a solution and the other an expression of the conditions of a problem of which the whole is a project. I have tried to establish an organic integrity to begin with, forming the basis for the subsequent working out of a significant grammatical expression and making the whole, as nearly as I could, consistent" (Wright 1908, p. 158).

5.6.3 Howe's Interpretation

Howe noted in 1980 that "with any building, Mr. **Wright** designed in plan, first and foremost. Then he moved (to) a section, and then the elevations were the result of the plan and the section. His buildings really were designed from the inside out … Mr. **Wright** would establish the grammar of the buildings from working on the elevations" (Lipman 1986, p. 25). Two messages in **Howe**'s note are important. One reveals **Wright**'s process, which is that plan came first, then section, and finally elevation. Perspective drawing is not mentioned as having been used in design. This confirms that perspective drawings are not used as means for developing concepts, but rather for communicating concepts. The other message is that **Wright**'s grammar of the buildings only refers to elevations. Combining **White**'s and **Howe**'s descriptions, a process diagram is shown in Fig. 5.9.

5.6.3.1 Scully's Interpretation

According to **Vincent Scully**, **Wright**'s processes of design had certain sequences (Scully 1960, p. 13):

1. His primary concern was **abstract**: first, usually, in the abstraction of the space, taking shape as it did out of his double will to embody its use and to form it into a rhythmically geometric pattern.

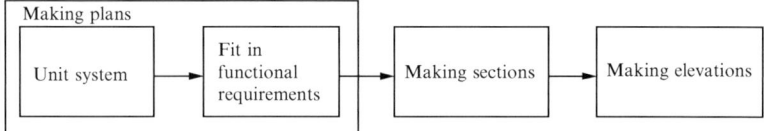

Fig. 5.9 Combination of Howe's and White's description of Wright's process (Source: Copyright Locke Science Publishing Company, Inc. Figure 6, p. 221, reproduced with permission. Chan (1992))

2. He wished both to enclose the hollow so created and to extend it or the expression of it to the exterior through the sculptural massing of the building as a whole. Sometimes, as in the earliest works of his several phases, a concern for the exterior massing may have preceded that for the interior space.
3. Having made his building visually integral in both its voids and solids, he then wished to build it of such materials and in such a way as to make it structurally integral as well. In some of his later projects, the structural principle may come first in the process, but when his work in general was studied, it is found that structural integration tended to come last at any stage in his development, and that he himself was most specifically pleased with any building when its structural rather than simply its spatial and sculptural aspects were intrinsic to the whole.

What **Scully** mentioned can be represented by a diagram to visualize **Wright**'s process of design, as shown in Fig. 5.10, which also incorporates **White**'s and **Howe**'s interpretations. In it, blocks stands for stages. The first stage was to develop an **abstract** of the design. This proceeding of developing abstraction at the very beginning can be confirmed by **Wright**'s writings. For instance, he said, "before the plan is a plan it is a concept in some creative mind… therefore conceive the building in the imagination, not on paper but in the mind, thoroughly— before touching paper" (Wright 1928, p. 49). Stages 2 and 3 together belonged to a plan-making stage in which a general shape was formed, in accordance with **White** and **Howe**. Stage 4 developed building form and sometimes (as **Scully** has implied occurred in the Prairie Houses period) happened before stage 3. Thus, two branches are noted after stage 2. Stage 5 is the period of integrating materials and structures into the building mass. In referring to **Howe**'s interpretation, it is inferred that this is the task that **Wright** would have pursued in the section-making stage. And, although **Scully** indicated that these stage-5 activities occasionally occurred at the very beginning of design in **Wright**'s later projects, this does not mean that sections were developed at the beginning. Instead, it indicates that an abstract idea of a unity of materials and structural methods was considered first in some instances, for example, the Johnson Wax Building (1936) and the Guggenheim Museum (1946).

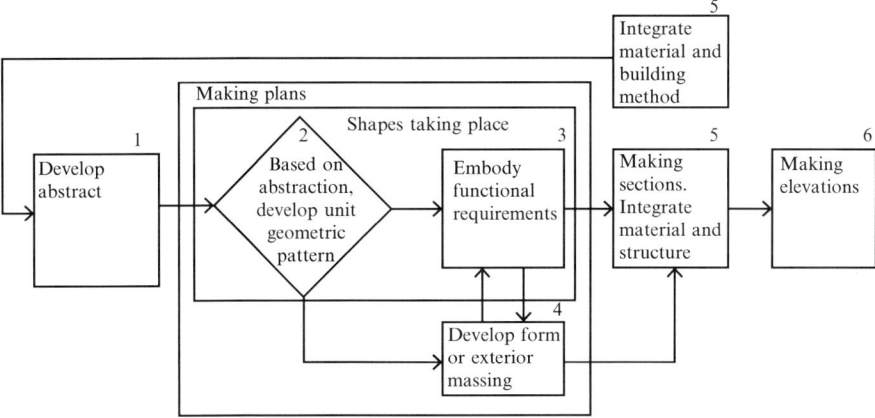

Fig. 5.10 Scully's description of Wright's process (Source: Copyright Locke Science Publishing Company, Inc. Figure 7, p. 222, reproduced with permission. Chan (1992))

5.7   Reconstruction of Wright's Design Process

A design process also can be explained as a series of design goals to be achieved. For each goal, there is a set of design constraints used for solution generation and testing (Chan 1989, 1990b). There also is a set of design principles that serve as a guide for supervising the direction of design. These principles remain unchanged throughout the whole process and are treated as global constraints. This concept, illustrated in Fig. 5.11, explains the design process in general. But **Wright** had special design methods in handling design constraints to accomplish a goal. Thus, in the lower part of the models in Figs. 5.11, 5.12, 5.13, and 5.14, design methods used at different stages are included to show what particular methods are applied at certain stages.

Wright had indicated that there were six factors he would consider in making plans, and the unit system and the grid system were the methods used to handle all of them. Meanwhile, he would achieve the design principles of horizontality, simplicity, plasticity, continuity, and the nature of materials. Fitting these details into the model, **Wright**'s example of making plans is shown specifically in Fig. 5.12.

Studies by **White**, **Howe**, and **Scully**, and **Wright**'s own writings reflect that **Wright**'s general design process can be classified into two approaches, associated with **Wright**'s early and late periods of designs. The first approach applies to **Wright**'s Prairie House period (see Fig. 5.13). In designing these projects, **Wright** would first conceive in his mind an abstract that was possibly an architectural program of the building. Then, on the basis of the abstract and considering the material and the structural method used, he would develop a grid system that best suited the whole purposes. Materials were determined by considering budget constraints.

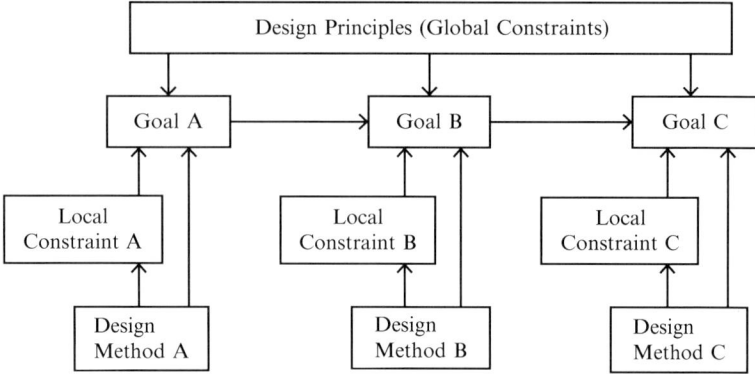

Fig. 5.11 A general model shows relationships between goal, constraint, and methods used in process (Source: Copyright Locke Science Publishing Company, Inc. Figure 8, p. 223, reproduced with permission. Chan (1992))

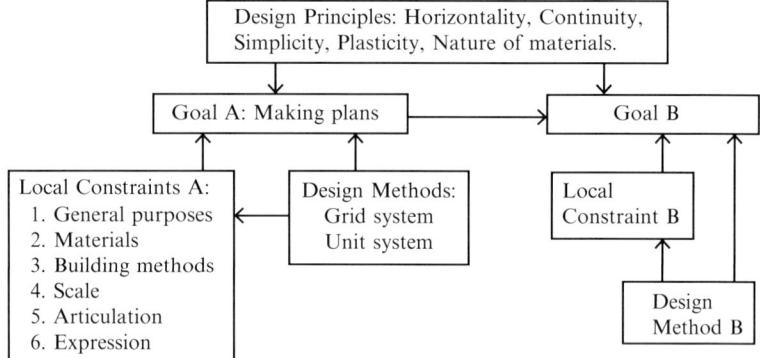

Fig. 5.12 Example of Wright's design process (Source: Copyright Locke Science Publishing Company, Inc. Figure 9, p. 223, reproduced with permission. Chan (1992))

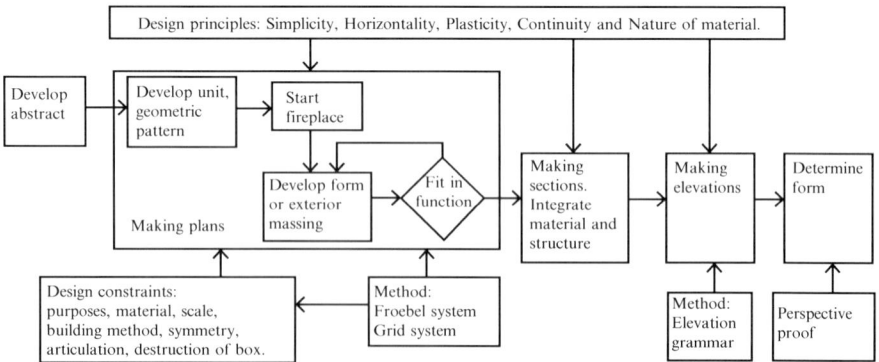

Fig. 5.13 Wrigh's design process in Prairie Houses (Source: Copyright Locke Science Publishing Company, Inc. Figure 10, p. 224, reproduced with permission. Chan (1992))

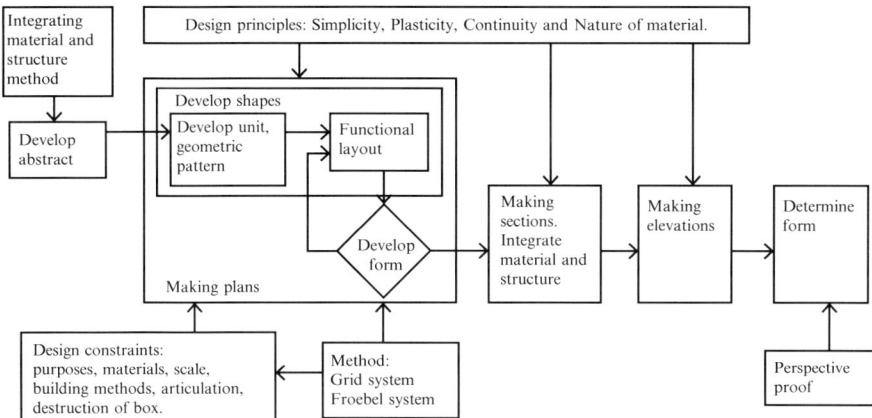

Fig. 5.14 Wright's design process in later period (Source: Copyright Locke Science Publishing Company, Inc. Figure 11, p. 224, reproduced with permission. Chan (1992))

If the budget was not tight, he would start working with brick and stone. But often he had to use wood and stucco. After the grid system had been decided, he started to make plans. **Wright** often referred to the fireplace as the "heart" of the house. And, for almost all of his Prairie Houses, the fireplace was located right in the center. Hence, it seems that **Wright** would use 'fireplace in the center' as a starting point in making plans. While making plans, the concern of mass was developed before the functional layouts were fitted in. Shapes were planned in alignment with grids. Then the whole building was integrated during the section-making stage. The elevation was the result of the application of elevation grammar on plans.

The second approach belongs to **Wright**'s design process other than and later than Prairie Houses (see Fig. 5.14). In this figure, the sequences in making plans are changed. According to **Scully**, **Wright** would concentrate on the functional arrangements first and develop the form later. The sets of design principles and design constraints are different from what he had used earlier. For example, horizontality and symmetry were not major considerations in his late projects. In the development of elevation stage, the elevation grammar may not have been the same as for the Prairie Houses. In some projects, the development of an integrated structural method and materials occurred at very early stage, as in the Johnson Wax Building (Lipman 1986). Hence, the overall form developed and the elevation generated did not share the characteristics of the Prairie Houses.

5.8   Matching Features with Design Methodologies

Results of how factors of design constraints, principles, methods, and procedures, which are design intelligence resulting from utilizing cognitive mechanisms of search, association, and memory recall (Chan 2008), do affect the product forms, and therefore, produce common features are analyzed in this section and summarized briefly in the last columns of Tables 5.3 and 5.4.

5.8.1 Results from Design Methods

5.8.1.1 Grid System

The unit system was the basic module of the whole composition. By using the unit and considering the nature of materials, building method, and scale, a grid system was developed. The grid used in designing the Ross House, for example, is shown in the upper part of Fig. 5.13. The use of a grid system had two effects. First, the grid system provided a norm for controlling the proportion of each part of the building and the relationships among parts. Second, the grid system integrated the nature of material and the construction methods. For instance, the nature of material determined the distance of the bay of columns. The construction methods determined both the dimension of bays and the size of columns. Overall, using a grid system not only addressed the various constraints of material, scale, proportion, and the construction methods, but also controlled the size and placement of each element in the plan.

Wright used the grid system method in almost every design throughout his career, even in his later period of **Usonian** houses. According to **Twombly**, **Wright** consistently used a two-by-four-foot horizontal module of grid. This grid system governed the entire plan in the drawings and on the floor, so that the contractor could easily locate doors and windows and reduce labor and waste because the materials were plywood that came in four-foot cuts. All the objects were centered on, aligned with, or related to the grid or its subdivisions (Twombly 1979).

The grid system was sometimes composed with rectangular blocks rather than squares, and these set up a tartan. The tartan effect resulted from the fact that certain lines in the grid, due to Wright's alignment of some major and minor elements in the design, received more emphasis than others. The tartan used in designing the Ross House is shown in the lower part of Fig. 5.15. **Manson** and **MacCormac** also have shown that approximate models of the Larkin Building, Unity Temple, **Ross** House, **Barton** House, and **Evans** House can be constructed with different patterns of grids and tartans (Manson 1953; MacCormac 1968).

The three-dimensional forms of the Larkin Building and the Unity Temple reveal characteristics of a possible usage of the Froebel system to generate blocks of geometric forms that symmetrically and massively lay upon the grid system. **Manson** has pointed out that the two-dimensional and three-dimensional schemes of the Froebelian exercises bear a distinct resemblance to **Wright**'s designs carried out from 1900 to 1910 (Manson 1958, p. 7). This indicates that the form generation can be the result of the method—the Froebel system of manipulating blocks on grids. This inference is challenged by **Van Zanten**, however, who argued that the Froebelian experience may be historically dubious. He pointed out that the influ-

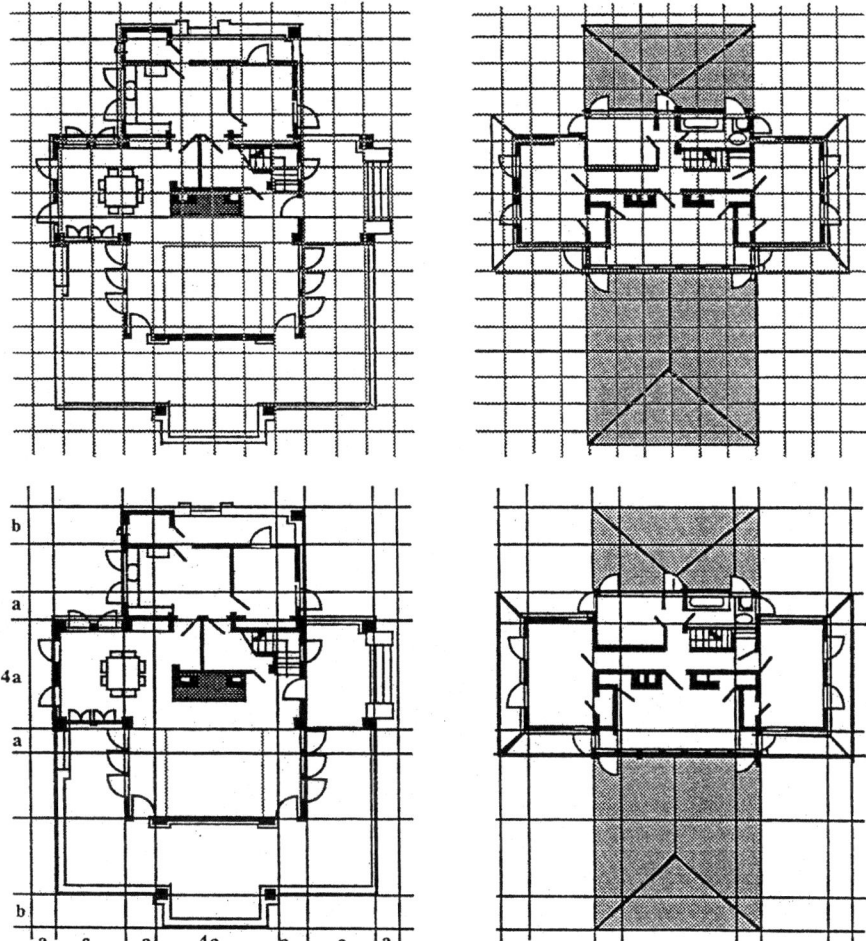

Fig. 5.15 The grid system (*above*) and tartan (*below*) of Ross House design (Source: Copyright Locke Science Publishing Company, Inc. Figure 12, p. 226, reproduced with permission. Chan (1992))

ence of the Froebel kindergarten blocks is doubtful in its validity and is a tremendously resilient myth (Van Zanten 1988).

5.8.1.2   Elevation Grammar

Wright explained that he had endeavored to establish a harmonious relationship between ground plan and elevation of the prairie houses he designed, considering the one as a solution and the other as an expression (Wright 1943). To him, the elevation was a response to the plan, and the elevation grammar he used is an invention of a personal idiom. Using this grammar, the resulting elevations showed some common and recognizable characteristics in his Prairie Houses. They all had a base, a plain wall, an odd number of casement windows and a low hip roof raised right above the window. The terrace, which is extended forward from the living room or entrance hall, always consisted of a base, a plain wall, and a coping. These characteristics did not appear in his non-residential designs because he did not apply the same elevation grammar as for houses.

5.8.2 Results from Design Principles

Three examples will serve to describe how a design principle yielded a particular feature. The first example relates to the principle of simplicity. **Wright** wrote in 1908, "a building should contain as few rooms as will meet the conditions which give it rise and under which we live, and which the architect should strive continually to simplify… Beside the entry and necessary workrooms there need be but three rooms on the ground floor of any house, living room, dining room, and kitchen, with the possible addition of a 'social office'…" (Wright 1908, p. 156). This explains that **Wright** adapted the principle of simplicity for the ground floor plan of his Prairie Houses. It turns out that, in most of his Prairie Houses, living and social quarters were located on the first floor, whereas bedrooms were arranged on upper floors.

The second example relates to the principles of plasticity and continuity. **Wright** had tried to develop a form of wood trim to embody this principle. He said, "I entirely eliminate the wooden trim. I did make it 'plastic', that is to say, light and continuously flowing instead of the prevailing heavy cut and butt carpenter work… the trim thus became only a single, flat, narrow horizontal band running around the room walls at the top of the windows and doors and another one at the floors" (Wright 1966, p. 36). Thus, these continuous bands of wood were results of the principle of plasticity and continuity.

The third example explains how a consistent application of a principle will yield a consistent characteristic of a form expression. In Prairie Houses, space was a horizontally moving entity, always in layers parallel to the earth. As discussed earlier, this was the result of the principle of horizontality. But this principle was not always appropriate for public buildings. **Wright** had to find and develop different principles for public buildings. Therefore, for the Larkin Building and for Unity Temple, **Wright** began to test the possibilities of space-in-motion up and down, as well as sideways (Blake 1960). It turned out that the vertical character of the Larkin Building and Unity Temple resulted from the application of the same principle—the principle of changing horizontal to vertical.

5.8.3 Results from Design Constraints

A conception of a particular constraint in a design yields certain features. Two examples provide an explanation.

5.8.3.1 Material as a Design Constraint

To **Wright**, different materials had different architectural expressions. For wood structures, **Wright** expressed the wooden skeleton and used light-colored stucco panels to fill in the surface between the darker stripping (**Wright**'s term for trim). This can be found in **Willitts House** and **Coonley House** (Scully 1960). For concrete, Wright provided three solutions. First, he expressed the corner pillars as heavy massive supports for the roof (Wils 1985), or in **Hitchcock**'s term, the broad corner piers (Hitchcock 1942). This solution appeared first in the Yahara Boat Club (1902), later in Unity Temple (1906), and again in a residential project of "A Fireproof House for $5,000" (1906). Second, roofs were flat slabs and cantilevered out above the windows. This solution was the common character of Unity Temple and "A Fireproof House for $5,000." Third, floors were strung between pillars to hold up the parapets. Sometimes the floors protruded for several feet, forming balconies that cantilevered from the wall without any visible form of support (Wils 1985). This solution was repeated in his later designs, the **Kaufmann House** (1936) and **Johnson Wax building** (1936).

5.8.3.2 Articulation as a Design Constraint

In **Wright**'s residential designs, each part of the building was either separate from the others or extended from the surface. The separation of a part from the others can be explained as each principal space, which was a block, occupying its own place on one axis. As a result, the living room, dining room, and kitchen together formed an L plan or a T plan. Along with the reception area (called the entrance hall from 1903 onward), the whole plan became a cruciform that can be seen as the result of articulating principal spaces. The articulation also can be explained as the extension of a part from the surface. This is found in his linear plan, in which the principal space protruded several feet to distinguish itself. Thus, the exterior forms projected internal spaces.

In his non-residential designs of the Larkin Building and Unity Temple, the staircases, resembling towers, were located in the corners. Wright wrote: "I worked to get that something into the Larkin Building, interested now also in the principle of articulation as related to that order" (Wright 1943, p. 151). His answer was to build the stair towers free of the central block and to push them into the corner to form independent elements for communication and escape, but also as air intakes for the ventilating system.[3] Thus the tower-like massive blocks resolved from the effort of articulation, and were applied again in Unity Temple. This result of articulation also made the inner court of the Larkin Building and the great hall of the Temple both extend forward and explore the full length outside. The function of each part of the buildings was well expressed.

These two examples of the material and articulation constraints explain: (1) how a constraint provides the generation of a form; (2) how the application of the same constraints yields the same resulting forms. By using the same constraints in different designs, the generated solutions embodied forms with similarities.

5.8.4   Results from Other Factors

5.8.4.1   Presolution Models

A *presolution model* is a design solution that is generated previously and is used on later designs (Foz 1972; Chan 1990b). The application of presolution models maintain similar forms across designs and, thus, constant features appear. In 1903, **Wright** designed the **Barton House** in Buffalo. The theme of Barton house was similar to the **Walser House** he designed in Chicago—reduction of the Prairie House formula to a tight, symmetrical plan with a side entrance and street elevation made up of a self-contained tripartite window for the living room on the first floor, a continuous strip of casements above, and the capping eaves. Manson pointed out that "the Walser-Barton parti was a scheme that **Wright** often used as the answer to the problem of fitting a Prairie House to a narrow lot and a limited budget; but, because neither of these conditions existed on Summit Avenue (where the Barton House is located), it is puzzling why he used it there except as an expediency, so as to get to work as soon as possible upon the main house" (Manson 1958, p. 140). Regardless of why the site condition did not affect the design solution, this example displays the application of presolution model, in this instance the similar plan arrangement, which provides similarities for the resulting forms. Also, several instances can be found where the arrangements of the principal spaces in floor plans are almost identical, for instance, the **Martin House** (1903) and **Horner House** (1908); the **Little House** (1903) and Barton House (1903); and the **Davenport House** (1901) and **Ingalls House** (1909) (see Fig. 5.1).

Another example of using a presolution model is the mushroom column Wright developed for the Johnson Wax Building of 1936. The section of mushroom column was initially developed for the Capital Journal Newspaper Building

---

[3] According to Einbinder, the solution locating stairs in the corners first appeared in the design of the **Larkin** Building. It is the result of a design constraint – escaping fire (Einbinder 1986).

Project of 1931. In the mid-1950s, Wright transformed this solution of a great mushroom column-filled space into two more schemes. The first one appeared in the initial version of the Freund Y Cia Department Store project in San Salvador, El Salvador in 1954. The second one was the 1955 design of an expandable tent-like headquarters for the Lenkurt Electric Company in San Mateo, California. The repetition of the same solution therefore generates similar forms (mushroom image) across designs.

5.8.4.2   The Use of Partial Forms

Different individuals have different tastes and favor certain geometric forms. For example, **Le Corbusier** preferred the free curve, whereas **John Portman** favored the cylinder. The use of a specific form in designs will mark an idiosyncrasy of the architect. Although the preference may change with time, the consistent application of the same form within a specific span of time labels the designer's style. For instance, the polygon used in **Wright**'s Bootlegged Houses (1889–1894) and the low hip roof in Prairie Houses were the most prominent and consistent forms applied in those two periods.

Polygon

A better example that explains the use of partial form to yield a particular feature is found in **Wright**'s preference for the polygon in the period before the turn of the century. According to Manson (1958), polygonal bay window and inglenooks were features that **Wright** was to incorporate again and again in his plans of 1890. From 1889 to 1894, **Wright** designed 17 residences (Storrer 1978) around the area of **Oak Park**, a Chicago suburb, while with the firm of **Adler** and. **Sullivan**. From studying the published floor plan drawings (Manson 1958; Steiner 1982) and photographs (Storrer 1978), it is found that the inglenook is used in three projects only—his own **Oak Park** House of 1889, **Blossom** House of 1892, and **Bagley** House of 1894. On the other hand, at least one bay window of polygonal shape (see Table 5.2) appeared in 15 of the 17 designs.[4] This 88 % occurrence demonstrates that a polygonal bay window was a significant feature during this period. **Wright**'s predilection for the polygon continues to be evident in his later Prairie Houses period, although not in every design project.

---

[4] **Blossom** House of 1892 is the only one without a polygonal shape. The other one of **MacHarg** House (1891) cannot be judged because no information is available.

Table 5.2 Houses designed by Wright from 1889 to 1894

Project	Date	Bay window location	Polygonal element	Source
Oak Park House	1889	Living Rm, entry Rm		Manson
L. Sullivan House	1890	Shown in photography		Storrer
Charnley sum House	1890	Shown in photography	Octagonal bay	Storrer
Charnley House	1891	Dining Rm		Storrer
MacHarg House	1891	<No plan available>		Storrer
Mc Arthur House	1892	Dining rm, parlor, living Rm	Polygonal bay	Manson
Blossom House	1892	<None>		Manson
Emmond House	1892	Reception, living Rm, terrace	Polygonal bay, roof	Manson
T. Gale House	1892	Living dining Rm	Polygonal roof, octagonal bay	Steiner
Parker House	1892	Living, dining Rm	Polygonal roof, octagonal bay	Steiner
Harlan House	1892	Library	Polygonal roof, octagonal bay	Steiner
Harlan House	1892	Library		Manson
A. Sullivan House	1892	Shown in photography		Storrer
W. Gale House	1893	Living Rm, Staircase		Manson
Wooley House	1893	Parlor	Polygonal dormer, bay	Steiner
Winslow House	1893	Living Rm		Manson
Bagley House	1894	Dining Rm	Octagonal library	Manson
Goan House	1894	Shown in photography		Storrer

Source: Copyright Locke Science Publishing Company, Inc. Table 2, p. 231, reproduced with permission. Chan (1992)

Roofs

Wright used gable roofs in several of his early Prairie Houses, for example, the **Bradley** House (1900), A Small House with Lots of Room in It (1901), **Hickox** House (1900), **Foster** House (1900), **Davenport** House (1901), and **Dana** House (1903). **Wright** mentioned in 1910, "a study of the drawings will show that the buildings presented fall readily into three groups having a family resemblance; the low-pitched hip roofs, heaped together in pyramidal fashion, or presenting quiet, unbroken sky lines; the low roofs with simple pediments countering on long ridges; and those topped with a simple slab... of the second type, the **Bradley**, **Hickox**, **Davenport**, and **Dana** houses are typical" (Wright 1941, p. 75). But the low hip gable roof with wide eaves appeared in most of his Prairie Houses. The earliest occurrence of this kind can be traced back to the **Winslow** House in 1893. **Manson**'s study indicated that "the most important innovation in the **Winslow** House is its roof. Here, in unambiguous form, that low, generous hip with wide eaves that is to become the keynote of the Prairie House makes its appearance" (Manson 1958, p. 62).

Neither **Wright**, nor any other source, has explained how this form (low and generous hip with wide eaves) was first generated. **Manson** (1958) inferred that it may have been suggested by the roof forms of traditional Japanese architecture. It also can be inferred that this solution, which precluded the use of attic space for anything, was partly suggested by **Wright**'s objection to the attic. In his autobiography, **Wright** expressed his disapproval of putting attics and basements into residences (Wright 1943). Regardless of the explanation, the repeated use of such a solution on most of his residential designs marks a prominent feature of the Prairie Houses. This is another example showing how the replication of presolution models manifests style. It also illustrates that **Wright** was fond of the particular roof form in his residential design, and that signifies his style.

Manson classified the designs of the 5-year period from 1889 to 1894 as **Wright**'s "Bootlegged Houses." If one would identify **Wright**'s style of this period, the preference for a partial form—polygonal bay window—marks an obvious feature, versus the low hip roof of his Prairie Houses.

Design Schemata

The design units (or architectural programs) used by **Wright** in Prairie Houses were quite similar across designs. Usually, they consisted of a living room, dining room, kitchen, servant's quarters, two or three bedrooms, a reception room, and a library (sometimes called a study room) or a music room, depending upon the client's interest and desire. **Wright** used similar topological arrangements in Prairie Houses. The locations of principal spaces (living room, dining room, kitchen, entrance, and servant's space) have certain similarities, and their interrelations have a certain regularity. For example, **Wright**'s prototypical space plan can be described as having a living room on the South, dining room on the West, kitchen on the North, servant's room further North, and fireplace at the center. The East was left for the entrance. If the entrance was a terrace, a T plan was formed. Here, the terrace was treated as a negative space that was open and without enclosure. If the entrance was a porch, or enclosed to form a reception hall, then a cruciform was shaped. Sometimes this reception hall or entrance hall was replaced by a study room or a library. In both the cruciform plan and the T plan, the location of kitchen and dining room were interchangeable. Typical plans found in such topological arrangements are shown in Fig. 5.16.

It is not clear whether **Wright** repeated the same design processes to reach the same layouts in the projects in Fig. 5.16, or whether **Wright** had certain fixed spatial layout **schemata**, so that each time, the instantiation of these **schemata** from memory caused the same topological layout. The term *schemata* explains many phenomena very well. For example, the central fireplace and the base (watertable) were constant features in Prairie House design. These two features can be seen as

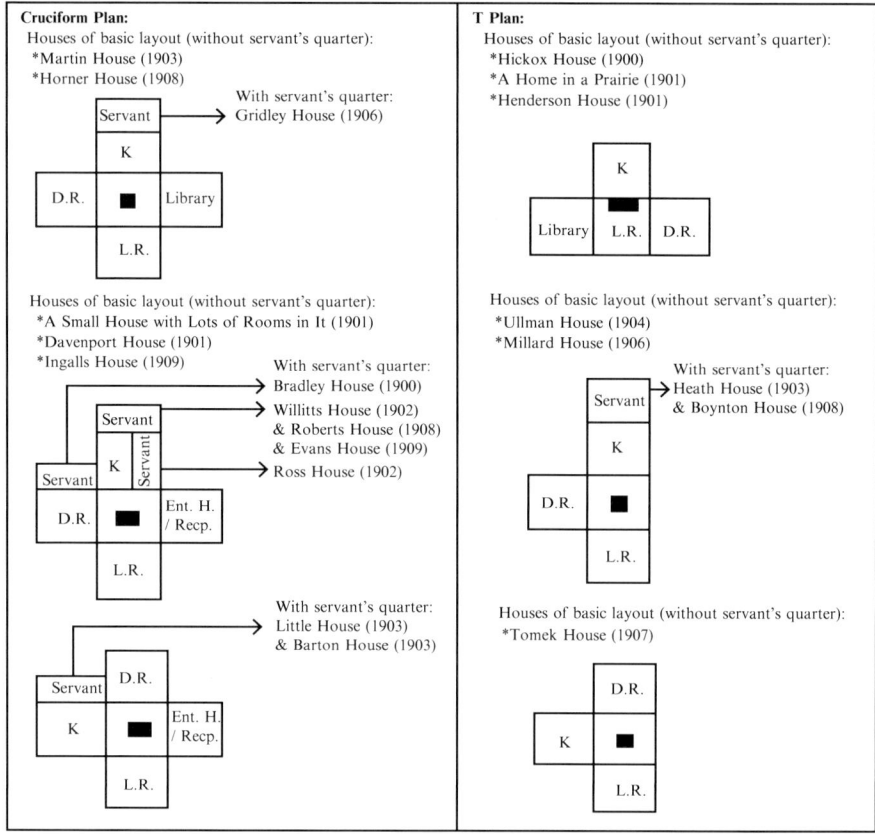

Fig. 5.16 A summary of Wright's typically topological arrangements (Source: Copyright Locke Science Publishing Company, Inc. Figure 14, p. 232, reproduced with permission. Chan (1992))

two constraint schemata, both described by **Wright**. In his writings, he referred to the fireplace as the heart of the house. He also indicated in his autobiography his desires, in building a new house, to "get rid of the attic, therefore the dormer… Next, get rid of the unwholesome basement, yes absolutely – in any house built on the prairie" (Wright 1943, p. 141). His schema on handling the basement was to get the unwholesome basement up out of the ground, entirely above it, as a low pedestal for the living portion of the home, making the foundation itself visible as a low masonry platform on the ground on which the building would stand (Wright 1943).

5.9   Conclusions

Examples collected in this chapter show how **Wright**'s work demonstrates how design-related activities (constraints, principles, methods, and procedures) influence the generation of certain common features by which an individual style is defined. Summaries of the common features in residential and non-residential projects, together with the causative factors discovered in **Wright**'s works, are listed in Tables 5.3 and 5.4.

This case study explains how consistent processes and constant applications of certain factors lead to the appearance of similar products. It also demonstrates that a product is a function of a process. A form that appears in products is caused by a factor in the design process. Thus, an individual style is identified either by the common features appearing in products or by the common factors operating in the

Table 5.3 Common features of Prairie Houses

	Common features	Result from
Plan	1. Fireplace at the center	
	2. Principle space (LR, DR, K Entrance) Each one occupies one end of an axis	Articulation (DC)
	3. No basement, no attic	
	4. Walls are extended to from terraces courts	
	5. A major shape in plan that is long and narrow	Articulation (DC)
Elevation	1. Overhang gable roof or low hip roof	Horizontal (DP)
	2. A ribbon of casement windows	Elevation grammar (DM)
	3. A base of the house	Elevation grammar (DM)
	4. Coping of terraces	Elevation grammar (DM)
	5. Horizontal elements (roof, wall, parapet, porch, terrace)	
Wall	1. Walls are screens	Horizontal (DP)
	2. Continuous horizontal band around the wall, above doors	Destruction of box (DC)
Material	1. Most houses are wood trimmed with white stucco	Plasticity (DP)
	2. Oak is the major material	Material (DC)

Source: Copyright Locke Science Publishing Company, Inc. Table 3, p. 234, reproduced with permission. Chan (1992)
Note: *DC* design constraint, *DM* design method, *DP* design principle

Table 5.4 Common features among three non-residential designs

	Yahara Boat Club	Larkin building	Unity temple	Result from
Plan	Simple block and rectangular shape	Stories are laid out around an inner court The inner court, lit by a huge skylight, located in the center	Galleries are laid around the auditorium court space Court is lit by skylight	Site condition (DC) (smoky air and noisy site)
	Symmetric plan	Symmetric plan	Symmetric plan	Symmetry (DC)

(continued)

Table 5.4   (continued)

	Yahara Boat Club	Larkin building	Unity temple	Result from
Elevation	Flat slab roof Cantilevered	Flat slab roof	Flat slab roof Cantilevered	Nature of materials (DP)
	A long band of windows		A band of windows	
	A continuous sill	A continuous sill	A continuous sill	Plasticity (DP)
	Building sits on an extended base of retaining walls		Has a solid base	
	Dominant horizontal character (horizontal roof plane and long side walls)	Vertical movement	Vertical movement	Horizontal/vertical (DP)
	Massive corner block	Massive corner block (Brick corner stair tower)	Massive corner block. (Corner stair tower)	Articulation (DC) Fire escaping (DC)
Material		Exterior and interior are both of brick, with floors and ceilings of concrete	Poured concrete	Budget (DC)

Source: Copyright Locke Science Publishing Company, Inc. Table 4, p. 234, reproduced with permission. Chan (1992)
Note: *DC* design constraint, *DM* design method, *DP* design principle

design processes. Findings and evidence collected in this research suggest that **Wright**'s Prairie Houses style can be defined as:

1. The repeated forms of a low hip roof, a band of casement windows, coping of terraces, extended terraces with a low parapet, a continuous band of sill, a continuous watertable, a symmetrical wing facade, planting urns, a massive brick chimney, and corner blocks.
2. The repeated absence of elements: basement and attic.
3. The repeated use of a fixed set of design constraints (e.g., material, scale, building method, articulation, symmetry, and destruction of box), and design principles (e.g., simplicity, plasticity, horizontality, and continuity).
4. The repeated use of the design methods of the grid system and use of elevation grammar.
5. The repeated use of a fixed sequence of design procedures (e.g., from **design abstract** to floor plan, section, elevation, perspective renderings, and to floor plan again, if necessary).

The case study shows that for different building types, some of the factors changed while others remained. For example, the grid system, the tartan system (design methods), articulation (constraint), and the sequences of procedures (from floor plan to section to elevation) were consistently used across building types (residential vs non-residential). Yet, the horizontality constraint was changed to the vertical in two examples of non-residential buildings. Thus, the application of different factors in design will generate quite different features. But, the repetitions of the same factors (causing forces), after all, define an individual style. Another factor determining style is a predilection for certain geometric forms, which also manifest style. For example, **Wright** was fond of overhanging low hip roofs, planting urns, and coping low terraces in residential designs and of a band of casement windows in both residential and non-residential buildings. His preference for these elements makes them obvious in form. All these forms together define **Wright**'s idiosyncratic style. It is all because **Wright** used a rich design language (including form, method, and expression) constantly across projects that the style manifested in his projects is a strong style: clear, dominant, and legible.

References

Ackerman JS (1963) Style. In: Ackerman JS, Carpenter R (eds) Art and archaeology. Prentice-Hall, Englewood Cliffs, pp 174–186

Ackerman JS (1967) A theory of style. In: Beardsley MC, Schueller HM (eds) Aesthetic inquiry: essays on art criticism and the philosophy of art. Dickenson, Delmont, pp 54–66

Akin O (1986) Psychology of architectural design. Pion, London

Blake P (1960) The master builders. Norton, New York

Brooks HA (1984) Frank Lloyd Wright and the prairie school. George Braziller, New York

Buffon MD (1923) Discourse on style. In: Cooper L (ed) Theories of style. Macmillan, New York, pp 169–179

Chan CS (1989) Cognition in design process. In: Proceedings of the 11th annual conference of the Cognitive Science Society. Lawrence Erlbaum, Hillsdale, pp 291–298

Chan CS (1990a) Psychology of style in design. Dissertation. Carnegie Mellon University

Chan CS (1990b) Cognitive processes in architectural design problem solving. Des Stud 11(2):60–80

Chan CS (1992) Exploring individual style through Wright's design. J Architect Plan Res 9(3):207–238

Chan CS (1994) Operational definition of style. Environ Plan B Plan Des 21(2):223–246

Chan CS (1995) A cognitive theory of style. Environ Plan B Plan Des 22(4):461–474

Chan CS (2000) Can style be measured? Des Stud 21(3):277–291

Chan CS (2001) An examination of the forces that generate a style. Des Stud 22(4):319–346

Chan CS (2008) Design cognition: cognitive science in design. China Architecture & Building Press, Beijing

Cleaver DG (1985) Art: an introduction. Harcourt Brace Jovanovich, San Diego

Connors J (1984) The Robie House of Frank Lloyd Wright. The University of Chicago Press, Chicago

Einbinder H (1986) An American genius: Frank Lloyd Wright. Philosophical Library, New York

Evans HM (1982) An invitation to design. Macmillan, New York

Finch M (1974) Style in art history. Scarecrow Press, Metuchen

Foz ATK (1972) Observations on designer behavior in the parti. Master's thesis, Massachusetts Institute of Technology

Goldstein KM, Blackman S (1978) Cognitive style: five approaches and relevant research. Wiley, New York

Gombrich EH (1960) Art and illusion: a study in the psychology of pictorial representation. Pantheon, New York

Gombrich EH (1968) Style. In: Sills DL (ed) International encyclopedia of the social sciences. Macmillan, New York, pp 352–361

Gombrich EH (1971) Meditations on a hobby horse. Phaidon, Oxford

Gombrich EH (1982) The image and the eye. Phaidon, Oxford

Hitchcock H (1942) In the nature of materials. Duell/Sloan and Pearce, New York

Kroeber A (1963) An anthropologist looks at history. University of California Press, Berkeley

Kubler G (1979) Towards a reductive theory of visual style. In: Lang B (ed) The concept of style. University of Pennsylvania Press, Philadelphia, pp 119–127

LaRue J (1970) Guidelines for style analysis. W. W. Norton, New York

Lipman J (1986) Frank Lloyd Wright and the Johnson Wax Buildings. Rizzoli, New York

MacCormac RC (1968) The anatomy of Wright's aesthetic. Architect Rev 143:143–146

MacCormac RC (1974) Froebel's kindergarten gifts and the early work of Frank Lloyd Wright. Environ Plan B Plan Des 1:29–50

Manson GC (1953) Wright in the nursery; the influence of Froebel education on his work. Architect Rev 113:349–351

Manson GC (1958) Frank Lloyd Wright to 1910. Van Nostrand Reinhold, New York

Messick S (1976) Personality consistencies in cognition and creativity. In: Messick S et al (eds) Individuality in learning. Jossey-Bass, San Francisco, pp 4–22

Meyer LB (1979) Toward a theory of style. In: Lang B (ed) The concept of style. University of Pennsylvania Press, Philadelphia, pp 3–44

Newton E (1957) Style and vision in art. Listener 57:467–469

Rapoport A (1969) House form and culture. Prentice-Hall, Englewood Cliffs

Rorick H (1975) The Frank Lloyd Wrighter. In: Negroponte N (ed) Reflections on computer aids to design and architecture. Petrocelli, New York, pp 49–60

Schapiro M (1962) Style. In: Tax S (ed) Anthropology today: selections. University of Chicago Press, Chicago, pp 278–303

Scott G (1980) The architecture of humanism. The Architecture Press, London

Scully VJ (1960) Frank Lloyd Wright. George Braziller, New York

Simon HA (1975) Style in design. In: Archea J, Eastman C (eds) Proceedings of the 2nd annual Environmental Design Research Association conference. Dowden/Hutchinson & Ross, Stroudsburg, pp 1–10

Smithies KM (1981) Principles of design in architecture. Van Nostrand Reinhold, New York

Sparshott F (1965) The structure of aesthetics. University of Toronto Press, Toronto

Steiner FH (1982) Frank Lloyd Wright in Oak Park and River Forest. Sigma Press, Chicago

Storrer WA (1978) The architecture of Frank Lloyd Wright. MIT Press, Cambridge, MA

Streich ER (1972) An original-owner interview survey of Frank Lloyd Wright's residential architecture. EDRA 3/AR8 conference, 2(13.10):1–8

Suedfeld P (1971) Information processing as a personality model. In: Schroder HM, Suedfeld P (eds) Personality theory and information processing. Ronald Press, New York, pp 3–14

Torossian A (1937) A guide to aesthetics. Stanford University Press, Stanford

Twombly RC (1979) Frank Lloyd Wright, his life and his architecture. Wiley, New York

Van Zanten D (1988) Schooling the Prairie School: Wright's early style as a communicable system. In: Bolon CR, Nelson RS, Seidel L (eds) The nature of Frank Lloyd Wright. University of Chicago Press, Chicago, pp 70–84

Wackernagel W (1923) Poetics, rhetoric, and the theory of style. In: Cooper L (ed) Theories of style. Macmillan, New York, pp 1–22

Weitz M (1970) Problems in aesthetics. Macmillan, London

White CEJ (1971) Letters from the studio of Frank Lloyd Wright. J Architect Educ 25:104–112

Whyte LL (1961) A scientific view of the creative energy of man. In: Philipson M (ed) Aesthetics today. World Publishing, Cleveland, pp 349–374

Wils J (1985) Frank Lloyd Wright. In: Brooks HA (ed) Writings on Wright. MIT Press, Cambridge, pp 139–145

Wollheim R (1979) Pictorial style: two views. In: Lang B (ed) The concept of style. University of Pennsylvania Press, Philadelphia, pp 129–145

Wright FL (1908) In the cause of architecture. Archit Rec 23:155–221

Wright FL (1925) Frank Lloyd Wright: the life work of the American architect. Wendingen, Santpoort

Wright FL (1928) In the cause of architecture. Archit Rec 63:49–57

Wright FL (1931) Modern architecture....Kahn lectures for 1930. Princeton University Press, Princeton

Wright FL (1941) Studies and executed buildings. In: Gutheim FA (ed) Frank Lloyd Wright on architecture. Duell/Sloan and Pearce, New York, pp 59–76

Wright FL (1943) An autobiography. Duell/Sloan & Pearce, New York

Wright FL (1953) The future of architecture. Horizon, New York

Wright FL (1954) The natural house. Horizon, New York

Wright FL (1966) Frank Lloyd Wright, his life, his work, his words. Horizontal, New York

Chapter 6
Creation of Style in the Design Process

A style could be seen as a cultural sign, a social phenomenon, a product symbol, or a way of doing things coming from some individual or group effort. Apart from examining style from outcomes of intentional purposes, this chapter focuses on how an individual style is shaped in design processes approached from the perspective of design cognition. Particularly, attention is on the schematic design stage, which is considered the most critical stage for a design project to be formulated before its final form is determined and constructed. The purpose is to explore how a style is developed in the design process. Fundamental concepts are based on the supposition that an individual style is identified by a set of common features created by a series of mental processes on managing design information. If products share many common features, then there should be many common processes of using similar information utilized in the process, and the style of the products will be strongly expressed and recognized. Therefore, the numbers of common features in products and similar operational factors in processes determine the degree of style, and a style can be defined and measured by the function of common features and factors.

Following this line of thought, a style can be explored through identifying the factors occurring in the design processes. This concept assumes that a style is a designer's distinctive personal and professional way of doing things resulting from a series of decisions made in the process to deliberate the pattern of expression. In architectural design, the way of doing things refers to certain procedures of designing. Theoretically speaking, a design consists of a series of design processes, each of which can be seen as an episode that is intended to achieve a design goal by utilizing design methods. These design methods may include manipulating mental images (Downing 1992; Chan 1997) for form generation, applying design rules to generate solutions, or utilizing design constraints to limit the memory search load. These fundamental components—mental images, design rules, design constraints, and design goal—are the means used to create tangible products. The tangible products are physical forms with certain levels of abstraction generated in the schematic design stage. The level of abstraction presented in a form depends on how far the design has been conducted and the level of details achieved. The less abstract a

© Springer International Publishing Switzerland 2015   197
C.-S. Chan, *Style and Creativity in Design*, Studies in Applied Philosophy,
Epistemology and Rational Ethics 17, DOI 10.1007/978-3-319-14017-9_6

design product is, the more detailed features will appear; the more features that appear in products, the stronger the opportunities to manifest a style.

Explained from the operational point of view, designers must go through a number of processes and utilize various methods to achieve certain design goals. If the design methods are repeated, then certain factors may appear and reappear despite their different design natures. The repetition of these factors in different design processes will generate similar features by which a style arises. A large number of similar processes will generate a large number of similar features, which will express the style more strongly. This defends the proposition that a style is a function of design processes. If the repetitive factors that correlate with the repetitious generation of features in design processes can be operationally defined and verified by scientific methods, then the driving forces of a style can be explained.

In order to justify the concept, empirical studies must be conducted by applying analytical and scientific methods to scrutinize a design process stepwise. One such method is **protocol analysis**, which collects cognitive data from design processes and analyzes the data to discover the cognitive mechanisms applied. This analytical method will yield insight into the processes by which designers solve design problems in different design projects but still produce the same style. As such, a case study on a practicing architect is conducted in this chapter to discover the driving forces of a style.

6.1   Methods for Exploring Design Processes

Protocol analysis, developed in the field of cognitive psychology, is a series of procedures for collecting verbal data and systematically analyzing the data to prove hypotheses on thinking patterns or cognitive mechanisms utilized for generating solutions or decisions. Details of utilizing protocol data to study cognitive phenomena in design, and methods used to analyze design behavior have been covered extensively in the literature (Cross and Cross 1995; Ericsson and Simon 1996; Eastman 2001; Chan 1990a, 2008). Research has also shown that protocol data provides robust evidence for the causal-effect and explanations of the phenomenon of concept formation (Dorst 1995). Thus, through the use of protocol analysis, it is possible to identify the repeating factors that appear across a number of design processes, which will reveal the story of the creation of a style.

A design process can be seen as an information processing sequence. When information processing for solving design problems occurs within the designer's mind, it is necessary to first comprehend the cognitive aspect of the design process. Then the explanations of how a style is generated will be meaningful and justifiable. In this regard, the study of intelligence and intelligent systems, with particular reference to intelligent behavior as computation (Simon and Kaplan 1989), becomes especially useful. The approach applied through the information computational point of view is to observe a number of a designer's processes as they occur, identify the repetitious features in products, then sort out the repeated factors to explain the underlying mechanisms involved in manufacturing the features that symbolize the style.

The ultimate purpose of this approach is to explore the mechanisms that produce a designer's individual style and to find whether these mechanisms have been repeatedly reused as generative forces. According to information processing theory, cognitive aspects might differ from person to person but not within the same person. In other words, the basic cognitive mechanisms utilized by a person should remain constant to be claimed as the driving forces of a personal style within a particular span of time during the designer's entire career life. Thus, if a number of protocol data sets are collected from one person within a period of time, it is appropriate to observe the generic patterns that manifest the style within this period.

6.2 Hypothetical Factors Generating Styles

Based on the hypothesis that a style is manipulated by prominent two- or three-dimensional entities (which are termed features) generated by design processes, it is possible to see that a style is a product of certain factors that occur in a design process. The process of architectural design is a unique thinking process that involves the use of logical reasoning, manipulation of images, two-dimensional or three-dimensional representation, and many other mental activities to reveal forms generated in the designer's mind. In general, there are two ways to elucidate design thinking. Seen from the view of information science, design problem solving has a logical and explainable sequence of processing information. From the fine arts perspective, design is to develop a piece of art with intuition involved to trigger the creation of beautiful features and forms occurring in the designer's mind. Nonetheless, to achieve a persuasive study on design thinking, efforts must be focused on how information is processed in the mind, which is symbolically described as a black box in the brain.

The information involved in design thinking has two key components: (1) symbolic representation of design knowledge and reasoning that moves the design stage forward[1]; and (2) iconic representation that portrays the mental image of the design forms. Within a design course, these components are intentionally manipulated and handled to achieve certain goals until the final product is accepted by both the designer and the client, or by the students and the instructor. Therefore, a design

---

[1] **Representation** has been explained in Chap. 2. In short, it means a set of conventions about how to describe a class of things. Knowledge representation in design signifies a set of conventions available for rationally describing design knowledge. The set of conventions can be seen and elaborated as a hierarchical network of symbols stored in the designers' memory, and each symbol in the network has attributes defining architectural objects or tectonics.

process can be seen as executing a series of design goals sequentially to arrive at an acceptable solution. This is the fundamental characteristic of problem solving. Within the processes, several factors can be seen as operators to move problem-solving stages forward and to identify the driving forces that create a style. Thus, ten hypotheses are proposed.

6.2.1   Personal Preferences on Certain Forms

Designers may have different tastes, unique preferences for certain primitive forms, or favor certain materials used or created in their past designs (Schapiro 1962). The primitive forms are geometric shapes and volumes used in design compositions. For instance, **Le Corbusier** preferred free curve in some projects, whereas **Frank Lloyd Wright** preferred polygonal bay windows in his Prairie House period (Chan 1992). Thus, the consistent application of the same form within a particular span of time labels a designer's style. Usually, different forms accommodate different functionalities. A designer may choose the way things are done, which is reflected in the form used. For example, a staircase can be spiral, rectangular, circular, or straight. If a circular staircase is used by the designer most of time, it marks the designer's idiosyncrasy, and its repetitious use signifies a style. Although preferences may change with time, the consistent application of the same preferred form within a specific time span labels the designer's style. *This factor of repetition of personal preferences on forms or materials sets up* **Hypothesis 1**.

6.2.2   Design Goals

A design goal accomplishes a particular design task. Different goals can be used as a means to differentiate design stages for describing different design activities. Within each goal, designers search for either the knowledge that provides solutions, or the sources of precedent solutions existing in memory. The knowledge for potential solutions could be seasoned knowledge obtained from learning, practicing, or from various information resources, whereas the precedent solutions existing in memory could be images from cases generated in previous designs.

In design, the process is usually guided by a goal sequence. Such a goal sequence not only defines the design strategy used in a design but also reflects the designer's general design method. Different designers might have specific ways of approaching a design and specific strategies for tackling design problems. If a designer always applies the same sequences of goals (or procedures) on different projects, then the products will exhibit certain similarities in results. As long as a pattern of design strategy repeats in designs, a style appears. Thus, **Hypothesis 2** *is that the sequences of goals, which refer to the design strategies, generate a style.*

6.2.3 Design Constraints

An experienced designer may have a larger knowledge repertoire.[2] But only a limited subset is utilized at different stages for different goals, and this sets up a design constraint. If a design constraint is applied to generate a schematic design solution, affiliated with many design units, and maintained across the entire design process, then it is a **global constraint**. According to previous studies (Chan 1990a), a characteristic of a global constraint is that it appears early in a design process to set limits on the design form. Because global constraints are parts of design knowledge, the retrieval of a global constraint from a repertoire stored in memory must result from choices that reveal a style. Some choices are exercised at the time of knowledge acquisition, some at the time of memory retrieval, and some after the successful retrieval. Thus, ***Hypothesis 3*** *is that the choice of global constraints generates style.*

Global constraints have profound influences on forms at an early design stage, because the order in which global constraints appear in a design determines: (1) the ways a designer tackles a design problem; and (2) the ways of approaching a problem. The timing of applying a particular constraint reveals its importance and necessity to the problem. If a designer applies the same priority order of global constraints to solve many problems, this order, of course, dominates the sequences of introducing global constraints into the problem and is the generative source of a style. Explained from another perspective, a design might have more than one global constraint considered and their priority of consideration would determine the character of the products. Because a design problem has indefinite solution paths, each path could lead to a final solution.[3] Therefore, a constraint helps limit the search effort in the problem space and defines the design solution path. If a designer always uses the same set of design constraints associated with a particular goal in designs, then the design solution paths will show some similarities and the products will possess resemblances. Thus, ***Hypothesis 4*** *is that a similar priority order of managing global constraints will generate similar design style.*

---

[2] Designers have a large set of design knowledge accumulated from experience. Generally speaking, design knowledge as explained in Chap. 2 consists of two major parts, declarative knowledge and procedural knowledge. Schemata are kinds of "knowledge representation" computer codes construed to show knowledge in design constraint format with design rules embedded. They usually are expressed in the form of **production systems**, which is an ordered set of processes called productions. Each production contains a paired condition and action. The condition part contains declarative knowledge, and the action part has a set of rules representing the procedural knowledge. Whenever a condition is satisfied, an action is executed (see Newell and Simon 1972, pp. 32–33.) In architectural design, the application of a recalled schema from memory at a particular design stage will make its associated rules available and a solution can be generated.

[3] Usually, a design problem contains an initial situation for the designer and is referred to as the initial state. A goal state is the stage at which the design problem has been resolved. The process of problem solving from initial state to goal state can be modeled as a series of transformations generating a sequence of problem states. A problem state is a particular stage in which a designer knows a set of things, and is referred to as a knowledge state. The various states that the designer can achieve are called problem spaces. The various ways of changing one state into another are symbolically called paths. A solution path is the path leading to the final solution (see Chan 2008).

On the other hand, if a constraint is applied to a few design units at a particular stage to resolve local problems, then it is termed a local constraint. In design, each goal stage has a particular set of design constraints developed for solution generation. A solution to a design problem is produced by imposing a constraint onto the process and searching for the associated knowledge from memory to meet the problem situation. Of course, designers must find and compose a form to satisfy a set of constraints. Although some constraints are given in the design problem by clients, the designer also must provide personal constraints to help organize problems and develop solutions. Furthermore, the generated solutions should be evaluated by applying other sets of constraints for finalizing the solution. Thus, **Hypothesis 5** *is that the common sets of global and local constraints applied for generating solutions and evaluating solutions among alternatives generate style.*

Design constraints are bound to design goals. Each design goal has its own specific design constraints. For instance, the constraints used for developing floor plans differ from those used for generating elevation plans. The change in goal order will influence the use of bound constraints and affects the form of final products. Thus, **Hypothesis 6** *is that maintaining the priority order of goals will sustain the priority of constraints for generating the same style.* In short, *the interaction of goal order and constraint order affects the generation of style.*

The knowledge in design constraints consists of spatial relationships among building elements, physical laws, structural and material properties of buildings, qualitative requirements of spaces, or energy efficiency. Since the repertoire of design knowledge stored in memory is in the form of **chunks** (Miller 1956; Simon 1974; Larkin et al. 1980), thus, the format of constraint could also be seen as a chunk with some knowledge or rules embedded. Design rules are procedural knowledge that tells one how to perform a task. For a design unit, there are a number of constraints applicable, and each has a number of ways to fulfill the constraint. If a designer applies the same rules to solve the same constraints for the same design units, resolutions will resemble each other and a style is determined. Thus, **Hypothesis 7** *is that repeatedly using the same constraint with the same set of rules may generate similar solutions, and thus a style emerges.*

6.2.4 Search Pattern and Order

The large sets of knowledge chunks and image schemata in memory become, metaphorically, a data ocean full of data items. The processes of searching for an appropriate item of information and applying it shape the search pattern. The search pattern can be a search for mental images, design rules, design constraints, or design goals. Different designers may have different search patterns. For instance, some designers rely more on the search for mental images and less on design rules or vice versa. Thus, the search pattern utilized in a design affects the generation of a style.

Search order (Simon 1975) is another factor in the design process that determines a style. Designers have certain sets of procedures to determine which design unit (space requirement), design constraint, or design goal to start with. Because the process of design is considered as the process of satisfying the design constraints (Simon 1969, p. 138), the first object taken into consideration will satisfy a specified set of constraints and yield a first satisfying solution. Any successive object taken into account will be based on the first solution to generate a further solution that also is satisfying. Thus, the order in which possibilities are examined will have a major influence on the solution and will ultimately cast an influence on forms. Therefore, **Hypothesis 8** *is that search order also serves as a driving force of an individual style.*

6.2.5   Presolution Models

Presolution models are developed with experience (Chan 1990a), in which a designer starts out with some constraints, operates a procedure, then arrives at a product. In his or her later designs, the designer simply copies his or her own product and, therefore, maintains the same style. For instance, Manson pointed out that the Walser-Barton House (1903) was a schema that **Frank Lloyd Wright** often used as the answer to the problem of fitting a Prairie House to a narrow lot and a limited budget (Manson 1958, p. 140). Thus, a designer could generate a solution and reuse it for later design to sustain the same style. Thus, **Hypothesis 9** *is that presolution models not only generate a style, but also maintain the same style.*

6.2.6   Mental Images

Mental image (Chan 1997) is another design constraint. Designers often mentally manipulate the spatial relations among fundamental geometric shapes and represent the product in graphic or model representations. In other words, sketches are tried out in the mind's eye before they emerge onto the drawing board. Whenever an architect is confronted by a design problem, his or her initial problem-solving process involves a visualization of potential solutions; that is, a mental image of something not real or present. It is a mental reproduction or imitation of the form of a physical object. **Gombrich** (1960) explained that one way a painter learns to paint a mountain is to go out to the field to observe a mountain. The process of observation would abstract the form of the mountain; then a pattern or a prototype is developed and stored in memory. Such a pattern not only has the form of a mental image, but also is the painter's concept of the mountain. The term "**schema**" is used to describe the pattern stored in memory. The formation of the schema makes a pattern available, and the painter is able to draw his own mountain, which may not be any particular mountain but the painter's concept. In a painter's memory, there are thousands of patterns of objects that form a data bank. Similarly, in the field of architectural design, a designer may have many images of architectonic elements stored in his or her memory to form a part of the image schemata. This notion shapes **Hypothesis 10** *that repetition of the same image may create similar features in design products, and a style is generated.*

6.3   An Empirical Study

A series of design tasks are conducted in the following example to verify the
proposed factors in the ten hypotheses that generate a style. Approached from the
problem solving point of view, it is assumed that a design task is a major design
problem uniquely defined by a **design brief**.[4] For instance, a simple residential
design brief may consist of a particular owner, a unique site, a certain set of constraints, a particular usage of the building, and a fixed set of space requirements
termed design units. Changing any of the settings will create a new design problem.
Yet, regardless of the changes made in the design brief, the proposed hypothetical
factors should remain across designs to generate constant forms that manifest a
style. The following experiments are designed accordingly to purposefully change
the brief for detecting the existence of the consistent factors in the design process
that generate an individual style.

6.3.1   Subject

The subject, who had won several awards and had published his design works in
Architectural Record, is an experienced architect. At the time this study was conducted, he had been practicing in the field for more than 25 years, owned an architectural firm in downtown Pittsburg, and was principally in charge of the firm's
design activities. The subject was selected from a list of nine architects (from the
Pittsburg area) who had won at least two architectural awards. The subject was
chosen after initial interviews scheduled randomly with the architects, and was the
first architect on the list who was willing to participate. Thus, the subject selection
process was an independent **random selection**.

6.3.2   Experimental Tasks and Procedures

The experiment was designed to mimic real design activities in a laboratory setting.
Design works in this experiment were carried out in a situation similar to that which
should take place in an ordinary architectural office or in a design studio in schools.
The subject preferred to work on design through pencil-and-paper sketch mode. He
brought his own drawing equipment, which included different sizes of drawing

---

[4]**A design brief** usually lists the design requirements and describes the nature of a design task,
which can be categorized as the owner's background (social, cultural, and economical), building type,
design issues (design constraints), site issues (climatic, contextual, or geographical conditions),
and space requirements (the quantitative and qualitative requirements). The term "owner" should
be referred to as users in a larger context or in a complicated building design.

Table 6.1 The systematic change of design sessions (A, B, C, and D represent different values of a variable)

Sessions	Owner	Site	Climate	Design units	Design issues	Budget	Result
#1	A	Ⓑ	A	A	A	A	Solution A
#2	A	A	A	A	A	A	Solution B
#3	A	A	A	Ⓑ	A	Ⓑ	Solution C
#4	A	A	A	A	Ⓑ	C	Solution D
#5	Ⓑ	A	A	A	A	A	Solution B
#6	A	A	A	A	A+	A	Solution B
#7	A	A	Ⓑ	A	A	A	Solution B
#8	C	C	C	C	C	D	Solution E

pens, scales, and a set of color pencils, to the laboratory. The experimenter provided the drawing paper.

The subject was invited to do eight design sessions. The design problems in the eight sessions were systematically changed. Five sets of problem-definition variables were used to set up the basic design problem. A change in any one variable delimited a new design problem, as those shown in Sessions 1, 3, 4, 5, 6, and 7 in Table 6.1. Session 8 had a totally different set of variables. The goal was to change the problem-definition variables for the purpose of identifying the invariant forces that appeared in each process.

The basic design premise was a simplified residential design of a one-bedroom house for a young, single, male professor. The flat site (70 ft by 110 ft) running from northeast to southwest was located in an urban area in Pittsburgh. It overlooked a landscaped park across a street and was surrounded by adjacent houses on the other three sides (site B, see Fig. 6.1). The weather in winter was cold, damp, and windy with winds from the southwest, and in summer was hot and humid with breezes from the northeast. The owner could only invest $70,000 and wanted views toward the park and to reduce the noise from the street side. Three design issues were required: cost limitation, view requirement, and noise reduction. Five design units were essential and mandatory: a living room, dining room, kitchen, bedroom, and bathroom. This fundamental brief served as the basic structure of the design project and was randomly selected first to balance out possible bias; thus, it was titled Session 1.

In Session 2, the site with the same size as Session 1 had been changed to a rural Pittsburgh area with the Allegheny River running through the back (see site A in Fig. 6.2). The property was accessible from a street in front, with adjacent houses to the left and to the right. Other conditions and requirements remained the same as in Session 1. Session 3 kept the same conditions as in Session 2, except that an extra study room was added and the budget was increased to $85,000. Beginning with Session 4, only one condition changed in each subsequent session (see Tables 6.1 and 6.2). For example, the design constraints changed to the requirements of privacy,

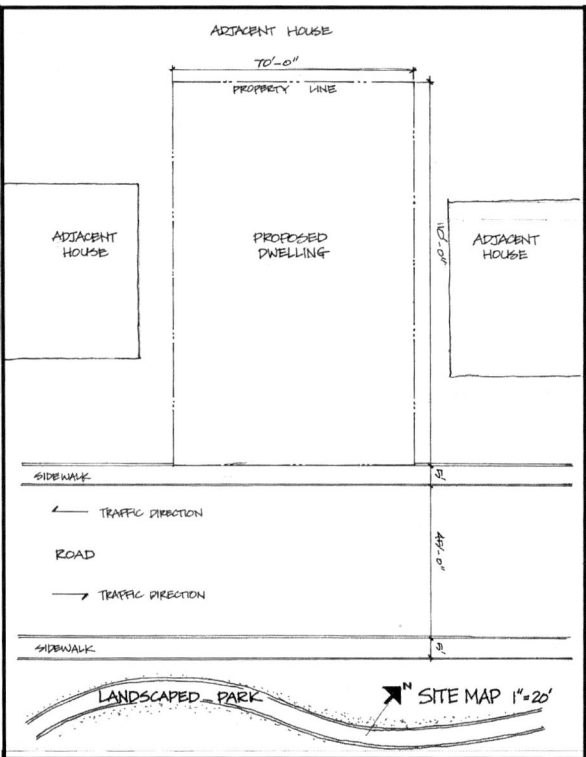

Fig. 6.1 Site B for Session 1

intimacy, skylight, and $140,000 budget limit in Session 4. The owner in Session 5 was changed to a retired executive of a company who loved music very much. In Session 6, the original owner insisted on having three design issues considered in the priority sequences of noise, budget of $70,000, and view to explore whether the imposed sequences would have changed the design results. In Session 7, the weather of the site was changed to a mild climate zone similar to the Seattle area that had no heavy snow in winter and was warm in summer. Winds came from the north in summer and from the south in winter.

An entirely new set of variables was given in Session 8. The owners were a middle-aged couple whose son lived in another city. The couple liked to entertain their friends. Occasionally, they would invite friends to their house for dinner and social gathering. The husband was a filmmaker, and the wife was a well-known professional watercolor painter. The site (80 ft by 95 ft), located in a southwestern rural area of Connecticut, sat on a gently sloped hillside near the edge of a forest (maple trees) that was northeast of the site (see site C in Fig. 6.3). The site also had

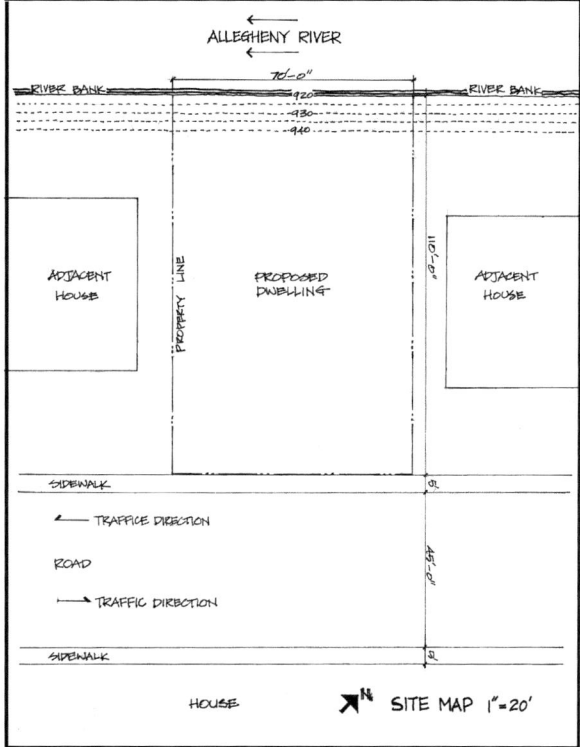

Fig. 6.2 Site A for Session 2

several maple trees on two sides and overlooked a town some distance south. The owner would invest $300,000 and would like to have a dining room, a kitchen, two bedrooms, bathrooms, a study room, a professional workshop for painting, and a two-car garage. All these design sessions were restricted to the residential building type to avoid dramatic changes in design approach.

Before the experiment, the subject was informed that each session was independent of the others and that he was expected to design in the way he ordinarily does. The eight sessions were conducted at different times to diminish the possibility of learning effects—which might cause the direct association of the results obtained from previous designs. The procedures for each session were similar. First, a design brief was shown to the architect. Then, he was asked to start designing with pencil and paper and to voice his thoughts throughout the study. The information provided in the design brief was reduced to a minimum to discover the kind of design information that was developed by the designer. The subject was not encouraged to ask questions, but was expected to make assumptions about certain design issues. At the

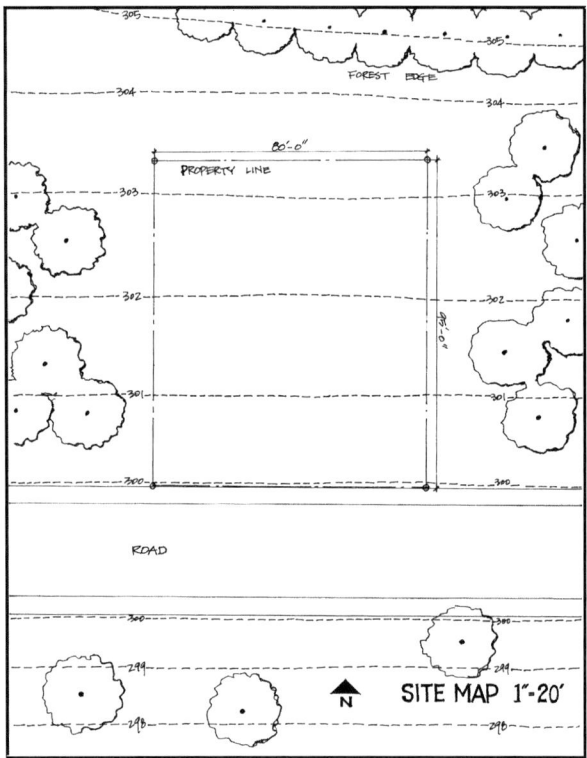

Fig. 6.3 Site C for Session 8

end of each session, the subject was required to finish a design product presented with floor plan, site plan, and elevation drawings. The processes were recorded by two sets of cameras. One was hung above and in front of the subject to capture the entire background; another was set to the side to focus on detail drawings. There was no time limitation, but the architect spent an average of 5 h to finish each design. Data were collected from verbal protocols obtained from each design.[5]

---

[5] It is essential to get enough data about the individual designer, the information he has, and how he is processing it. The method is to use the high output rate of verbal behavior as data which are termed verbal protocols. The process is to instruct the designer to report verbally everything he thinks about as he works on the design problem.

Table 6.2 List of the eight design sessions

Session 1	*Session 2*
Owner: A single male Professor	Owner: A single male Professor
Site: *B in Pittsburg urban area*	Site: A in riverside
Climate: Pittsburg weather	Climate: Pittsburg weather
Design units: BR, LR, DR, K, bathroom	Design units: BR, LR, DR, K, bathroom
Design issues: Noise, view, maximum cost	Design issues: Noise, view, maximum cost
Budget: $70,000	Budget: $70,000
Session 3	*Session 4*
Owner: A single male Professor	Owner: A single male Professor
Site: A in riverside	Site: A in riverside
Climate: Pittsburg weather	Climate: Pittsburg weather
Design units: *BR, LR, DR, K, bathroom and study room*	Design units: BR, LR, DR, K, bathroom
Design issues: Noise, view, maximum cost	*Design issues*: *Privacy, intimacy, skylight, and maximum cost*
Budget: *$85,000*	*Budget*: *$140,000*
Session 5	*Session 6*
Owner: *A retired executive*	Owner: A single male Professor
Site: A in riverside	Site: A in riverside
Climate: Pittsburg weather	Climate: Pittsburg weather
Design units: BR, LR, DR, K, bathroom	Design units: BR, LR, DR, K, bathroom
Design issues: Noise, view, maximum cost	*Design issues in sequence*: *Reduce noise, $70,000 budget consideration, view*
Budget: $70,000	
Session 7	*Session 8*
Owner: A single male Professor	*Owner*: *A middle-aged couple*
Site: A in riverside	*Site*: *C in Connecticut*
Climate: *Seattle weather*	*Climate*: *Connecticut weather*
Design units: BR, LR, DR, K, bathroom	*Design units*: *2-BR, LR, DR, K, bathroom, study room, workshop, garage*
Design issues: Noise, view, maximum cost	*Design issue*: *maximum cost, open and sunny interior space*
Budget: $70,000	*Budget*: *$300,000*

6.4   Experimental Results

The architect finished the eight design sessions within 2 months. As shown in Table 6.1, there were five sets of drawings (Solutions A–E) generated in Sessions 1, 2, 3, 4 and 8. Because drawings of Session 3 were not adapted due to video failure, there were four sets of drawings {1, 2, 4, 8} used for data analysis. Design activities occurring in each session are briefly explained in the following.

1. Session 1: Within 5½h, the architect completed this design and evaluated Solution A (see Fig. 6.4) as a satisfactory one.
2. Session 2: Although the site was changed, the architect recalled and applied the design Solution A developed in Session 1. His strategy was to modify the site plan to accommodate Solution A. After working for 1½h, he discovered that he was stuck on the applied solution and decided to take a break. His strategy was to walk away whenever he was stuck and return to the drawing board at another time. A week later, he continued the design and spent 4 h in developing a new Solution B (see Fig. 6.5).

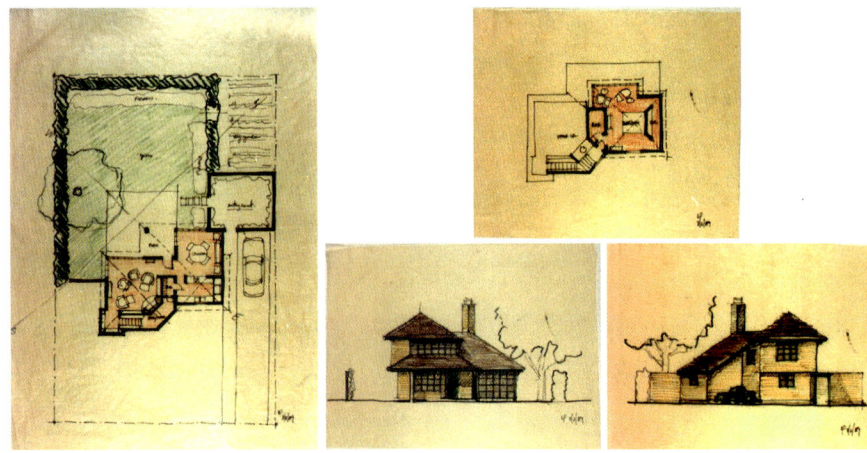

Fig. 6.4 Plans and elevations of session 1 (Source: Reprinted from Chan (2001). Figure 3, p. 332, Copyright (2001), with permission from Elsevier)

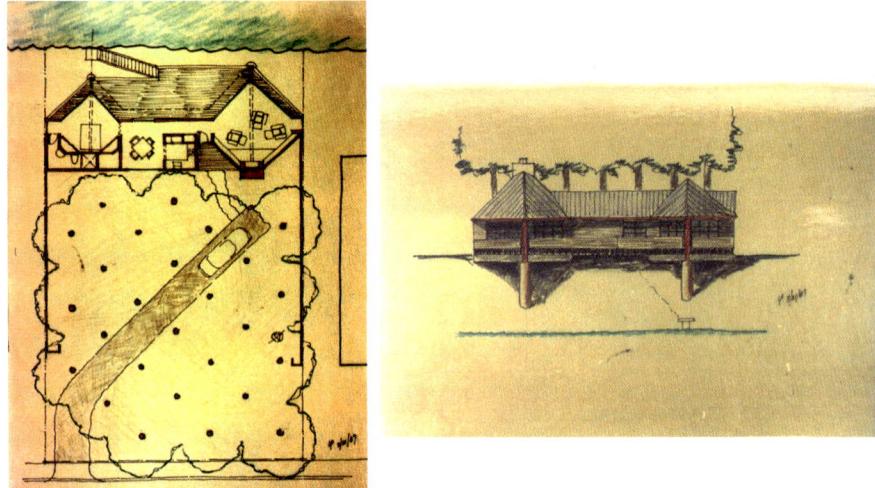

Fig. 6.5 Plan and elevation of session 2 (Source: Reprinted from Chan (2001). Figure 2, p. 331, Copyright (2001), with permission from Elsevier)

3. Session 3: A study room was added to the set of design units. The architect recalled Solution B in Session 2 and did several modifications to fit the study room into the context of Solution B. Three hours later, he completed a Solution C (see Fig. 6.6), which was to add a lower level unit under the living room portion to serve as the study room. This lower level is accessible through a spiral staircase located at the east end. This solution fit perfectly and elegantly with the

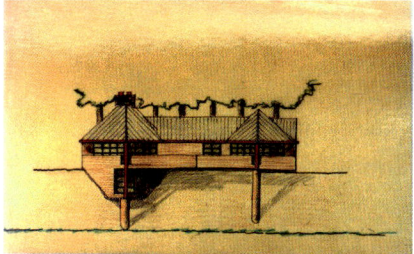

Fig. 6.6 Floor plan and elevation of Session 3 (Source: Reprinted from Chan (2001). Figure 6, p. 336, Copyright (2001), with permission from Elsevier)

previous solution in terms of structural, functional, and site context considerations. The architect called this solution a remodeling of Session 2 without having major changes in plans and elevations. Unfortunately, a machine failure occurred in the first 2 h of the experiment. Consequently, one video lost audio signal and could not be used for data analysis. However, in the following experiments, an extra sound recorder was installed specifically for audio recording.

4. Session 4: The architect worked on this session for 2 h until he obtained a final conceptual scheme. Because of physical fatigue, he decided to come back a week later to continue working on the refinement of the scheme. In the follow-up sessions, he spent 3 h finishing the whole design, called Solution D (see Fig. 6.7).

5. Session 5: In this session, the owner of the project was changed to a retired executive who loved music. The architect recalled Solution B of Session 2 and compared the solution with the design requirements. He indicated that Solution B fitted the new owner well. In particular, the bookshelves in the living room area developed in Solution B could be used for shelving musical instruments. Therefore, after 20 min of evaluation and reflection, he decided not to generate a new solution. Protocols quoted in the excerpts, the subject indicated that:

   "I don't see anything different." "The rooms are the same; the only difference with this person is that instead of books, he is into music. And he is retired, so he may spend a little more time here. But music is his love, and he wants the view, and this was his…he would buy this house from the other guy, he ought to buy the professor's house that we designed in the second one."

6. Session 6: The design order was imposed in this session. The architect evaluated Solution B against the given design constraints and indicated that the solution satisfied the constraints well. He remarked that if this session had come first, he would possible have generated a totally different solution, but in this session he was not motivated enough to pursue a new design. Thus, this session stopped. The protocol data showed that:

   "This is the same program as Session number two, except that … the rooms are the same, this (site) is the same. The only thing he said here… and the budget is the same, and it supposed to be views, and all you have done is to prioritize the

Fig. 6.7 Plans and elevations of Session 4 (Source: Reprinted from Chan (2001). Figure 4, p. 333,
Copyright (2001), with permission from Elsevier)

> *noise, the dollars, and the views…Well, I might come back to this very same
> thing (of Solution B), and say … all he has done is to come back and say that he
> wants to prioritize three issues. I am not sure how that would change (Solution B).*"

7. Session 7: The climatic condition was different in this session and winds were
   from a different direction to the site. Yet, the architect recalled Solution B again
   and evaluated the design requirements against the solutions. He indicated that

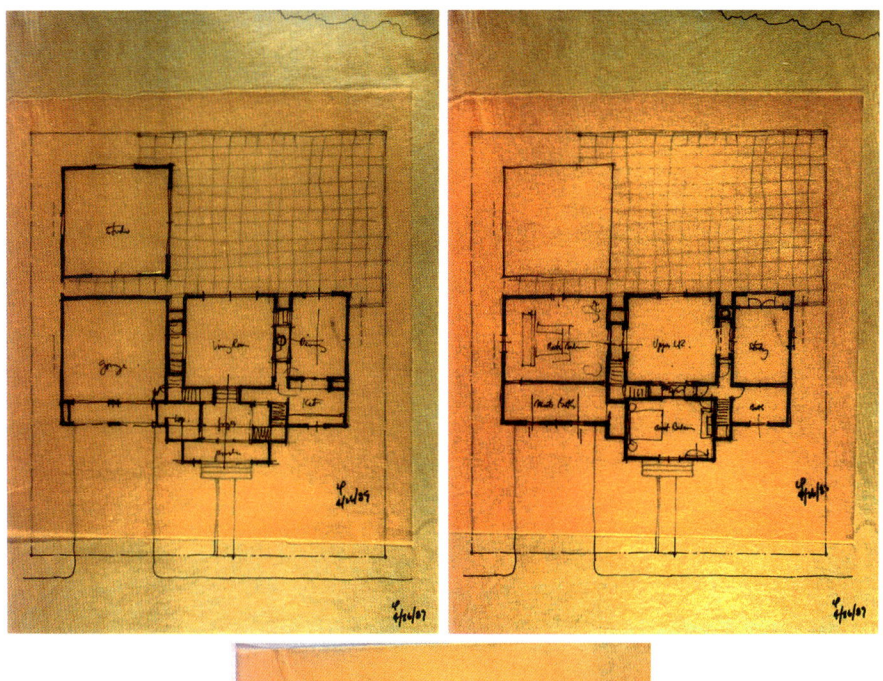

Fig. 6.8 Plans and elevation of Session 8 (Source: Reprinted from Chan (2001). Figure 5, p. 334, Copyright (2001), with permission from Elsevier)

there was no conflict between the solution and the requirements at all. Therefore, no new solution was provided. The subject says:

"I don't see any major climatic information that would create a significantly different orientation of this house (Solution B). I really don't." "I think the only thing I could conceive here is that the … if you really want to do a very sophisticated shading here, you might want to do some changes of the overhang here (along the north side of the deck). Nothing else."

8. Session 8: The architect was excited about this totally different design program. He spent 4 h developing a final scheme of Solution E (see Fig. 6.8).

Methods used for analyzing protocol data were approached from seeing the process as goal driven, which means that sequential events occur in the process to accomplish certain design intentions. This approach has the following steps as explained extensively in other literatures (Newell and Simon 1972; Chan 1990a, b, 2008). First, the architect's verbalizations were converted into written forms. Then the transcripts were divided by episodes, which are discernible segments of behavior in attaining a design goal. A design goal is identified by tracing a series of actions attempting to solve one design unit, which is one of the required living spaces specified by the design brief, or by identifying a particular recognizable intention under which a group of design units or an intentional design unit is to be resolved. Afterward, the transcripts of the design procedures were coded by a special format of **production systems** (Newell 1973; Valkenburg and Dorst 1998) to represent the design knowledge used to produce a design unit. This could: (1) identify the design constraints applied within a particular goal in solving a particular problem; (2) detect the associated **procedural knowledge** and **declarative knowledge** used by the architect; and (3) scientifically discover the elements that generate the common features and determine whether they do occur across design sessions.

6.5   Data Analysis

As discussed in this chapter, the phenomenon of repetition begins with the premise that a style is verified by the repetitious factors in design processes generating similar features in products. Results obtained from data analysis on this experiment show that there are, indeed, some repetitions occurring in the products and throughout the entire design processes, including similar features, design procedures, constraints, pre-solutions, and mental images.

6.5.1   Repetitious Features Shown in Products

After carefully examining the elevation and floor-plan drawings, there are common features found in Sessions 1, 2, 4, and 8, which are summarized in Table 6.3. These common features are not evenly distributed among sessions. Their distribution in the table, however, reflects the phenomenon of the degree of similarity. For example, in elevations, only one feature of horizontal board siding appeared in all four designs. But, one feature is insufficient to stand for a style. Thus, these four designs are not stylistically similar to each other, and the architect's style is not strongly expressed in the elevations of four designs at this schematic design stage. Applying the similarity measurement formula (see Chap. 4) to the data, it is found that the similarity between Sessions 1 and 2 is 5, which compares with similarities of -3 and -5 for the other sessions; therefore, Sessions 1 and 2 have the strongest resemblance of the designs generated. In contrast, Session 8 has only one feature in common with Sessions 1 and 2 and is dissimilar to these two designs.

Table 6.3 Common features appearing in four designs

	Session			
Feature	1	2	4	8
Elevation				
Horizontal board siding	+	+	+	+
Grid-pattern full-height window	+	+	+	-
Double-pitch roof	+	+	−	−
Exposed column	+	+	−	−
Brick chimney	+	+	−	−
Circular metal chimney	−	−	+	+
Corner plate	−	−	+	+
Similarity to session 1		$5-0-0=5$	$2-3-2=-3$	$1-4-2=-5$
Floor plan				
Bedroom with walk-around closet in the back	+	+	+	+
Sink in kitchen facing a window	+	+	+	+
Sink in kitchen facing a window	+	+	+	+
Corner windows	+	+	+	-
Staircase around the living room	+	−	+	−
Entrance next to the kitchen	+	+	−	−
A corner fireplace	+	+	−	−
An enclosed inner court	+	−	−	+
Kitchen in front of dining room with views through	+	−	−	+
Skylight	−	−	+	+
Symmetry	−	+	+	+
Similarity to session 1		$5-3-1=1$	$4-4-2=-2$	$6-2-2=2$

The same situation of less features also occurred in floor-plan drawings where only two common features were shared by all four designs. As shown in Table 6.3, the number of common features shared by any two sessions ranged from −2 to 2, which is considered low. Applying the similarity formula to measure the similarity of floor plans among sessions, results do not show a strong stylistic tendency. For instance, calculating the similarity of Session 1 to Sessions 2, 4, and 8; the scores are 1, −2, and 2 respectively. In comparison with **Wright**'s similarity score, the architect's style in this experiment is not very expressive in both floor plans and elevations; and thus is a weak style. The meaning of weak refers to the quantity but not quality of shared features.

6.5.2   Observation on Repetitious Utilization of Design Goals and Procedures

In organizing protocol information for data analysis, verbal data can be organized into groups representing actions of accomplishing goals. Similarly, a design process can be broken down into stages corresponding to goals. In the data, a goal is sometimes indicated by a statement or a spoken sentence with multiple actions following. Sometimes a goal is not explicitly expressed verbally by the subject, but can be distinguished by intentions of the proceeding actions and the executed tasks. Usually, the tasks and affiliate consecutive moves can be coded aggregately as one episode representing one goal. The method of grouping data into an episode can be explained by the following example.

In Session 1, the architect said: *"Ohm…well, let's see what have we resolved? I think if we have a plan organization that would be nice."* This statement indicated that the forthcoming goal was to work out a plan organization. Before this statement, the architect worked on a site sketch to develop a scheme for carrying out a well-planned site scenario. After this statement, the architect started developing primary floor plans. Thus, the goal statement and the subsequent moves (identified from protocol statements) constitute one episode, which is called "develop floor plans," a name that describes the accomplished tasks. Any episode executing the similar task of developing floor plans in other design sessions will bear the same title. This explains the methods used to organize goals for data analysis.

Sequences of design goals, in fact, reveal a designer's strategy of managing a design process to finish a design. Without having a goal sequence, a design would not move on effectively to achieve a promising solution. In a design studio, an instructor could teach the same design method, but students eventually will develop their own methods, which are different from the ones of the instructor and fellow students. Regardless of the variations among individuals, a designer may have different strategies available under different design conditions and situations. After years of practicing, an experienced architect develops a unique routine while designing a specific building type, and a set of fixed procedures should exist and be applied.

For instance, the tasks in the episode of "determining the building size" in Session 1, were: (1) to set up several assumptions for determining the cost of the site development and based on the budget constraint to determine the construction method; (2) to set up assumptions to decide the average unit cost per square foot; (3) to divide the construction cost by the unit cost to obtain the dimensions of the house; and (4) to resolve the room sizes. This methodological sequence signifies a strategic procedure that were found in Sessions 4 and 8. Thus, the episodes that carried this sequence in Sessions 4 and 8 also were labeled as "determining the building size." Diagrams shown in Figs. 6.9 and 6.10 were used by this method to represent the processes. The identified design constraints applied at each stage (or procedure, or episode) were also included in the figures. Because an episode represents an intensive effort to accomplish a specific task, the progress of episodes reflects the general method of creating a design. We can observe that the architect had a standard routine, representing his general design method, appearing in four designs.

a   **b**

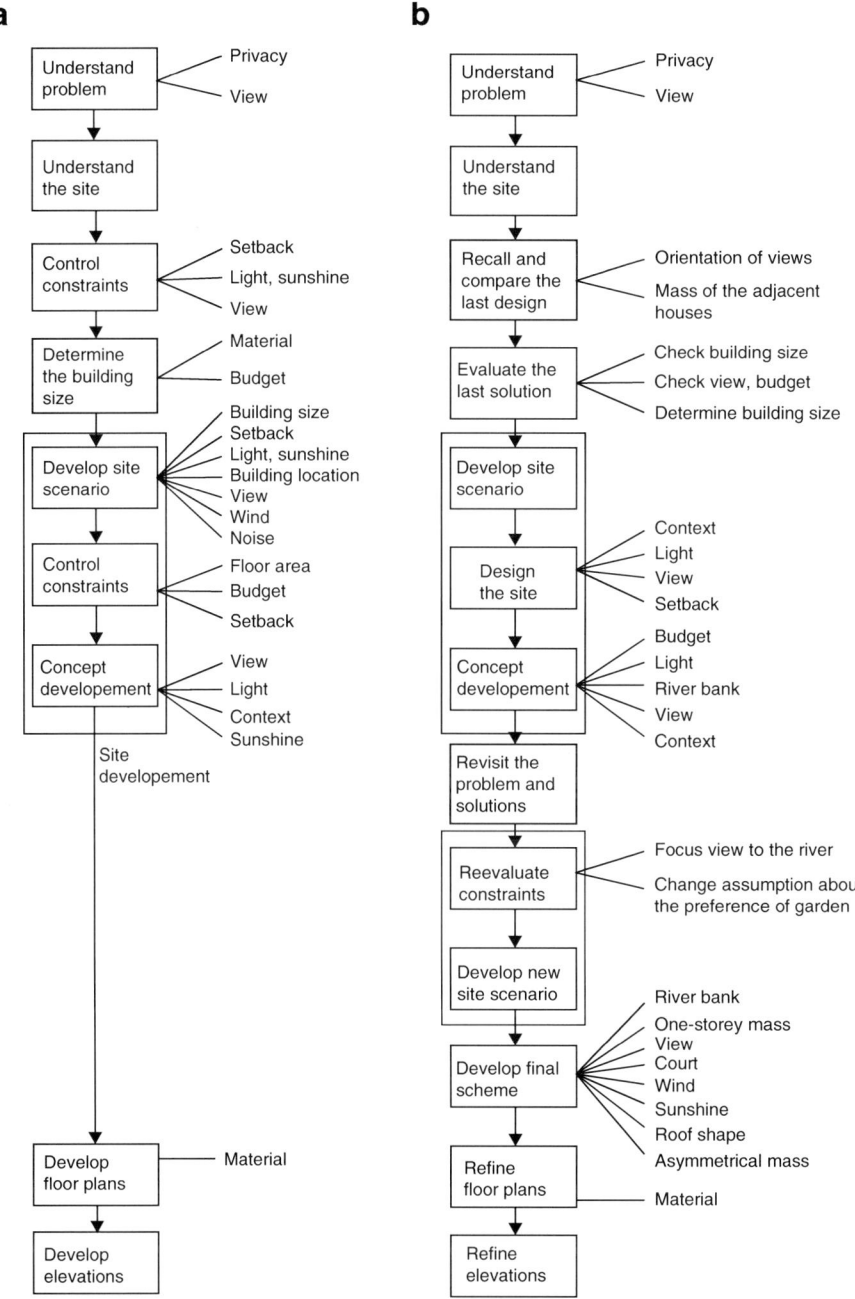

Fig. 6.9 Design goals and their sequences of Session 1 (**a**) and 2 (**b**) (Source: Copyright Pion Limited, London. Figure 5, p. 403, reproduced with permission. Chan (1993), Websites: www. pion.co.uk, www.envplan.com)

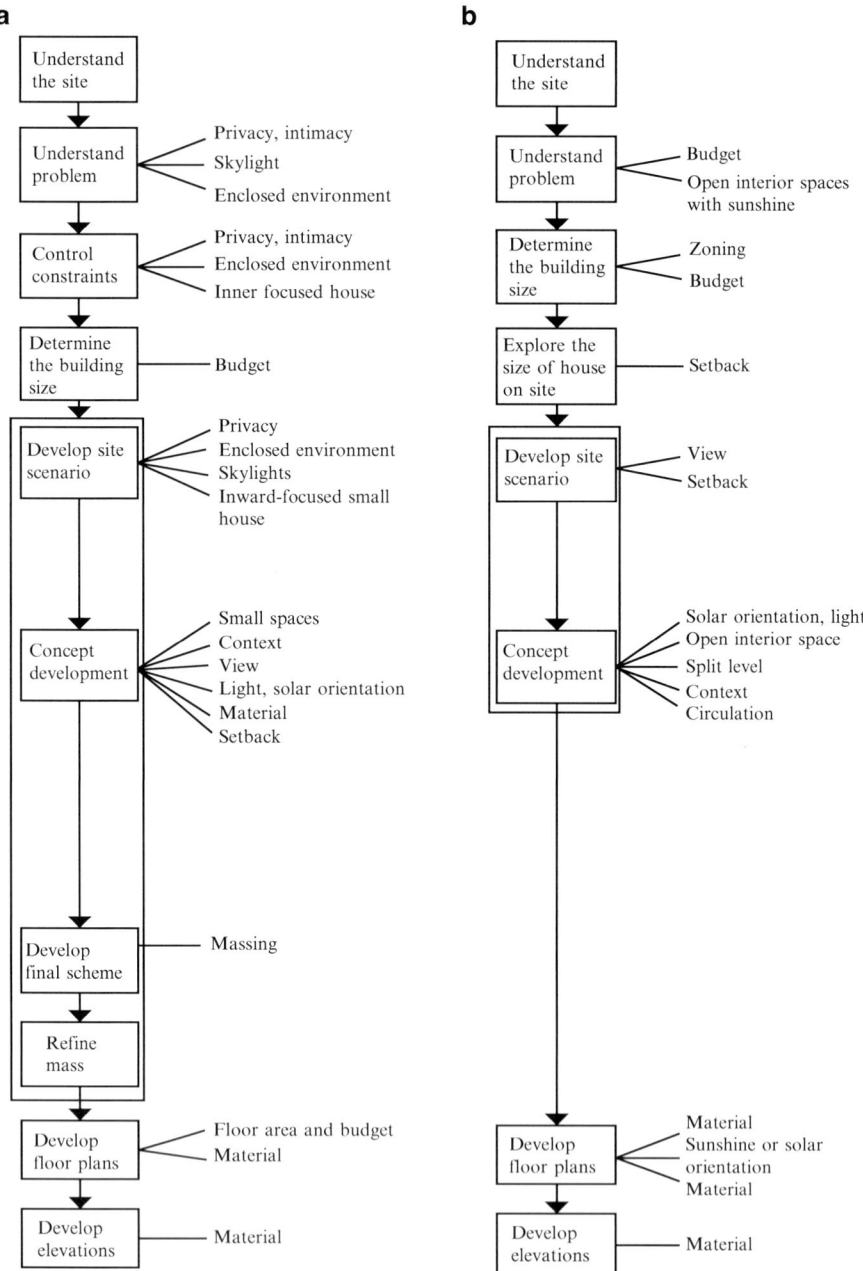

Fig. 6.10 Design goals and their sequences of Session 4 (**a**) and 8 (**b**) (Source: Copyright Pion Limited, London. Figure 6, p. 404, reproduced with permission. Chan (1993), Websites: www. pion.co.uk, www.envplan.com)

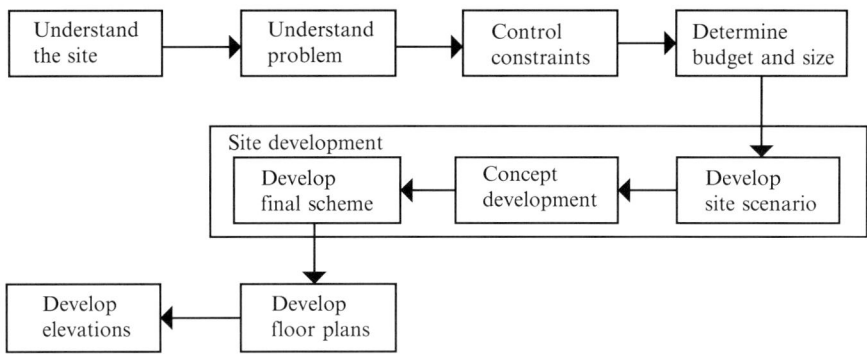

Fig. 6.11 The general design methods used in four sessions (Source: Copyright Pion Limited, London. Figure 7, p. 405, reproduced with permission. Chan (1993), Websites: www.pion.co.uk, www.envplan.com)

In short, the architect's strategies were to understand the site issues and design problems first. Then, based on the gathered visual information through his reading of the site map, a set of important design constraints was formed in a list format, which could be seen as the global constraints; **semantic solution** was developed to resolve these addressed global constraints.[6] Afterward, the architect determined the construction budget upon which the total floor area (dimensions) and the room sizes were decided. Then, based on the semantic solution of the constraints, the architect developed a graphic design scenario that had correspondence to the site, and served as a guide for the upcoming design procedures. Next, the generated design scenario went through a series of refinements with several alternatives and selection of a satisfying one to produce a workable scheme. Through an evaluation of the alternatives, the final version of the design was embodied through the refinements of floor plans and elevation drawings. Such a series of design goals and their specific order shown in four design sessions can be generalized through a diagramed flow as shown in Fig. 6.11, which also signifies the design method used by the architect. In Fig. 6.11, the stages of site scenario development and the concept development both related to the site planning, and were grouped as the "site development" stage, presented from the macro level.

6.5.3   Repetitious Constraints Used in Designs

Protocol data showed that there were global constraints repeatedly used for solution generation and evaluation, namely, sunshine/light, view, context, and budget. The **sunshine/light constraint** was used to determine the orientation of rooms. For

---

[6]The term "**semantic solution**" refers to a solution that is very abstract and is expressed mostly in verbal terms. The subject in this experiment always developed the first overall idea (solution) of the building in the form of a design scenario that managed his developed design issues. The solution at this stage was very flexible, schematic, and open-ended.

instance, in Session 1, it was the factor creating an outdoor space next to the living area at a very early design stage (before the building size and the construction budget were determined). In Session 2, it determined the geometrical configuration of the living room and the bedroom. In Session 8, it was applied to locate the painting studio. All these occurrences took place at the same stage of concept development (see Figs. 6.9 and 6.10). Moreover, sunshine/light also was the major constraint used to determine the locations of living room, dining room, and kitchen at the later concept development stage in Session 1. These examples indicate that the sunshine/light constraint is an important factor in spatial arrangements of living units, which share some similarities of internal spatial layout corresponding to their orientations.

The **view constraint** was unique for it provided views for the design units toward an inner court, and set a view path from the kitchen through the dining room to the outdoor viewed object. For instance, it was applied in Sessions 1 and 8 to generate the final solution scheme and in Session 4 to create two alternative solutions for decision making. All these solutions shared a common feature of an enclosed inner private court located at the backyard of the buildings. Furthermore, the view constraint, together with the sunshine/light constraint, yielded a collective feature in Sessions 1, 4 and 8, that living room and dining room were located at the back of the house, providing views to the inner court. Similarly, data also showed that the view and sunshine/light constraints were combined to determine the southeastern corner of the kitchen location in Sessions 1, 4, 8, and in one conceptual scheme at early design stage in Session 2 (see the protocol data listed on the inferred rules). Thus, the geometric location of the kitchen had similar results shown in Sessions 1, 4, and 8.

The **context constraint** refers to architectural context, which is the relationship between the designed building and the surrounding buildings. In the early design stage, the 25-foot setback from the street side and 5-ft setback from the property line was used to define the building location on the site plan. The context constraint also was used for generating solutions in Sessions 1, 2, and 4. For example, in Session 2, the subject said: "*I think the mass of these adjacent houses once again is an assumption unfortunately. But, I think it is a reasonable one. Ohm…I don't think that there will be a house twenty-five by twenty-five squares with these kind of side lines, it wasn't a plain old Pittsburgh type of big four-square house. So, I might not want to put this house out here with where these houses are, because it's going to look pretty bizarre and over-powered by them. Actually, what I might want to do is …to build a house that is very much farther back on the edge, toward the edge, so that it got… really capitalize the view.*"

The **context constraint** also was used for evaluating temporary solutions in Sessions 2, 4, and 8. For instance, it was used in Session 2 to evaluate one of the alternative solutions, which positioned the house toward the river with a living room on the left and a kitchen on the right, and the house was accessible by a long driveway. The architect indicated that because the adjacent houses were located near the river, the volume of the house was overpowered by these adjacent houses, and the solution was abandoned. He explained: "*Course here is another problem. Boy! There is this little one-story living room sitting next to this big bulk over here. The problem is that these other big houses are hanging on the back of this (river) cliff, too*" (Session 2).

In Session 4, the **context constraint** was used to evaluate the final scheme. The architect said: "*I guess I was thinking about how this would sit next to all these other big houses (draw a perspective of the adjacent house), I mean it's going to sit between some of these massive houses. And it's going to be a little piece here (draws the house), and another big house down next to it (draw another adjacent house). So…and…and this driveway, it looks like a little toy there, and this sort of…Ohm… So, what I think I need to do, is to pick up where I was, which was trying to resolve the mass, vis-à-vis the plan.*"

In Session 8, one of the alternative solutions had a U-shaped massing; the scheme occupied the full width of the site, with a painting studio located on top of the garage. The architect evaluated this solution and said: "*Course I am not sure that you want this (house) to be too close to the property line. Maybe, that means some-body else is going to build as close as that.*" These examples demonstrated that the context constraint defining the relationships with other buildings was constantly used to generate and evaluate solutions in these design sessions.

Lastly, the **budget constraint** showed a specific method of applying an average unit-cost-per-square-feet to divide the given budget for determining the room sizes, dimensions of the house, and the materials to be used. Methods applied were explained in the previous section. After all, this sequence on considering the budget constraint represented a strategic procedure that had been used in design sessions 4 and 8.

On the other hand, a set of **local constraints** with limited set members repeated in the protocols of four sessions was found, which included **view and material** together appearing at the floor-plan and elevation refinement stages (see Figs. 6.9 and 6.10). View was used to determine the window location and sizes, and the relation between kitchen and dining room. Material was used to determine the shape of the chimney and fireplace. For instance, in Sessions 1 and 2, the architect developed a brick chim-ney with a rectangular, massive form, and in Sessions 4 and 8, the architect used a metal chimney that created a circular shape. The material consideration was also used to determine ground level pavements on the yard. For example, in Session one, the architect said: "*There is nothing but stepping stones across the garden and that is what the small courtyard that is paved.*" In Session 2, he indicated that "*maybe there are some stepping stones that connect to this (entrance), or a pathway.*"

Data showed in this experiment that the architect did have precise rules for applying the constraints. For example, the light constraint had two components of morning and afternoon light. The morning light was used: (1) to determine the kitchen location at the southeast quarter of the site in Session 1; (2) to locate the kitchen in the southeastern site, and later, to test the bedroom solution in Session 2; and (3) to locate the painting studio in Session 8. The afternoon light was applied to generate overhangs and to test the living-room solutions in Session 2. Relevant protocol segments are quoted in the following and the associated rules could be converted into the formula shown in Table 6.4.

- In Session 1, a solution generated by Rule 1: "*But this was the kitchen you could see someone come in and you see out into the garden and you get some sun in the morning.*"

Table 6.4 Inferred rules in sunshine/light constraint schema

Light schema:
<Light> (<X>)
Rule 1: If <Light-source>=morning light/sun
Then <Opening-location>=eastern surface
and the axis from <Light-source> to <Opening-location> is not intercepted
Rule 2: If <Light-source>=afternoon light/sun in winter
Then <Opening-location>=southwestern surface
and the axis from <Light-source> to <Opening-location> is not intercepted
Rule 3: If<Light-source>=morning light/sun and <X>=Kitchen
Then <X-location>=eastern part of the site
and <Opening-location>=eastern surface
Rule 4: <Light-source>=afternoon light/sun and <Opening-size>=a large view window
Then <X> gets <Overhangs>
Rule 5: If <X>=Kitchen
Then <Light-source>=morning light
Rule 6: If <X>=bed room
Then <Light-source>=morning light
and <Light-source>=afternoon light/sun
Rule 7: If <X>=Living room
Then <Light-source>=afternoon light/sun

Source: Copyright Pion Limited, London. Figure 10, p. 412, reproduced with permission. Chan (1993), Websites: www.pion.co.uk, www.envplan.com

- In Session 2, a solution generated by Rules 1, 3, and 5: "*Or the house is going to get some morning sun in here (kitchen). So that is why kitchen was on this (eastern) side of the other house.*"
- In Session 2, a solution tested by Rules 1 and 6: "*And the bedroom would be in this little spot here. It is not going to get any morning sun. Ohm, it is not going to get any morning sun.*"
- In Session 8, a solution generated by Rules 1 and 2: "*Well, I guess the position of the studio really depends on whether this artist would like the light in the morning. If you put it over here [western site], you can get a morning light [from east], and you can give her morning light, and that can be some late afternoon light [from west] … I am going to assume she doesn't work [elsewhere], that she prefers to be over here [on the site], where she gets eastern light in the morning, and maybe you do some skylights [above] there [in the south of the room], something to give you a little [from south] during the day, and maybe more private in this [north] side. Yeah, I think that is what we are going to do for her.*"
- In Session 2, a solution generated by Rule 4: "*The real problem is going to be the low afternoon sun from across the river. It is going to be pretty fierce. It could really hit you. Wait, somewhere sun…afternoon actually starts at three o'clock or so. It is going to be very low. You might need some kind of big overhang on this side or something.*"

- In Session 2, a solution tested by Rules 2 and 7: "*In the winter, sun is going to set, you will get some afternoon sun over into this room (living room). I think this could be nice.*"

The rules that explicitly define the contents of a constraint are the main factors creating identical design results (solutions). Described in brief technical terms, parts of the light/sunshine constraint can be represented by schemata representation as shown in Table 6.4.[7] The name of the schema is symbolized by <light>, and the following <X> signifies the working design unit. The entire rule is expressed by an "if-then" clause representing the format of a **production system** (see footnote 2). In this notation, design actions (procedural knowledge) are displayed on the right-hand side of the system, and the left-hand side represents declarative knowledge. Embedded in the light/sunshine constraint is another rule that prohibits any interceptions (Rules 1 and 2 in Table 6.4). These examples explained that rules were applied to determine the kitchen, studio, and overhang locations, and to evaluate the bedroom and living room solutions, but not the dining room or bathroom designs. It was inferred that these rules were not applicable to these design units. Although the same rules (including Rule 3) of morning light determined the kitchen and openings in Sessions 1 and 2, they were not applied in Session 4 because they contradicted the inward-focused space requirement. Yet, the architect was aware of having not used this light constraint to determine the orientations and pointed out that: "*It is interesting that this house has no…It is just a sort of inward looking. I haven't really been thinking about the orientation of the rooms. This spatial issue (inward focused) of the room, just sort of … got me into the massing of the building.*"

Another example of design rules was from the global constraint of view that was used to determine the room openings in Session 1, the orientation of the house in Session 2, and the massing in Session 8. The corresponding protocol data is quoted in the following and the inferred rules are shown in Table 6.5.

- In Session 1, a solution generated by light Rules 1, 3, and 5 (Table 6.4); and view Rule 1 (Table 6.5): "*Ohm, I guess, the morning sun is over here, I tend to…I want to have my…cause this (house) is also tiny. You could see right through here. I mean you could do something that will have living room, kitchen, dining room, like that (visual link), and kitchen will get sun and be nice to be in the kitchen and*

---

[7] The notion of the light schemata differs from the notion of pattern language developed by Christopher Alexander in 1977. Pattern language is a method of generating building designs. It is an approach which constructs concepts for a building by combining patterns for smaller parts of the building. "The elements of this language are entities called patterns. Each pattern describes a problem which occurs over and over again in our environment, and then describes the core of the solution to that problem, in such a way that you can use this solution a million times over, without ever doing it the same way twice" (Alexander et al. 1977, p. x). A pattern can be selected from past experience or from precedents that seem most appropriate for the new project. Then a pattern is developed by combining smaller patterns to represent a summary of the building element and to show the overall idea of the building. This method of combining patterns to solve a design problem differs from applying design constraints and their associated rules for solution generations.

Table 6.5 Inferred rules in view constraint schema

View schema:	
<View> (<X>)	
Rule 1: If <Object> is a member of (river, forest, park, town scene)	
Then sight should be continuous from (X-location) to (Object-location)	
and (Opening-location be continuous from (X-location) to (Object-location)	
and (Opening-size)=large view window	
Rule 2: If <Object> is a member of (river, garden, forest, park, town scene)	
and <X>=Kitchen	
and dining room is between <Object> and <X>	
Then <Opening-location>=boundary wall	
and <Opening-size>=large view opening	
Rule 3: If <Object> is a member of (river, garden, forest, park, town scene)	
and <X>=living room	
Then <Opening-size>=full height from floor to ceiling	

Source: Copyright Pion Limited, London. Figure 11, p. 413, reproduced with permission. Chan (1993), Websites: www.pion.co.uk, www.envplan.com

see someone to come in the car and I guess the kitchen won't be a bad place to be able to see this view by standing there."

- In Session 2, a design scenario generated by Rule 1: *"This house with the setback here (near the river) could have a view of the river on one side and a view of the garden made on this (front) side. And, probably still have a little vegetable garden over here (front yard) all the way. So this house would be different in that it would have a view on both sides. I mean a large view potentially. Ohm… could have a large view."*

- In Session 8, a design scenario generated by Rule 1: *"Ohm, I guess there is view in this direction over to the town. And here is the forest, which we know is already be here, and one of the nice thing about that is if you look at it, it is always going to be…if you look at that more, it is always going to be lit. I like that, I think what we could do is to build, a scheme of some kind of a space, the house is sort of around the courtyard, doesn't look in this direction (west) necessary, or in this direction of view (east) … with blinds on it. Look out in that direction (north), has some kind of space here (in the center), outside space, and somewhere here (southeastern corner) is the garage."*

The light/sunshine and view constraints were also applied locally to determine the window openings and their locations. For instance, in Sessions 1 and 2 the architect used the light constraint to set up window locations for the bedroom and kitchen, and applied the view constraint to determine window openings in kitchen and staircase. The protocols were:

- *"Might be nice if there is a window up here, to let some lights coming round this chimney, Ohm, with some morning light fall into that room. It's really northeastern light"* (Session 1). (Windows generated by light schema.)

- *"I guess in this case, the morning sun comes into the kitchen, and the bedroom up here will get some morning sun. And get the window over there"* (Session 2). (Windows generated by light schema.)
- *"Well, it is just to have some potential high window there (on the kitchen wall section). Course you may want some other windows to look out through these pines, the pine forest here (put a window below the high window). Be nice from the…be nice from the dining room and kitchen, the kitchen has view windows to look out into the pines. And maybe when you are sitting in the dining room, there is a … a good-size view out through the pines"* (Session 2). (Windows generated by view schema.)
- *"Stairs that come up here, and so some kind of landing here, and at that landing, it could be a window on this side of the landing with a view into the park"* (Session 1). (Windows generated by view schema.)

The same Rule 2 in the view constraint had been used in both Sessions 1 and 8 to create a visual path from the kitchen through the dining room to the viewed object, and consequently created similar features (see Figs. 6.4 and 6.8). The protocols indicated that:

- *"Actually, this kitchen is more focused on the garden than the last one. The last one really, the last one looks across the, across the dining room, and the same thing could happen here. It could have a (visual) path over here (dining room). You could look out to the river. So, that could be nice"* (Session 2). (A solution generated by view schema.)
- *"And, so, kitchen could be set up so that…you could look out there. You could look out across the dining room"* (Session 8).

Another example of the applications of the same rules appeared in the use of Rule 3 (see Table 6.5) in the view constraint to determine the window pattern in the living room in Sessions 1, 2, and 4. The architect indicated that:

- *"Well, I bet this one (window) would come down to the floor. Need one place for the window to come down to the floor in the living room, and that could be nice"* (Session 1).
- *"Get in here (living room), you can see these view across. The view has been kept in form of the special, the special, like in the living room, it is all view. You can see everywhere"* (Session 2).

These results demonstrated that the same set of design rules indeed had been repeated on a global level to determine the kitchen location as well as on a local level to determine window openings and patterns. Some interesting findings are: (1) the resulting form generated by a design rule can be changed in later design stages (local wide) by other rules from other constraints; and (2) the same rule may not be applicable to the same design unit in every design. Therefore, forms can be changed and modified, and the same forms created by the same rule are sometimes difficult to maintain the constant features in every design projects. It is because that forms might be unpredictably changed by other rules at any time.

6.5.4 Observation on Repetitious Presolutions Used in Designs

A presolution model is a well-defined, concrete two- or three-dimensional form representing a design solution generated in previous designs. It can be recalled from memory, modified, reapplied, and saved to memory again for future use. It is analogous to the terms of a **case** in the **case-based reasoning**,[8] a **parti** (Foz 1972), or a **priori concept** (Kant 1998). The architect in this study applied several presolution models to generate design scenarios and to resolve design details. The first example showed the application of previously generated solutions for developing a design scenario, which was the conceptual scheme of the final design solution. In Sessions 1, 2, and 8, the architect consecutively recalled houses he designed or observed before and used them as cues for developing concepts of scenarios. The following protocol statements explain this phenomenon.

- *"Kind of reminds me of a house that I did a long time ago, which was a renovation. Although the house doesn't, the plan doesn't, but the notion of the house to a garden, that was a wonderful little idea. That house, it was all roof on the garden, the whole idea was related to a kind of massing roof house that plan to be sort of…It had a mass roof came down to the different positions along the edge. It was…and then it had a wonderful dormer (window) for bedroom overlooking the garden. As I recalled I did a green roof, a really green roof…But this house (current design) could…might have a…could have some fun with this house, reminds me another house that I once did…actually in Pittsburgh, is a house that had two faces. It got a large public scale face that was keeping a scale with the houses that around it. And then in the inside, on the inside but it wasn't deep, it was one room. It was on the corner and it was basically three stories high on the front and in the side, but was only one story high in the middle, because the…let's see…the courtyard was here, the living room was like this. Ohm, maybe it was two stories inside but ran like this, the garage was up here so it really appears almost like a three and half stories house on the street side, but was one room deep, and had this great big slope roof drop into the courtyard here and we have trees. It was not really dissimilar to this kind of site plan, Ohm, and its relation to the scale, had a big scale in the outside and little scale in the inside. Ohm"* (Session 1).
- *"Ohm, reminds me of a house that I once stayed at in a summer. Basically, it was buried in a pine tree forest on an edge of a lake, except that was a horrible house. But it was a beautiful scene sitting in pine trees. Although it didn't get a lot of sun over. You could always look out through the pine trees to wherever the sun was"* (Session 2).
- *"And I know there is a little house that I did. Ohm, this is the program of a house that I already did once. Also in Connecticut. Maybe it books better than this. That was very open too…exactly what you are talking about (open interior spaces). I just made an addition to that house fifteen years later"* (Session 8).

---

[8] **Case-based reasoning** relates to the comparison of a new situation to existing situations, selecting appropriate architectural solutions from memory and adapting those solutions.

The second example related to the formation of presolution models and the usage of them in later designs. For instance, a skylight was required in Session 4. The architect, actually, did not like skylights in his professional practice. He mentioned that: "*I wonder whether he wants skylights. I don't like skylights. I mean I think skylights can be a problem. They let in uncontrolled sun, and it is hard to control, even in control, that the skylights would leak, heat buildup on top of the house, I am just not big on the value on skylights*" (Session 4). Regardless of the preconception of the skylight, the architect did respond to it. The skylights he developed in Session 4 were located at the top of the perimeter of the living room. A month later in Session 8, the skylights solution was applied again to solve the sunshine constraints for the living room. The configurations of these skylights in both Session 4 and Session 8 were similar.

The third example of using a newly generated solution in later designs was found in the design of a walk-around closet in the bedroom. This solution was developed in Session 1. The protocol transcriptions were: "*Well, obviously, the nicest thing in this (bed) room would be to have a bed (six feet six) here, and maybe this (space) is big enough to…Maybe it is big enough to do some closets back here. Oh, Yeah, I got that, that is not a bad idea, there is a wall here. Perfect, a low element here, a headboard of the bed, and got the…here comes the closet, it is walk in. It is a walk-around closet, and you can walk around either side, and left no door, cheap, and this does not necessary go up to the ceiling, but it might…because this might be where that skylight (dormer window) piece on above, could be there to light up there…Ok, that solves that*" (Session 1). The similar walk-around closet behind the bed solution was again used in Sessions 4, and 8. It also was used in design sketches in Session 2, but was finally modified in the final scheme with a similar shape (see Figs. 6.4, 6.5, 6.7, and 6.8).

The replicated presolution models of either the recalled previous solutions from memory or the generated new solutions (e.g., skylight and walk-around closet) that were utilized across designs did demonstrate the concept of knowledge learning phenomena. In short, designers would apply their learned knowledge for problem solving or generate a new knowledge and gradually alter its structure to solve new problems. The applications of the repeated presolution models created similar configurations and forms in the final products. This use of a past solution as a presolution model to solve later designs provides evidence that the repetition of presolution models generates recognizable features that manifest a style.

6.5.5 Observation on Repetitious Mental Images Used in Designs

Spatial ability relates to the cognitive skills of: (1) perceiving the visual world; (2) translating and assimilating initial perceptions; and (3) mentally recreating spatial aspects of the visual experience. Such an ability, which is critical in design and is capable of developing certain mental images of objects and keeping track of orientation of objects (Shepard and Metzler 1971), has been described as a source of intelligence (Gardner 1983). Thus, designers should have strong spatial ability, good visual perception, and good memory of image. An expert designer, in particular, should have more mental images stored in memory than do novice designers (Chan 1997). The stored body of mental images includes previous design solutions, primitive shapes and some forms that fit certain functions. On the other hand, a designer may prefer certain shapes and primitive forms and apply them as visual vocabulary in design for a period of time and later change to other forms. For instance, **Michael Graves** used a heavy column and a wedge shape on top or reversed the composition in several projects during the 1970s (see Fig. 6.12). The feature of this composition is very simple and easy to recognize; it is a strong signature of Graves' style in the 1970s.

Similarly, the architect in this experiment used images from his own experience as analogy to work out concepts and details for solving a design problem. For instance, in determining the dimensions of a round dining table for eight people, the architect recalled the table he owned and applied its measurements to determine the desired table size. Later on, these measurements again provided information for the dimensions of the dining room. The protocol recorded: "*I think, this dining room will be nice with a round table. Ohm...so, if he wants...Oh, will be nice to be able to seat eight people, eight people on a round table...well, let's see, my own table at home is square and that is four feet six, eight inches dome you get eight, and five six, and so you have four feet (of spare space) to get around, and then five feet six table and four feet, you are going to be up to a fourteen feet (dining) room, fourteen by fourteen is...what is that? Sixteen, fifty, four, hundred-and-ninety-six...that is two-hundred square feet, that is a way over (the size of 180 square feet on the checklist), Oh, no, not too bad, well, let's try that, you could always refine that, now let's make it thirteen feet square*" (Session 1).

Another example of plugging in a visual image from common experience to design was found in solving the spatial relationship between the dining room and the kitchen. The architect indicated that: "*Actually, maybe this kitchen, dining room relationship could be looked like my own. Where the kitchen is on the back here, it has counter that you overlook to the dining room, you can talk to the people, which it has place to look out to the garden to what I have, maybe that works*" (Session 1). Observed from data, the dining room was always located to the north of the kitchen. This spatial relationship did appear in the final scheme of Sessions 1 and 8, and it also appeared in two alternative solutions developed in Session 2 as shown in the following protocol segments:

- "*Actually, this kitchen is more focused on the garden than the last one. The last one really, the last one looks across the, across the dining room, and the same thing could happen here. It could have a (visual) path over here (dining room). You could look out to the river. So, that could be nice*" (Session 2).
- "*If this is the living room, then we had something like we had the last time. Then we had dining room over here. And we came into the kitchen over here. And we had this little thing (counter) down here, and this is stair. Oh, this might make much sense*" (Session 2).
- "*And, so, kitchen could be set up so that...you could look out there. You could look out across the dining room*" (Session 2).

Warehouse Conversion, Guest bedroom, 1977.   Schulman House, Fireplace sketch, 1976.

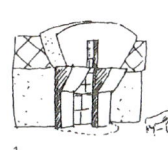
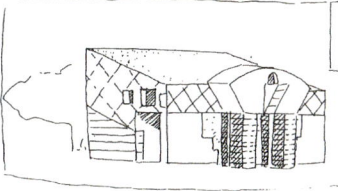

The primitive features.   Abrahams Dance Studio, preliminary studies, 1977.

Fargo-Moorhead Cultural Center, 1977-1978.   Crooks House, Fireplace detail, 1975.

Fig. 6.12 The primitive forms used by Michael Graves in the 1970s (Source: D. Dunster (1979). Images were reproduced by permission of 1979 Architectural Monographs and Academy Editions)

The mental image might also come from previous solutions.[9] If the image of it was not clear during the design, the solution may not be available or workable. For

---

[9] The architect verbally described the location, the date, and the form of the images of his early design works. In some instances, he also sketched the images.

instance, in developing the roof form in Session 1, the architect used a vague image from experience but could not resolve the problem. He indicated that: *"This is going to like this, and this is something funny in the roof, happens like that (corner part). Ohm…it is a piece of roof. I don't know how to do that piece of roof, that's a funny piece of roof. I think I have seen that kind of roof before. But I don't understand the geometry there. But, I think properly done that might work"* (Session 1).

It is also shown in the protocol data that a visual image from past designs was used as an instant solution followed by a limited modification. The image of a gate in an inner court in Session 1 and an image of a staircase in Session 2 were two examples. The architect said that:

- *"Maybe it will be nice if you come through this gate. It reminds a little wall that I did in a house at Long Island. Where you came from a parking and went through a, probably it was not a gate, but a gate with an opening in it, but it created…and then there is nothing but stepping stones across the garden, and that is what the small courtyard that is paved"* (Session 1).
- *"I have to figure out the way for that spot right there, and up for the stair to go downhill to a little thing overhang on the…overhangs on the edge of the water. Well, this reminds me a stair that I once did. It went back like that. Pretty bizarre. I went and saw it recently"* (Session 2).

Other than the mental images used for solving spatial and functional relationships among design units, the architect also used a U shape with articulation in the curve on floor plans in this study, as shown in Fig. 6.13. In sum, all repetitious features found on the floor plans and elevation drawings together with the studied

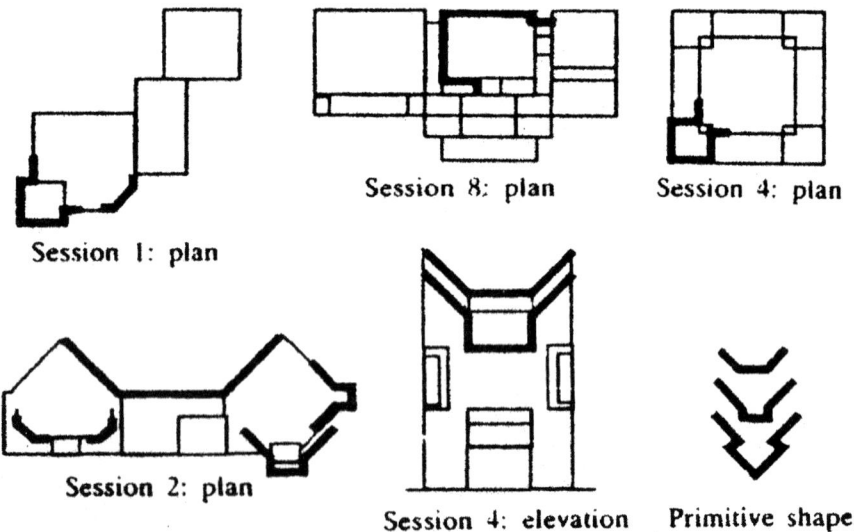

Fig. 6.13 Primitive forms (Source: Copyright Pion Limited, London. Figure 8, p. 408, reproduced with permission. Chan (1993), Websites: www.pion.co.uk, www.envplan.com)

Table 6.6 Repetitions of forms and their causes

Features	Session 1	Session 2	Session 4	Session 8	Forces
Elevation					
Horizontal broad siding	+	+	+	+	_Constraint_
Grid pattern, full-height window	+	+	+	−	_Presolution model_
Double-pitch roof	+	+	−	−	_Presolution model_
Exposed column	+	+	−	−	_Image_
Brick chimney	+	+	−	−	_Image_
Circular metal chimney	−	−	+	+	_Image_
Corner plate	−	−	+	+	_Image_
Plan					
Bed with walk-around closet in the back	+	+	+	+	_Presolution model_
Sink in kitchen facing a window	+	+	+	+	_Presolution model_
Corner windows	+	+	+	−	_Presolution model_
Staircase around LR	+	−	+	+	_Image_
Entrance next to kitchen	+	+	−	+	_Constraint_
A corner fireplace	+	+	−	−	_Constraint_
An enclosed inner court	+	−	−	+	_Constraint_
Kitchen in front of DR with views through	+	−	−	+	_Constraint_
Skylight	−	−	+	+	_Presolution model_
Symmetry	−	+	+	+	

causes that shaped these forms are summarized on Table 6.6. These forms and the primitive geometry were easily recognizable and identifiable in the final products; they had direct influence on the final appearance. Thus, the favored forms served as signatures to mark the architect's personal style and are direct factors in generating a style.

6.6   Discussion

This chapter intended to establish and test the operational concepts of style through design studies. The concepts emphasized are that: (1) repeated forms are used to identify a style; and (2) repeated factors in processes creating the repeated forms are the driving forces in generating style. Those forces are assumed to be the designer's seasoned knowledge together with the design constraints applied at each design stage, the fixed order of design goals, presolution models, and the designer's favored forms.

Table 6.7 A summary of the subject's style

Constant features	Factors generate the features
In plans	
1. Walk-around closets	Rule and presolution model
2. Centralized kitchen sink, window, and wall	Rule and presolution model
3. Staircase around the living room	
4. Entrance next to the kitchen	View constraint
5. Corner windows	Rule and presolution model
6. Symmetrical disposition	
In elevations	
1. Horizontal board siding	Material (local constraints)
2. Grid-pattern full-height windows	Presolution model
In design processes	
1. Consistent design methods	Constant goal order
2. Light and solar zone	Global constraint
3. Context	Global constraint
4. Enclosed and private inner courtyard for views	Global constraint
5. Discrete spatial arrangement	Personal preference
6. Primitive shape	Personal preference

Source: Copyright Pion Limited, London. Table 4, p. 422, reproduced with permission. Chan (1993), Websites: www.pion.co.uk, www.envplan.com

Results obtained from analyzing four design sessions showed that: (1) there were repetitions of forms in plans and elevations; (2) some processes were apparently consistent; and (3) relationships existed between these forms and processes. Interestingly, some relationships between processes and forms could be clearly traced from and supported by protocol data, whereas some could not (see Table 6.7). For the replicated forms that had no evident generating process, data showed that the architect simply drew the forms without verbalizing the causes. This is because the architect retrieved the images of those forms from memory and applied them directly into the design.

Many examples obtained from the study results indicate that using presolution models, primitive forms, and constraints has a direct impact on forms. Therefore, these variables are regarded as **direct factors** in expressing a style. There also were instances of repeated processes for achieving design goals, but this consistency did not generate any immediate influence on final forms. There were, however, certain constraints constantly bound to certain goals. For instance, the architect mainly applied the budget constraint to achieve the goal of determining the building size, and applied the constraints of context, light, view, and budget to accomplish the goal of developing the site plan. The change of the design goal order changed the sequences of design constraints and, consequently, changed the design method and the resulting form. Thus, design goal order is regarded as an **indirect factor** contributing to the formation of an individual style.

Table 6.8 Summary of the tests of the hypotheses

Hypotheses	Results
H1: Personal preferences on forms or materials	True (direct factor)
H2: Order of goal sequence	True (indirect factor)
H3: Choices of global constraints	Not applicable
H4: Priority order of global constraints	False
H5: Common set of global and local constraints	True (direct factor)
H6: The interaction of goal and constraint orders	Goal order dominates constraint order (indirect factor)
H7: Rules in constraint schemata	True (direct factor, local >global)
H8: Search order	True (indirect factor)
H9: Presolution model	True (direct factor)
H10: Mental image	True (direct factor)

A summary of the ten tests of hypotheses is listed in Table 6.8. Because the experiments were designed to systematically test the hypotheses and to be executed in a laboratory setting, the background conditions of some sessions were set in a straightforward way, for instance, the change of the owner in Session 5 to a retired executive who loved music. It is possible that the choice of this condition is arbitrary to some degree to discourage the search for novel solutions. If the condition were more sophisticated, for example, a professional musician who played a special kind of instrument instead, it might require the subject to seek a new design. Also, the change of design condition in Session 6 to set up some design priorities might not mean that a design order is imposed. Therefore, it could be the reason that the subject did not consider this issue seriously.

Regarding the subject's design behavior displayed in this experiment, one might argue that a majority of contemporary architects might use the context, light/sunshine, view, and budget constraints; however, how can one tell how much of the subject's behavior reflects personal style as opposed to schooling, training, and, in general, common practice? It is possible that a group of architects from the same region accepts the same training, schooling, and common practice and, therefore, shares the same design constraints. If this were true, the same constraints would create some similarities of forms that could be used to define a regional style or a group style. But it still is possible that there are individual differences in the use of the same constraints. The differences may exist in: (1) the design knowledge (i.e., design rules) embedded in the constraints; and (2) the method of implementing these constraints.

Another interesting point is whether certain design features that appear more than once across designs reflect individual style as opposed to fashion or standard practice? Some features could be standard commercial products used in practice—for instance, the horizontal board siding in this experiment. Observed from protocols, it is because the subject chose to use the commercial product for generating the wanted features (i.e., horizontal lines on elevations) in the final form that the choice of using siding manifested a style. The protocols explaining the choice situations are:

"Oh, he is probably not going to be able to afford anything other than stained aluminum siding, or some kind of hard board siding. I don't think we could use something like stucco" (Session 1).

"Oh, no, this is going to be the siding. I think it's going to be some horizontal siding, natural, maybe it's cedar. Not rough wood but" (Session 2).

"Siding, horizontal siding, beveled siding probably. Actually, really, probably the wrong kind for Pittsburgh...much more New England material. But it had the same sort of scale as brick. These houses next door are probably brick. And depending upon how we paint it, turning it into something else" (Session 4).

One also might question what would have been different with other architects in the general design procedures as shown in Fig. 6.11. The differences can be explained by another protocol study of design problem solving (Chan 1990a). The general procedures used by the architect in that study were to understand the tasks, do the site organization, develop a scenario, solve some image units, initiate space layouts, determine room sizes, generate space organizations, and then refine floor plans and elevations. Thus, the differences are: (1) the architect in the previous study controlled the global constraints at the time of understanding the tasks; and (2) he did not use **algorithms** to determine the building size. **Frank Lloyd Wright**'s design process in his Prairie Houses style is another example. **Wright** would begin a design from an abstract, proceed to the development of a unit system to fulfill the material, scale, and structure constraints, then develop forms of exterior massing to fit functions, and finally develop sections and elevations. Therefore, based upon these studies (Chan 1990a, 1992), it is suggested that there are differences among individual designers in their approach to design; the differences are especially obvious in the development of design concept before generating floor plans.

The style generated by the architect in this study can be seen as a weak style (Chan 1995) that compares to the works created in his design firm. In other words, if many strong and significant features appear in products representing a style, this style can be claimed as a strong one compared with the one with fewer and less significant features. In this study, the architect spent an average of 5 h for each design. The products are very schematic, and fewer features have been added into them; thus the style resulting from the short-term effort is regarded as a weak style. It should be noted that the term "weak" style at this point signifies the quantity, not the quality of the style.

Finally, **seasoned knowledge** is a very influential factor in determining the generation of the pattern of an individual style. It is a body of specialized, expert knowledge that a designer acquires through education and practice. If a designer is interested in energy-related issues, he or she is likely to apply more energy-related knowledge for making design decisions. The design forms would reveal more lighting-related features. For instance, an experienced architect who participated in another study on design process showed such a tendency. This architect has been practicing for more than 25 years and is teaching energy-related courses and studios in an architecture department at a university.

The second study used to test the concept of seasoned knowledge in this chapter was on a kitchen design. As with the previous study, the architect was required to think aloud the entire time he was designing. The given design problem was to design a kitchen for a new house with four bedrooms and a two-door garage in Ames, Iowa. The budget was $225,000. Several design constraints were required: (1) the client wants views; (2) the design should include at least one entrance and window; (3) reduce the street noise as much as possible; (4) the total square footage of the kitchen is between 200 and 350 square feet; and (5) the whole family loves a colorful appearance and good materials for the kitchen facilities. Again, the entire design process was videotaped.

The architect spent 90 min to finish the kitchen design, as shown in Fig. 6.14. Because he is an expert in energy-related issues, his design reflected such a character on determining the orientation (see the sketch) and the materials applied on

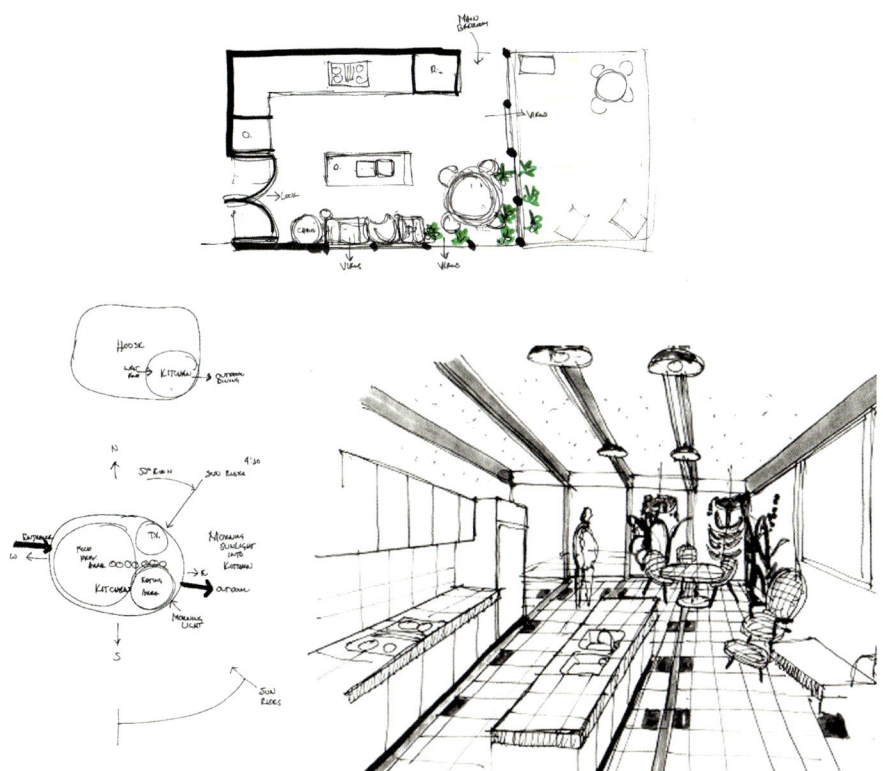

Fig. 6.14 Sketch, floor plan, and perspective drawings of a kitchen design (Source: Reprinted from Chan (2001). Figure 10, p. 343, Copyright (2001), with permission from Elsevier)

elevation (see perspective drawing). His verbalization obtained from the protocol data, as cited in the following, indicates his seasoned knowledge.

- *"If this is my overall kitchen bubble (diagram)…If I orient this within the house on a southeastern corner, where this would be north, east, west, and south; if I orient this within the bigger area of the house, and this kitchen is located in this area here (0:05:41) … then in the winter when the sun rises in the southeastern sky about 53° east of south, I'm a passive solar architect, so I just happen to know these dimensions"* (0:06:06) …
- *"I will always have morning light into this kitchen. Now I am going to say one thing that is at variance between the old Iowa farmer and a more modern household is that a farmer may have liked to get up at the crack of dawn to go to do his farm chores. No one else seems to want to do that. And therefore if you come into the kitchen early in the morning, if there is some light coming in it just makes it a far more cheerful situation"* (0:06:43).
- *"Now in the summer, the sun rises 53° east of north at about 4:30 in the morning. What this means is that this kitchen is in the southeastern corner. I will always have morning sunlight … coming into the kitchen"* (0:07:13).
- *"And also that in the summer time in the late afternoon, when it's very hot, the sun is beating on the house, having sun beating into the kitchen is not necessarily a good idea. So this is a perfect orientation. I'm going to orient this kitchen in the southeastern location of the house, and I'm going to be making some assumptions here that it would be nice to be able to lead out to an outdoor eating area on the eastern part of this house. And that probably the entrance to the kitchen. A major entrance coming from within the house is probably on the west end. Now if that's the case and I like that in concept"* (0:08:24).

If an architect is interested in structure and specializes in truss, for example, his or her design would have more features on truss structure. Seasoned knowledge is a long-term factor that builds throughout a long effort of cultivation of interest and expertise. This phenomenal aspect is more apparent in works by expert designers who have long been devoted to the profession than the works done by students in design studios. Individual style appearing at the studio design stage tends to be more apparent in the selections of media and mode for presentation style.

The concept developed in this chapter formulates a theory about individual style. The theory can be applied to differentiate a bad style from good ones. A good style is judged by the characteristic context that is generated by the topological relationships among features. Theoretically, a poor topological relationship is determined by: (1) the disproportion of features and among features (see the example shown in the upper right drawing of Fig. 6.15; the upper left drawing is the original living room facade of Wright's Little House design in 1903); (2) poor aesthetic expressions (as shown in the lower left corner drawing in Fig. 6.15); and (3) the violation or conflicts of functional requirements (as shown in the lower right drawing in Fig. 6.15). These could be the consequences of having poor quantity and quality of design knowledge, image, methods, and design goals. A poor topological relationship will cast a bad characteristic context, which is regarded as a bad style.

Fig. 6.15 Examples of bad or poor styles (Source: Reprinted from Chan (2001). Figure 11, p. 345, Copyright (2001), with permission from Elsevier)

6.7 Conclusions

The elaborated theory of individual style provides a notion about the understanding of a style that can be applied as a theoretical basis for studio teaching. For example, the theory can be used to help instructors and students identify students' design tendencies reflected in the design processes, and thus provide the basis for modifications that can improve their design skills. For instance, a better understanding of design constraints will improve design knowledge. More design experience will accumulate more available presolution models. Investigating various existing and precedent geometric forms will increase the ability and opportunity for creating new forms. Exploring algorithms of solving design problems will improve the ability of achieving design goals. A comprehension of the processes that generate a style will enhance the possibility of changing an individual style yet maintain certain characteristics and diversify design ability. On the other hand, if an instructor repeats similar critiques or routinely focuses on certain design aspects, then students may digest the same set of information and reapply it again in designs. If there are repetitions of replicating the learned knowledge in designs, the design constraints, goals, methods, and preferred images might reoccur cyclically. Consequently, some individual (student) styles will be generated within a group (studio) style.

This study was a first step in establishing a systematic process by which the factors of generating styles could be studied, and the findings gathered from this study support the assumptions. In other words, it was hypothesized that certain behaviors determined an architect's style; observation confirmed that this architect did, in fact, exhibit those behaviors. However, it was difficult to establish that the products of those behaviors were definitely the elements of the architect's style or that his style was recognizable. If several experts in style could examine this architect's works and identify the elements by which his style can be accurately defined (particularly, the weak styles), then it would strengthen and contribute to the reliability and validity of this research. In summary, a style results from executing fixed sequences of design goals (design method), applying fixed sets of constraints (design knowledge) at each goal stage, and exercising preferred presolution models and primitive forms (images). Based upon the observations made in this research, the study of an individual style should be approached from both the common-features and design-process points of view.

References

Alexander C, Ishikawa S, Silverstein M et al (1977) A pattern language, towns, buildings, con-
    structions. Oxford University Press, New York
Chan CS (1990a) Cognitive processes in architectural design problem solving. Des Stud
    11(2):60–80
Chan CS (1990b) Psychology of style in design. Dissertation. Carnegie Mellon University
Chan CS (1992) Exploring individual style through Wright's design. J Archit Plan Res
    9(3):207–238
Chan CS (1993) How an individual style is generated. Environ Plan B 20(4):391–423
Chan CS (1995) A cognitive theory of style. Environ Plan B 22(4):461–474
Chan CS (1997) Mental image and internal representation. J Archit Plan Res 14(1):52–77
Chan CS (2001) An examination of the forces that generate a style. Des Stud 22(4):319–346
Chan CS (2008) Design cognition: cognitive science in design. China Architecture & Building
    Press, Beijing
Cross N, Cross AC (1995) Observations of teamwork and social processes in design. Des Stud
    16:143–170
Dorst K (1995) Analyzing design activity: new directions in protocol analysis. Des Stud
    16(2):139–142
Downing F (1992) Conversations in Imagery. Des Stud 13:291–319
Dunster D (1979) Michael graves, architectural monographs 5. Rizzoli, New York
Eastman C (2001) New directions in design cognition: studies of representation and recall. In:
    Eastman C, McCracken M, Newsteller W (eds) Design knowing and learning. Elsevier,
    Amsterdam, pp 147–198
Ericsson KA, Simon HA (1996) Protocol analysis: verbal reports as data. MIT Press, Cambridge,
    MA, pp 48–62
Foz ATK (1972) Observations on designer behavior in the parti. Master's thesis, Massachusetts
    Institute of Technology
Gardner H (1983) Frames of mind: the theory of multiple intelligences. Basic Books, New York,
    pp 170–204
Gombrich EH (1960) Art and illusion: a study in the psychology of pictorial representation.
    Pantheon, New York
Kant I (1998) Critique of pure reason. Cambridge University Press, Cambridge
Larkin JH, McDermott J, Simon DP, Simon HA (1980) Expert and novice performance in solving
    physics problems. Science 208:1335–1342
Manson GC (1958) Frank Lloyd Wright to 1910. Van Nostrand Reinhold, New York

Miller GA (1956) The magical number seven, plus or minus two: some limits on our capacity for processing information. Psychol Rev 63:81–97

Newell A (1973) Production systems: models of control structures. In: Chase WG (ed) Visual information processing. Academic, New York, pp 463–526

Newell A, Simon HA (1972) Human problem solving. Prentice-Hall, Englewood Cliffs

Schapiro M (1962) Style. In: Tax S (ed) Anthropology today: selections. University of Chicago Press, Chicago, pp 278–303

Shepard RN, Metzler J (1971) Mental rotation of three-dimensional objects. Science 171:701–703

Simon HA (1969) Sciences of the artificial. MIT Press, Cambridge, MA

Simon HA (1974) How big is a chunk? Science 183:482–488

Simon HA (1975) Style in design. In: Archea J, Eastman C (eds) Proceedings of the 2nd annual Environmental Design Research Association conference. Dowden/Hutchinson & Ross, Stroudsburg, pp 1–10

Simon HA, Kaplan CA (1989) Foundations of cognitive science. In: Posner MI (ed) Foundations of cognitive science. MIT Press, Cambridge, MA, pp 1–47

Valkenburg R, Dorst K (1998) The reflective practice of design teams. Des Stud 19(3):249–271

Part III
Creativity

Chapter 7
Development of Studies in Creativity

Broadly speaking, creativity is the ability to create meaningful new ideas, forms, sounds, methods, performances, and interpretations. It implies that the creator's mind is non-conformist with the freedom of action. It also is a phenomenon that shows that some people can generate more beautiful, usable, and effective new things and marvelous new ideas than others. Such a phenomenon could be fine-tuned into aspects of invention and innovation, which have been broadly discussed in the field of engineering. Invention is the "creation" of a product or introduction of a process for the first time that has never been made before. Such a product or process is new, novel, and without precedent. For example, the original phonograph created by Thomas Edison in 1877 was a big invention that had not been seen before (see Fig. 7.1).[1]

Innovation is the "implementation" of a new thing. It occurs when a person improves on or makes a significant contribution to an existing product, process or service. It is about the practical application of new inventions into marketable products or services. Thus, innovation is only the first step in a longer and risky process of invention. Most innovations are, in fact, created by making improvements to existing things. IPod, the portable music device, was an innovation for it improved upon the idea of SONY's Walkman device. There are few totally new inventions, for novelty is an essential part of an invention, but novelty is not essential for innovation. The home phonograph developed in 1886 is an innovation by Edison, improving his original tinfoil phonograph, shown in Fig. 7.1, to wax cylinder sound input, shown in Fig. 7.2. Whereas Emile Berliner's change of input media in 1892 from phonograph cylinder to gramophone record with five inches of disc (Fig. 7.3) is another innovation. Of course, the compact disc and

---

[1] The history of the Edison cylinder phonograph and its cylinder and disc record devices can be seen online at: http://memory.loc.gov/ammem/edhtml/edcyldr.html and http://www.edisonphonology.com/cylinder.htm. Detailed illustration of the device could be found in Britannica Kids page at: http://kids.britannica.com/elementary/art-90615

© Springer International Publishing Switzerland 2015
C.-S. Chan, *Style and Creativity in Design*, Studies in Applied Philosophy,
Epistemology and Rational Ethics 17, DOI 10.1007/978-3-319-14017-9_7

Fig. 7.1 Edison Tin Foil
phonograph, 1877 (Source:
http://memory.loc.gov/
ammem/edhtml/edcyldr.html.
Accessed 10 Oct 2013)

Fig. 7.2 Edison home
phonograph, 1899 (Adapted
from Norman Bruderhofer/
Wikimedia Commons, http://
commons.wikimedia.org/
wiki/File:EdisonPhonograph.
jpg. Accessed 10 Oct 2013)

Fig. 7.3 A Berliner's disc
phonograph, 1907 (Adapted
from Norman Bruderhofer/
Wikimedia Commons, http://
commons.wikimedia.org/
wiki/File:VictorVPhonograph.
jpg. Accessed 10 Oct 2013)

digital recording are other innovations following the line of the phonograph. So, there is more innovation than invention.

In patent law, invention has been interpreted and regarded as principally a mental activity. An inventor is the one who conceives a patented invention. Conception is defined as the formation of a permanent idea of the operative invention in the inventor's mind, as it would be applied in practice. After inventors have conceived their inventions, they may use the services, ideas, and aid of others to perfect their inventions without losing their right to a patent.[2] However, when people talk about the "capability" or act of making something unique, novel, and original, be it an invention or innovation, then it relates to creativity. It is the ability of the mind to recognize a new pattern, or generate a new idea to serve as the basis for creating a new product. Therefore, creativity is the operation of cognitive factors generating the acts of invention and innovation. Such cognitive operations have been explored in design professions and artistic creations. Likewise, studies in creativity have been approached from various perspectives with different attentions in different fields including business (Amabile 1999), music, drama, art, design (Dorst and Cross 2001), English (Olson 1984), engineering and technology (Rubinstein 1975), Mathematics (Schoenfeld and Herrmann 1982), physics (Larkin 1980), philosophy (Hausman 1984), information and communication technology, scientific discovery (Kaplan and Simon 1990), gifted pupils (Gardner 1983), psychology (Sternberg 2010, 2012), and education (Isaksen 1988), to name just a few. Because creativity has benefits for individuals and society as a whole, a great deal of research has focused in the past 30 years on personality and creativity, and assessment of creativity for the purposes of improving human creation.

Most publications concentrate on studying the characteristics of creativity within a specific field. Notions discussed in this book on creativity are approached from a different perspective. Firstly, in this chapter, historical development of the concept of creativity, literature on creativity, and methods used to study creativity across various fields are broadly reviewed. All related findings are summarized to provide a comprehensive picture of the cognitive nature of creativity. In the following chapter, the cognitive aspects of creativity, related cognitive mechanisms triggering creativity in the areas of design, and the cognitive relationships between creativity and individual style occurring in the design process are further discussed through a number of case studies. As commonly agreed, creativity is essentially part of the problem-solving process (Wallas 1926; Torrance 1971; Mumford et al. 1991; Runco 1994) and creative ideation allows the individual to be flexible and innovative (Flach 1990; Runco 1986). The theory of this book is centered on these two points.

7.1 Historical Overview

Unlike the concept of style, creativity was not identified as a human ability in ancient times. But, as reported by Ryhammer and Brolin, the development of original products and new ideas is a particularly human characteristic (Ryhammer and Brolin 1999). New ideas must come from somewhere. In the ancient Greek period,

---

[2] "Invention is Not Innovation and Intellectual Property is Not Just Like Any Other Form of Property: Competition Themes from the FTC's March 2011 Patent Report" by Henry C. Su can be found on line at: http://www.americanbar.org/content/dam/aba/publishing/antitrust_source/aug11_su_7_26f.authcheckdam.pdf

Fig. 7.4 Stone sculpture of the nine Muses (Source: Jastrow/Wikimedia Commons, https://commons.wikimedia.org/wiki/File:Muses_sarcophagus_Louvre_MR880.jpg. Accessed 10 Oct 2013)

the word *inspiration* was introduced to explain where ideas came from. It was believed for centuries that a higher power produced new ideas. In Greek mythology, the Muses were the high power for inspiration in science, literature, and the arts, providing sources of knowledge (see Fig. 7.4). The Muses were seen as mediating inspiration from the Gods (Dacey 1999). Creative works were only recognized in poetry and within the poet who created it. For art, as Plato indicated, was the imitation of nature. Because nature was perfect and subject to laws and rules, so artists should discover laws and make art-works by rules. Thus, there was no freedom of action and creativity was not discussed in or connected to arts by the Greeks. The artist was a discoverer, not a creator.

After the Roman conquest of Greece in 146 BC, Greek art and sculpture had big influences on Roman art. In the Roman period (27 BC – 476 AD), art was recognized as work similar to poetry that possessed the characteristics of imagination and inspiration. *Creating* and *creator* were words used to describe the generation of art. But, there is no record of the masters of Roman art, and no signed works have been found. Greeks recognized aesthetic qualities and wrote extensively on artistic theory. Roman art was more decorative and indicated status and wealth. Art was not the subject of scholars or philosophers. However, in sculpture, facial expressions in Roman busts were more emotionally realistic and physically accurate than sculptures in the Greek period (see Fig. 7.5). Roman culture in sculpture provided its own thematic content, such as political and military figures, as well as the pantheon of gods. In architecture, Rome assumed an artistic expression of its own. The inventions of concrete, the arch, the vault, and the dome allowed buildings to be constructed on a larger scale; for instance, the Coliseum, the Pantheon, and the Aqueduct Bridge (see Fig. 7.6). In the Christian era of the late Roman Empire, wall paintings and mosaic ceilings were prevalent, but full-sized sculptures and panel paintings faded out due to religious reasons.

In the Middle-Ages (roughly the sixth through the sixteenth centuries), the medieval art of the western world grew out of the artistic heritage of the Roman Empire

Fig. 7.5 Bust of Emperor
Claudius (Jastrow/Wikimedia
Commons, http://commons.
wikimedia.org/wiki/
File:Claudius_Pio-
Clementino_Inv243.jpg.
Accessed 10 Oct 2013)

Fig. 7.6 The Aqueduct of Segovia, Spain (Bernard Gagnon/Wikimedia Commons, http://commons.wikimedia.org/wiki/File:Aqueduct_of_Segovia_08.jpg. Accessed 10 Oct 2013)

and the iconographic traditions of the early Christian church. It covers over 1,000 years of art in Europe, the Middle East and North Africa. During this era, the ancient view persisted that art was not a domain of creativity. Special talent or unusual ability by an individual was the manifestation of an outside spirit that served as a means of conduit (Albert and Runco 1999). The general concept of creation was to call for an external creative genius connected to the sacred or the divine (Dacey 1999), and this "genius" was associated with mystical powers of protection and good fortune. Ideas of art were given to the artisan by the genius to do God's work on Earth (Boorstin 1992; Albert and Runco 1999). So, creativity was a matter

Fig. 7.7 Giotto's painting, The Kiss of Judas, in the Scrovegni Chapel, 1305 (Source: Public domain, http://commons.wikimedia.org/wiki/File:Giotto_-_Scrovegni_-_-31-_-_Kiss_of_Judas. jpg. Accessed 10 Oct 2013)

of divine inspiration, and the conception of creativity was the religious one provided by the Book of Genesis. Creativity in the Judeo-Christian tradition only applied to religion; human beings were not considered to have the ability to create something new or serve as a cause of creation until the Renaissance. Also, the term *creativity* was not the same then as it is used in modern sense. Thus, from this religious interpretation of expression, art was a craft; it was making, not creating, objects. The most famous painter of late medieval art is Giotto (1266–1337). His paintings at Scrovegni Chapel, 1305 (see Fig. 7.7) have been recognized for representing the medieval (or early Renaissance) painting art.

In the Renaissance (fourteenth to sixteenth centuries), concepts of creativity started to change and the notion of art moved away from the imitation of nature and from the conduit for the divine towards individual expression. The source of inspiration and its artistic expression were created by human beings. The human mind was popularly thought to act independently of natural law and to be capable of almost any achievement, if compelled to exert itself by a will that had a power of initiation. Originality, insight, creative genius, and the personal feeling described in arts were valued in the Renaissance period. Creativity was shown in work generated by the abilities of great men.

In the Enlightenment period (approximately the seventeenth and eighteenth centuries), philosophers were able to move towards a concern with imagination, individual freedom, and society's authority in human affairs. The Enlightenment introduced an interest in science, the promotion of religious toleration, and a desire to construct governments free of tyranny. Individuals had their own right to explore

Fig. 7.8 Saint Jerome in His Study by Jan van Eyck, 1442 (Source: Public domain, http://commons.wikimedia.org/wiki/File:Jan_van_Eyck_-_St_Jerome_-_WGA07621.jpg. Accessed 6 Feb 2014)

their world without divine guidelines or intervention. The idea of research emerged. Science and scientific thinking started to take form as the preeminent instruments of discovery and models for thinking about the physical world (Albert and Runco 1999). This led to scientific research on the study of creativity.

Starting from the late nineteenth century, scholars began to investigate what fostered creativity and the possible contexts that affected creativity. The first scientific study was done in 1869 by Francis Galton on exceptional talent or genius (Galton 1869). He defined a *man of genius* as a man endowed with superior faculties and achievements acknowledged in the wider public arena. He collected 300 families containing nearly 1,000 eminent geniuses and analyzed 415 of them on whether their genius was an inherited natural ability. He hypothesized that natural ability was biologically inherited. The geniuses, who were at the top end of the social scale, became the leaders and eminent achievers of their generation. Genius level was directly connected to a higher level of prominence and greater level of presumed natural ability. Galton studied the accomplishments achieved among groups of judges, politicians, commanders, writers, scientists, poets, musicians, painters, and divines, and collected data to prove that genius clustered in "Notable Family Lines." For instance, he selected 42 painters whose work dated from 1370 to 1707, with the

Fig. 7.9 Dutch men-o'-war and other shipping in a calm, by William van der Velde, 1665 (Source: Public domain, http://commons.wikimedia.org/wiki/File:Willem_van_de_Velde_II_-_Dutch_men-o%27-war_and_other_shipping_in_a_calm.jpg. Accessed 10 Oct 2013)

earliest artist Jan van Eyck, 1370–1441 (see Fig. 7.8), and the latest William van der Velde, 1633–1707 (Fig. 7.9). Among them, half of the population of the data possessed eminent relations, some of them being direct kinsmen. He evaluated the chances of eminent people having eminent relations, considered the closeness of biological connection, and the magnitude of achievement of the eminent parent. Results of his findings showed that the more famous the parent and the closer the blood tie, the greater possibilities of having of those chances for eminence (Simonton 1994). He concluded with two interesting points: (1) no man could achieve a very high reputation without being gifted with very high abilities; and (2) few who possessed those very high abilities could fail in achieving eminence (Galton 1869, p. 43). Thus, the natural ability of genius was endowed by birth, but the achievement of eminence could be affected by other factors.

In the early twentieth century, similar methods on studying genius were used to explore the human intelligence of great men. **Catharine Cox** recorded the childhood and adolescent achievements of 301 great men and women who lived between the years 1450 and 1850, including writers, musicians, painters, philosophers, scientists, statesmen, and religious leaders. The purpose of these studies was to characterize a group of young geniuses with respect to certain mental traits and to estimate what their IQs would be at the age of their achieved accomplishments. For instance, the IQ score of the development of Michael Faraday (1791–1867), an English chemist and physicist at age 17 was 105, and the score of development from age 17–26 was 150. Methods used were to gather data of psychological significance of groups of eminence and average attainment through exhaustive analyses of the individual members, and to base upon these data, statistical studies of certain aspects of the psychology of these two groups. Then, perform comparative studies on the original data and the statistical data. From the findings, Cox concluded that those great men of the past with the highest IQs had achieved higher eminence than those of lower IQs, and those with higher IQs had more variety in their achievements (Cox 1926). She indicated that there were two leading factors in producing a man and making him what he was: the endowment given at birth, and the environment into which he comes. In other words, natural ability was not only endowed by birth, but also by nurture.

In psychology, the focus on the comparison of genius in the early twentieth century gradually shifted to the investigation of the creative side of intelligence, most significantly after 1950 (Guilford 1950; Isaksen 1988). In the 1950 meeting of the American Psychological Association, Guilford (1950) argued in his presidential address that the area of creativity was an understudied but essential field for research. He explained that scores on intelligence tests cannot be taken as a ranking to indicate the superiority of some people, or groups of people over others. For instance, the most creative people may have lower scores on a standard IQ test due to their approach to the problems, which generates a larger number of possible solutions and some are original. Thus, studies in creativity moved forward and achieved a lot of scientific findings. A number of handbooks were published, including one that focused on individual differences (Glover et al. 1989), one that provided the most comprehensive review (Sternberg 1999), and a wide ranging volume across many domains that introduced the history, theories, and functions of creativity (Sternberg and Kaufman 2010). An online "Creativity Research Journal" was established in 1988 as well. Thus, as acknowledged by scholars, the year 1950 was a turning point in the research on creativity. Even though studies in the early twentieth century mainly explored the creative process (Wallas 1926; Rossman 1931; Spearman 1931; Patrick 1937), extensive studies after 1950 examined creativity through scientific methods for understanding the mental processes shaping creative thinking and the factors facilitating creative production.

After studies were advanced, scholars recognized the importance and significance of human creativity. It has been argued that creativity is a habit that becomes a way of life that one regularly utilizes so that one is hardly aware one is engaging in it (Sternberg 2012). Such a habit refers to an acquired behavior pattern that is a part of the cognitive activities applied by human beings. Because creativity is a phenomenon of cognitive operation that could be performed by everyone, various creative phenomena performed by different individuals have been further categorized into four levels (Kaufman and Beghetto 2009). These four levels of creativity are eminent creativity found in Picasso, Beethoven, Darwin, and others of that level (called Big-C), everyday creativity found in nearly all people (called little-c), creativity that is inhered in the learning process (called mini-c), and the developmental progression beyond little-c that represents professional-level expertise in a creative area (call Pro-c). Simply speaking, creativity of Big-C is the famous master level. Little-c is the everyday life creativity used in everyday problem solving. Mini-c is the creative interpretation on new findings during learning. Pro-c is the creativity of the persons who are moving toward the master level but are not yet recognized by the public. The focus of design creativity covered in this chapter is on exploring the cognitive performance that achieves the level of and above the professional.

7.2 Current Theoretical Development

In 1963, Ross Mooney proposed four significantly different approaches to the exploration of creativity, including specific aspects of creative environment, product as an outcome of creativity, creative process, and the person doing the creating (Mooney 1963). These approaches have been applied by scholars, who have yielded a number of findings. If we put these four approaches into a formula, the creative process and the creative product are the criteria of creativity, the creative person is the fundamental predictor in the equation, and the environment is the modifier and the stimulus situation through which the inner creative processes are activated (Taylor 1988). Other than defining creativity from these four dimensions, methods of measuring creativity have also been explored recently. Following this equation, all research conducted in areas of personal psychology, social psychology, cognitive psychology, education, and engineering could be generally categorized into groups that investigated: traits that made creative people creative (personality); environmental conditions that promoted creativity (stimulation of creativity); the process of creativity (cognition), and; assessments used to evaluate creativity (measurement). Each of these groups of investigations are reviewed in the following.

7.2.1 What Traits Make Creative People Creative?

Much research supports the importance of certain personality attributes for creative functioning (Barron 1969, 1988; Feist 2010). Among them, some studies have focused on exploring dispositions of prominent creative persons correlating either positively or negatively with creativity. Positive characteristics include strong motivation, endurance, intellectual curiosity, deep commitment, independence in thought and action, strong desire for self-realization, strong sense of self, strong self-confidence, openness to impressions from within and without, attracted to complexity and obscurity, high sensitivity, and high capacity for emotional involvement in their investigations (Brolin 1992). Negative personality traits include dogmatism, conformism, narcissism, frustration, resilience, elation, hypomania, and affect tolerance[3] (Shaw and Runco 1994; Eisenmann 1997). A distinctive set of attitudes categorizing life-long, high level, and creative achievement has been identified as having self-control, sustained hard work, determination, and perseverance (Dacey

---

[3] Affect tolerance is defined as the ability to respond to a stimulus which would ordinarily be expected to evoke affects by the subjective experiencing of feelings, rather than by an apparent non-reaction response such as impulsive behavior, somatic dysfunction, or personality disorganization (Sashin 1985).

and Lennon 2000). Even the studies on personality have been criticized as having been too narrow with focus only on eminent and productive persons or on special groups that cannot represent creativity of all people, but Eysenck has argued that studies over the years of creative individuals have agreed with the results (Eysenck 1997). He explained that positive personality traits are central to becoming a creative genius (Eysenck 1995).

Among the discussed traits, much of the literature emphasizes motivation as a key variable for creativity (Amabile 1997; Sternberg 2012; Sternberg and Kaufman 2010). There are two different types of motivation—extrinsic and intrinsic. If a motivation comes from outside a person, from some reward system offered by the environment for example, it is extrinsic. Extrinsic motivation can lead short-term creation. After the incentives disappear from the environment, then motivation is weakened and people do not find their work interesting. If motivation comes from passion and interest, it is intrinsic motivation that calls an internal desire to create things. Intrinsic motivation is the most critical variable for creativity. Creators will be most creative when they are motivated mainly by the interest and passion and not by external pressures. When they are intrinsically motivated, they will engage in the work for the challenge, satisfaction, and innovation (Amabile 1999). In most cases, the intrinsic, task-focused motivation is the driving force that constantly motivates creators' devotion to solve a problem or to work on artistic creation.

The famous painter, Vincent Willem van Gogh (1853–1890) is a good example. He began to draw as a child and continued making drawings throughout the years leading to his decision to become an artist. He started to paint in his late twenties and completed many of his best known works during the last 2 years of his life. He attended the Academie Royale des Beaux-Arts in Brussels in 1880, studied anatomy and rules of perspective and modeling. The training provided him with some knowledge, as he explained in a letter, "you have to know just to be able to draw the least thing" (Tralbaut 1969). He began to paint in oil during the summer of 1882. In 1886, he studied painting at Ferman Cormon's studio, one of the leading historical painters of modern France. Based on these data, we could say that after more than 30 years of practicing drawing, learning basic painting techniques from professional institutes, and working with experts, he accumulated good painting expertise. Through imaginative thinking, he did generate special techniques in his art works. For instance, he wrote: "I should like to paint portraits which appear after a century to people living then as apparitions. By which I mean that I do not endeavor to achieve this through photographic resemblance, but by means of our impassioned emotions…using our knowledge and our modern taste for color as a means of arriving at the expression and the intensification of the character." Yet, without intrinsic motivation to paint landscapes, portraits, self-portraits, wheat fields and sunflowers for 30 more years, he would not have been able to achieve the eminent level of creativity. One of his intrinsic motivations could be explained by his writing on describing his portrait painting, "the only thing in painting that excited me to the depths of my soul, and which makes me feel the infinite more than anything else."[4]

---

[4] These quotations are cited from Wikipedia at: http://en.wikipedia.org/wiki/Vincent_van_Gogh. Retrieved August 2, 2013.

7.2.2 What Conditions of Society Affect Creativity and the Creative Environment?

Environment is a context that provides incentive to promote and maintain creativity. Creativity is not the product of single individuals, but of social systems making judgments about an individual's created products. Even though creativity has been studied and recognized as a mental process, it should also be seen as recognition by the culture of an environment and judgments by the social aspect of an environment. The judgments rely on past experience, training, cultural biases, current trends, personal values, and idiosyncratic preferences (Csikszentmihalyi 1999). As such, the environment, or the place that the creator is in, has a number of factors affecting creativity. A creative environment would be the situation in which the creative processes are stimulated and sustained through to completion. As explained by Feldman, there are certain instances in which social/cultural realities determine the possibility or lack of possibility for developing creativity in a given field (Feldman 1999). In some environments, people who are creative are treated as though they are mentally ill.

Environment is also a factor that influences creativity. Fashion design serves as one example. Fashion design, categorized as an ill-defined problem task, as handled by big design companies could have different approaches than design by individual licensed designers. Designers in big design companies have certain layers of considerations that must be attended to under the constraints that they have connections with retail outlets and manufactures. They have to collect and analyze information about current trends in color, fabric, styling, and market sales of the current season and predict future trends (Sinha 2002). In their design processes, these fashion designers should not only integrate the design function into all the related business processes, but also integrate the design methods and decisions to the design vision within the company mission and strategy (Cooper and Press 1995). Therefore, creative thinking in big companies has more design constraints and limitations sensitively connected to the environment than free-lance fashion designers or even other design professionals. In consequence, either the environment manipulates design thinking, or designers creatively manufacture an atmosphere to affect the environment.

7.2.3 What Are the Cognitive Processes Engaging Creativity?

Generation of creativity could be explained as a segment of thinking that is a cognitive part of the problem solving process, or as a series of segments conducted consciously, to create an identifiable new and novel product. Explained from the series of segments point of view, it could be the cognitive stages completed sequentially for a period of time to come up with a surprising solution. Explained from a segment of thinking point of view, it is the operation of usual cognitive mechanisms applied in the process, and because of the utilization of such a mechanism at the right time within the right problem context, a creative product is generated. All these cases of cognitive conducts are attributes of the cognitive processes causing creativity.

The notion of a series of segments that occur in thinking has been described by **Graham** Wallas (1926). According to his famous creative process model, there are five stages of preparation, incubation, intimation, illumination and verification involved in generating a creative thought or a scientific discovery. Preparation is to investigate the problem in all directions. Incubation is to think about the problem in a not-conscious manner. Illumination is the happening of the happy idea, and verification is to test the idea and reduce it to exact form. For intimation, Wallas explained that he found it convenient to use the term 'Intimation' for that moment in the Illumination stage when our fringe-consciousness of an association-train was in the state of rising consciousness which indicated that the fully conscious flash of success was coming. It is the phenomenon one feels at the moment they get the feeling that the solution is coming. Thus, among the five stages, the intimation stage had been argued in psychology as a sub-stage, lacking evidence of actual existence (Hayes 1981). Yet, most scholars believe that creativity comes from a long time of preparation on a special problem task, and constantly working on the problem through a number of new approaches, until a unique solution is accomplished and recognized as a creative one. This process model provides a very general explanation of creativity that occurs in major problem solving processes, scientific discoveries, or even in everyday problem solving.

In cases of utilizing a cognitive mechanism to generate creative products, a number of cognitive procedures and functions were discussed in Chap. 2; but, cognitive processes that lead to creativity in solving everyday life plus design problems done by scholars are covered in this chapter. For instance, studies on general creative thinking for everyday life problems have approached recognizing creativity as a problem-solving capacity (Wallas 1926), as an associative process (Spearman 1931), as think in analogies and metaphors (Lakoff 1993), as constructing processes of mental representation (Carroll et al. 1980; Eastman 2001), and as activities of problem finding and solving (Ryhammer and Brolin 1999) to name just a few. Major concepts in cognitive processes leading to creativity could be categorized into the following processes:

- **Knowledge accumulation process**

  As recognized in a number of studies, in order to achieve the master level of creativity, expertise on domain specific knowledge is essential. It will take time to fully develop a skillful repertoire to reach the master level. Hayes (1981) indicated that 10 years of intensive preparation in the field of concern is necessary to become a master performer in a domain including chess, the arts, sports, and sciences. Gardner (1993) argued that 10 years effort should be on actively experimenting and exploring rather than simply learning from standard protocol. In writing, Kaufman and Kaufman (2007) found that there is 10 years of time between the contemporary fiction writers' first publication and their peak publication. For classical composers, Simonton (2000) suggested that it takes 10 years to learn the fundamental mechanics of a field, but it also takes further time to accomplish the eminent level of a domain, especially for the domains that require some artistic expressions and considerations (Martindale 1990).

- **Association and restructuring process**
  The source of creativity comes from how we uniquely apply knowledge to solve a problem. It has been argued that various heuristics are searched and used by effective problem solvers (Newell and Simon 1972; Hayes 1981). As more heuristics are established from experience, the problem solver is more effective in solving the problem. Thus, scholars who see problem solving as a major and central aspect of creativity often use techniques based on the heuristics that allow people to effectively apply available expertise (Mumford et al. 2003). On the other hand, knowledge is built up by association. Associationism (Thorndike 1911) is the belief that possessing more associations improves creativity, and creative ideas are generated incrementally. Scholars, who see associational mechanisms as an important aspect of creativity, are more likely to apply imagery techniques in training (Gur and Reyher 1976). Imagery provides another format of knowledge representation that has a strong connection to the information stored in memory. Psychological attributes, such as greater access to unusual associations and thoughts or increased motivation in pursuing unconventional or risky ventures, appear to be related to creativity (Barron and Harrington 1981).

  Regarding the restrictive process of re-organizing knowledge to solve problems differently, Gestalt psychology (see footnote 1 in Chap. 2) has proposed some notions (Duncker and Lees 1945; Wertheimer 1959). Theories in Gestalt psychology implied that creative thinking or insight involves incremental processes of knowledge or quick restructuring of ideas in creative discovery. The restructuring relates to the redefinition or reorganization of the problem situation. If a designer can make novel associations and constructs the problem structure unconventionally to create novel and useful forms, then creativity happens. In some cases, the problem structure would have gone through a series of modifications and revisions in order to reach a satisfactory solution.

- **Learning process**
  Creativity might happen in the process of learning. When we learn a new concept or make a new analogy or metaphor, a creative insight is experienced and a new personal knowledge is developed. This is similar to everyday (mini-c) creativity in the learning process, which is defined by Beghetto and Kaufman as the novel and personally meaningful interpretation of experiences, actions, and events (2007). This phenomenon is the cognitive process based on using existing chunks of knowledge in memory to construct personal knowledge and understanding for coping with a particular sociocultural context that a person encounters (Moran and John-Steiner 2003). This could also be explained as knowledge assimilation— not only generating a new knowledge (Piaget 1967), but also generating it creatively. In design, this could be the learning process of formulating a new

design concept from interpreting the existing design product or feature generated by other designers. It also explains how innovation is generated in the conceptual design stage. If the learned knowledge is applied in a new design project, then a creative design result might be generated or innovated, and creativity happens.

- **Search process**

  Eysenck argues that intelligence is essentially the search process to discover neo-genetic solutions, and bring together different ideas from memory to produce new answers to problems. Such search processes are guided by explicit or implicit ideas of relevance. There are individual differences in the definition of relevance. Some people are overinclusive in their thinking, having a rather wide conception of relevance, whereas others have a narrow or a more conventional conception. If the individual whose thought processes are overinclusive with a larger sample of ideas for the search process, then he or she would make possible the production of unusual, novel, and creative ideas. This thinking style is the basis of creativity (Eysenck 1993). In this notion, relevance is the closeness of information that connects to the question at hand. If we think knowledge is stored in memory and connected by associations, then the concept of relevance is equivalent to association in some sense. Creators must build up links among association or think divergently to get needed information. Particularly, psychological attributes, such as greater access to unusual associations (through search) or increased motivation (personality) in pursuing unconventional or risky ventures, appear to be related to creativity (Barron and Harrington 1981). Thus, the cognition of search and personality of motivation are parts of the multi-faceted nature of creativity. In design, functional requirements are the most important consideration in solving a design problem. If designers can search for new and novel forms to satisfy the major functional requirements, then creativity is manifested from the search efforts.

- **Reasoning**

  Problem solving is a complex cognitive function. It sometimes involves decisions to be made either from deductive reasoning or induction reasoning. Other times it needs to make conclusions from given information. When problem solvers are making assumptions for deduction or induction to figure out the problem situation, they are analyzing the problem context and analytical thinking is in process. When judgments must be applied, problem solvers must combine evidence and assumptions to reach conclusions, which is synthetic thinking. Such analytical thinking and synthetic thinking are similar to what were termed analytic and synthetic skills by Sternburg (1988) as parts of intellectual skills.

  However, such intellectual skills are, indeed, the rationales used to guide the cognitive moves for solution generation and should be categorized as logical thinking components. Among them, there is another format of abductive reasoning (Peirce 1997), which is to formulate hypotheses inductively from observing certain events, looking for new data references, challenging accepted explanations, inferring possible new forms and new functions, and contemplating consequences. Abductive reasoning plays an important role in daily life, and is at the heart of creativity. It is the notion of looking at a set of seemingly unrelated components.

Abductive reasoning usually starts with an incomplete set of observations, develops a likeliest possible explanation for the set (Peirce 1998), and creates the kind of everyday decision-making that does its best with the limited available information at hand. Most of time, the information is often incomplete. This phenomenon also is what happens in design. Sometimes the unrelated components might make different and unconventional association to generate unconventional design. Such a method of making unrelated associations or links is what is called divergent thinking style (Guilford 1950, 1967; Torrance 1962, 1966; Runco 1991), but the problem solver must make some judgments after the link and interpret the linked results.

- **Mental representation**
  Mental representation has been discussed and explored more in solving domain specific problems than everyday problems, for instance in art (Gombrich 1960), architecture designs (Eastman 2001; Goldschmidt and Porter 2004; Visser 2006; Chan 1997, 2011), computer engineering (Korf 1980), psychology (Shepard 1974; Kosslyn 1975; Greco 1995), scientific discovery (Kaplan and Simon 1990), and writing (Thevenot and Oakhill 2006), to name just a few. It is because everyday problems (or the same problem that occurs many times), could be resolved by: (1) heuristics developed from experiences; or (2) solutions already created in the past and recalled from memory, modified to fit. Most daily problems are likely to be routine problems, for which solutions might become standardized and automatic. Thus, mental representation is not critical in this regard.

  In solving domain specific problems or any new and strange problems, appropriate representation is necessary. Without correct representation, a problem cannot be solved. The classic mutilated checkerboard problem explained in Chap. 2 is a good example. In design, it is extremely important to use a correct internal representation that matches the external representation. Needless to say that if a unique mental image is found to serve the representation for the design, the resulting design product would be a unique and creative one.

7.2.4   How Is Creativity Assessed?

Creativity is multi-faceted and many variables are interrelated. The interaction between components of the equation and environment needed for creative performance is complex. Any changes of one variable can affect other variables simultaneously. It is impossible to identify the major variables that dominate creativity due to the individual differences among creators. All the variables should be treated as equally important in magnitude and their order in the assessment scale. Thus, in order to capture the nuances of the complex nature and individual differences between creators, the use of multiple tests of different criteria has been proposed for assessment.

For example, divergent thinking tests including the dimensions of fluency, flexibility, originality, and elaboration have been used as measures of creativity. Fluency is the ability to produce a larger number of ideas, while flexibility involves producing a large variety of ideas. Elaboration is the ability to develop and embellish an idea, whereas originality is the ability to produce ideas that are unusual, not banal or obvious (Guilford 1950, 1967; Torrance 1962, 1966; Runco 1991). Such methods of multiple measurements have been termed *psychometrics* and further developed as the Torrance Tests of Creative Thinking (Torrance 1974). Following Guilford's divergent thinking model, Torrance Tests assess the number of responses generated in the test (fluency), how many categories are shifts in responses (flexibility), the uniqueness of responses (originality), and the refinement of responses in the test to measure the extent of divergent thinking for testing students' creative thinking in learning. To the present, the Torrance Tests remain the most widely used assessment of creative talent (Sternberg 2006).

These creativity tests have been criticized for measuring intelligence related factors rather than creativity, and the subjects could be easily affected by external circumstances. Scholars also agree that both Guilford and Torrance Tests concentrated on divergent thinking as the basis of creativity and devised tests that emphasized the assessment of divergent thinking. Therefore, they have been challenged because their procedures would only measure creativity on study as opposed to creativity in daily life. But, Bachelor and Michael (1997) have proven that the tests have potential to assess creative thoughts. Along this development, Sternberg proposed a general theory of wisdom, intelligence, and creativity synthesized (WICS) for testing creativity (Sternberg 2006). His theory asserts that creative people are willing to pursue ideas that are unknown or out of favor but have growth potential. They would pursue their creation publically and move on to the next new or unpopular idea after their creation is publically announced.

7.2.5   How Can Creativity Be Trained?

It has been argued that by fostering student's creativity in the classroom, they will improve their skills in their domain but also be able to identify and establish a framework for their own lives in the future (Annarella 1999). Thus, creativity training has been developed for kindergarten students (Meador 1994), elementary school students (Castillo 1998), high school students (Fritz 1993), and college students (Daniels et al. 1985); or in the areas of educational administration (Burstiner 1973), marketing (Rickards and Freedman 1979), and engineering (Basadur et al. 1986; Clapham and Schuster 1992).

One example, given by Baer (1996), describes creativity training. He developed thinking exercises for poetry writing. In his studies, he uses two groups of student participants. One group is trained by standard language arts, whereas the experiment group is trained by image construction. The image construction session asks students to invent words or descriptions of things that suggest other things, which is the divergent thinking style. Results in his study showed this domain specific training approach, in comparison with the control group, resulted in more creative products for poems but not for the embedded stories. Therefore, teaching students to exercise divergent thinking improves creativity in image construction.

Divergent thinking is not the only component of creative thought. Other models have also been proposed as a training set to provide a more complete description of the processes in thinking. Mumford et al. identified eight core processing operations including: problem construction or problem finding, information gathering, concept search and selection, conceptual combination, idea generation, idea evaluation, implementation planning, and action monitoring (Mumford et al. 1999). This model provides a fairly coherent description of creative thought on solving ill-defined problems. It shows the promise of possible implementation of potential solutions by combining and organizing relevant concepts. Along with this proposed model, a number of experiments have demonstrated that training in the processes of problem finding (Getzels and Csikszentmihalyi 1976; Rostan 1994), conceptual combination (Mumford and Gustafson 1988; Baughman and Mumford 1995), idea evaluation (Basadur et al. 2000) do relate to both creative problem solving and creative performance. Furthermore, checklist and feature listing techniques have been proposed and trained for enhancing creative problem solving (McCormack 1974; Clapham 1997; Scott et al. 2004).

7.3 Operational Definition of Creativity in Design

Between 1930 and 1960, more than 60 definitions of creativity were found in psychological literature; by now, the number is even greater. Taylor (1988) collected these definitions and classified them into classes representing major approaches utilized by scholars, which are listed below for reference. These classes of definitions have been explained by modern terminologies and expanded further to the major definitions given in the 1980s.

- Creative perception related definitions are on the meaning of recombination of Gestalt (see footnote 1 in Chap. 2) or restructuring of perceptions. The representative definition of this approach is the process of destroying one gestalt in favor of a better one (Wertheimer 1959).
- Creative process related definitions focus on the process that generates innovative end products. A typical one is the process which results in a novel work that is accepted as tenable or useful or satisfying by a group at some point in time (Stein 1953). Torrance defines creative thinking as the process of sensing difficulties, problems, gaps in information, missing elements, something askew; making guesses and formulating hypotheses about these deficiencies; evaluating and testing these guesses and hypotheses; possibly revising and retesting them; and finally communicating the results (Torrance 1988).

- Creative expression related definitions concentrate on aesthetic or expressive considerations, which emphasizes self-expression. Any self-expression is regarded as creative. For example, the process of change, of development, of evolution in the organization of subjective life is creative process (Ghiselin 1955).
- Creative personality related definitions are on psychoanalytic or dynamic aspects of creativity, which define creativity in terms of interactional strength ratios of the identity, personal ego, and superego (Kubie 1958).
- Creative intelligence skill related definitions focus on the ability of solution thinking. For instance, Guilford defined creativity on problem solving as the ability to develop information out of what is given by stimulation (Guilford 1959).
- Creative knowledge related definitions focus on cognitive operation. For instance, creativity has been defined as the addition to the existing stored knowledge of mankind (Rand 1952).
- Creative responses definitions are on everyday life tasks. A representative definition is that a response will be judged as creative to the extent that (a) it is both a novel and appropriate, useful, correct, or valuable response to the task at hand; and (b) the task is heuristic rather than algorithmic (Amabile 1983a).
- Creativity in social and cultural aspects defines creativity as any act, idea, or product that changes an existing domain, or that transforms an existing domain into a new one (Csikszentmihalyi 1996).

These different groups of definition represent studies done by scholars across various fields up to the 1960s (Taylor 1988) and beyond. In 1969, Torrance approached from the problem solving perspective with the definition that creativity has the processes of sensing a problem, searching for possible solutions, drawing hypotheses, testing and evaluating, and communicating the results to others. Such a process model largely covers the entire creation process. Furthermore, at the concept ideation stage, the cognitive activities including the generation of original ideas, developing a different point of view, breaking out of the mold, recombining ideas, or seeing new relationships among ideas are also discussed (Torrance 1969). After 1980, a consensus among researchers had formed that individual creativity should involve the multiple components of cognitive ability, personality traits, thinking style, motivation, knowledge, and the environment (Amabile 1983b; Woodman and Schoenfeldt 1989; Eysenck 1993). Following this trend, scholars began to realize the complexity and multi-faceted nature of creativity. Efforts moved towards exploring the differences between eminent talent and everyday creativity, and separately defining them.

On the eminent side, studies are on talent performance of creativity, which is the Big-C level. Definition of Big-C creativity is "the achievement of something remarkable and new, something which transforms and changes a field of endeavor in a significant way…the kinds of things that people do that change the world" (Feldman et al. 1994). Another definition used to explain high level of accomplishment is through a person's capacity to produce new or original ideas, insights, restructurings, inventions or artistic objects, which are accepted by experts as being of scientific, aesthetic, social, or technological value (Vernon 1989). These definitions apply to highly recognized and extremely talented masters with extraordinary achievements.

On the other side, ordinary people do possess the ability to be creative for solving everyday problems, which is the little-c level. Definitions of little-c could be explained by the following two examples. Craft (2000) defines it as the kind of creativity which guides choices and route-finding in everyday life, or what I have come to term little c creativity. Little-c creativity involves being imaginative, going beyond the obvious, being aware of one's own unconventionality, being original in some way. It is not necessarily linked with a product-outcome. Another definition given in the field of education on little-c is that creativity is the application of knowledge and skills in new ways to achieve a valued goal. To achieve this, learners must have four key qualities: the ability to identify new problems, rather than depending on others to define them; the ability to transfer knowledge gained in one context to another in order to solve a problem; a belief in learning as an incremental process, in which repeated attempts will eventually lead to success; the capacity to focus attention in the pursuit of a goal, or set of goals (Seltzer and Bentley 1999). This definition explains little-c creativity from the problem solving point of view.

Regardless whether it is eminent creativity or everyday creativity, as long as the result generated by the creator (or the problem solver) is novel, new, and functional, creativity is there. This could be seen as the overall generic and essential definition of creativity that applies to all levels. Therefore, the commonly accepted core characteristic of creativity is that the production of an idea or product is both novel and useful (Barron 1955; Mumford 2003). Such a common denominator of new, novel and functional production of cognitive activities should also apply to the fundamental nature of design creativity.

However, if we see design creativity from the human cognition point of view, then design involves different reasoning, knowledge, and most importantly aesthetic value in the creative processes. For exploring the driving forces of creativity in design, cognition should be specified and the definition could be further defined operationally as: *the particular actions of consciously operating knowledge through some reasoning to generate a design idea that has a certain functional, aesthetic, and marketable value; and that resulting production is new, novel, beautiful, and accepted by public*. Such actions of creation can also be utilized as an indicator of a person's problem solving skill and ability for generating new things. Based on this definition, creativity in design can be distinguished from the products side (recognition) and explored from the associated cognitive actions side that generates these products (identification).

It has been assumed that we can identify whether or not a person is creative by observing his or her behavior or discovering what his or her products are (Taylor 1964). By the same token, we could recognize design creativity through design products and processes. Products are physical forms. Processes are continuous actions of thinking that are invisible. Yet, we could well construe design creativity as a continuous internal process in design action and make the intangible action, through visible media, to be tangible. Thus, approached from the product side, creativity in a designer's product should be judged by comparing it with other similar products to see whether the generated product is a novel, innovative, functional, appropriate, and valuable one. Taking **Wright**'s Prairie Houses as an example, features of the horizontality and deep pitched roofs were new forms that not only were elegant in maintaining the horizontality representing the nature of prairie but also functioned to keep the rain away. Similarly, the narrow cuts on concrete walls by **Ando** to bring sunshine and light in, or let another piece of wall penetrate through (see Chap. 8), was also aesthetically and functionally new and creative. Therefore, these generated features do signify the creative character that had not been created by other designers.

Approached from the process side, after new and creative features are identified, a designer's creativity should be examined and judged as to whether the methods used or processes applied by the designer to create these features were new and novel. If the products were generated from an innovative and valuable cognitive process, the designer, of course, was creative. Such a study could be done through examining master designers' design projects to get a better understanding of design creativity.

7.4   Methods of Creativity Data Collection

As explained in Chap. 2, there are four methods of studying design thinking: interview, case study, questionnaire survey, and protocol analysis. The fundamental issue of any data collection methods applied is that the method should validly and scientifically prove any proposed research hypotheses on thinking. In the studies on particular aspects of creativity, methods applied by scholars have different concentrations and procedures which could be summarized into the following types. These summaries should provide an overall understanding on data collection methods used by scholars for evaluating the different levels of creativity, for studying the cognitive processes of creation in various fields, for exploring the thinking processes of generating creativity from different approaches, and for finding methods of assessing creativity.

- **Data collection for studying genius**

  Studies on the nature of creativity have focused early on the natural ability of genius (Galton 1869), and human intelligence of great men (Cox 1926). Both studies collected data on a great number of famous creators in various fields within a range of time and compared the differences between them through statistical analysis. For instance, Galton's selection of genius relied upon the judgment of historians and others. Data is collected from a biographical dictionary.

  Such approaches on studying talented creators to understand intelligence have prevailed for a century. Similar methods have been applied to study creative genius for defining achievement acknowledged in the wider public arena

(Simonton 1994, 2004). The methods used to select creative geniuses, by modern standards, include choosing eminent classical and opera composers whose works have lasted centuries (Simonton 1977, 1998), winners of the Pulitzer Prize in fiction, or people who have entries in the Encyclopedia Britannica longer than 100 sentences (Kaufman and Beghetto 2009), etc.

Another example is the social psychology experiment done by Csikszentmihalyi. He is interested in finding out those who express unusual thoughts, who experience the world in novel ways, and who have changed culture in some important respect. From 1990 to 1995, he contacted 275 subjects for interviews, and videotaped interviews with 91 exceptional individuals who have made a difference to culture and were older than 60 years of age. Among the participants, 14 were Nobel Laureates. The studies generated many research findings and conclusions. He also clarified the differences between talent and genius (Csikszentmihalyi 1996).

- **Study creative process** (interview, protocol data, and lab documents)
Creative process is the sequence of thinking on solving a problem or on creating an object. Data used to study the process should be records of continuous thinking activities. In psychology, methods on collecting thinking data have been done through interviews, to ask problem solvers (or designers) through specific questions about a problem they solved before. The problem solvers (or designers) would recall what they had done on solving the problems in the past. In forms of introspective reporting, designers may report the information that they have inferred or otherwise generated through tape recording. Such a method is **introspective verbalization**, and the data from the person might be reflective after the fact (Ericsson and Simon 1980; Chan 2008).

The other method is to ask problem solvers (or designers) to report what they have done right after the process has been completed and tape record the verbal data. This method of having the problem solver retrieve the trace of a process just completed is **retrospective verbalization** method (Ericsson and Simon 1980). Since the event was just completed, mental activity data would not yet be forgotten and is retractable. This method, therefore, is more reliable than the introspective verbal technique.

However, recall after the fact might not be 100 % accurate. Thus, **concurrent verbalization** was developed, which is to collect mental data at the time the problem solver is attending to it. It has been demonstrated that concurrent reporting reveals the information representing the sequence heeded by the problem solver without altering the cognitive process they report (Ericsson and Simon 1980, 1996). Data collected from this method is more reliable for constructing a realistic model of problem solving, for exploring the causal-effect of creative operation, or for studying creative mental processes. The most suitable method for gathering concurrent verbalization data is protocol analysis method through video recording. Details of such a method can be found in a number of publications (Ericsson and Simon 1980, 1996; Chan 1990). Similarly, this method was used in studying individual style in architectural design covered in Chaps. 4 and 6.

In studying the creative process of a scientific discovery or a great technological invention, it is difficult to collect protocol data due to the fact that the event might have happened in the past and data is missing, or it might take a very long time to think aloud for a long period of a problem-solving task. But, other methods of collecting lab documentation have been applied for scientific discovery. Lab documents are the records of lab activities. In the late nineteenth and early twentieth century, Thomas Edison kept detailed notebooks that documented activities in his labs. Such a lab documentation method has been followed by most of the national laboratories. The notebooks provide a remarkable (opportunity) window, enabling scholars to study how new ideas developed, from conception to reality. For example, Edison's invention of phonograph can be traced from the first drawing of the talking machine developed in November, 1877 to the first phonograph announced on December 6, 1877. Further down the line, a series of more advanced products of phonograph over the following years could be tracked to understand the sequential thinking and development of the same line of products.

- **Modeling thinking processes** (survey, interview, computation, and protocol analysis)

  In studying thinking, data collected from methods of survey, interview, and protocol analysis have been mixed to generate descriptive models to scientifically analyze patterns of creativity through diagrams. For example, computational modeling is an approach used to simulate operations that generate the same sorts of ideas produced by creative humans. This technique became the most influential approach after the 1980s. Its application, for instance, was on using computational and compositional schemata to model artistic creativity without explaining what cognitive processes these schemata would entail (Gardner 1982).

  In the science of management domain the creative process had been explored to identify factors that enhance creative management (Simon 1985). In the area of artificial intelligence, the creative processes of scientific discovery have also been examined through combining computer programs with the scientist's lab data to simulate his/her own discovery processes (Bradshaw et al. 1983; Kulkarni and Simon 1988). The **BACON** program used a series of rules and heuristics to rediscover **Kepler**'s third law of planetary motion (Langley et al. 1987).

  Other studies in the field of neuroscience (Martindale 1995), cognitive science (Finke et al. 1992), social psychology (Amabile 1983b), and social-cultural studies (Sternberg 1988; Health 1993; Gardner 1993) demonstrated that creativity is a series of mental processes that generate novel ideas leading to a novel solution, or to make discoveries to improve human knowledge. Particularly, experimental methods and protocol analysis has been utilized extensively to study creative cognition (Finke et al. 1992) and the processes of gaining the insight required to solve a problem (Kaplan and Simon 1990).

- **Assess creativity** (experiments)

  Creativity can be tested through conducting psychological experiments to understand and develop a specific theory. For instance, it was assumed that the ability to cope with relative novelty is a key aspect of intelligence (Sternberg and Gastel 1989). In an experiment, 50 people were given a sentence-verification task. Subjects were asked to verify a set of subsequent statements within either a factual presupposition or counterfactual presupposition. The purposes were to explore under what circumstances people's creativity would be triggered. It has been found through analyzing the data that when people subtract non-novel from novel presuppositions, their response time is significantly higher than subtracting from non-novel stimulus. Other studies on gathering experimental data to understand design creativity have been used to explore design representation (Carroll et al. 1980).

7.5 Conclusions

Reviews of these studies on Big-C, little-c, and everyday creativity could come to a general description that creativity has different levels for within-individual and between-individual creativity. For within-individual creativity, we could develop different types of creativity, depending upon how exciting ideas are moved forward from minor replications to major restructuring or reshaping in thinking (Sternberg 2006). For creativity between individuals, the different level of creativity could be achieved depending on how the environment is supportive, the individual differences on knowledge and intelligence, and the timing of the creation. If two similar inventions (or innovations) are created by different individuals at the same time, it might have to be settled by a copyright dispute. Yet, if the person understands the problem well, thinks better, has enough knowledge and cognitive skill, motivates to persistently generate a novel creation, with a supportive environment they will achieve a high (talent or eminent) level of creativity.

Combining the studies on cognition in Chap. 2 and everyday creativity in this chapter, three important generalizations about the character of creativity from the cognitive point of view are emerging, which can serve as a conceptual framework for the study of design creativity: (1) creativity emerges from the interaction of many types of high level mental processes of visualization, memory retrieval, reasoning, and motivation, which encompass special combinations and patterns of the same cognitive processes seen in other noncreative endeavors; (2) different domains have their own unique patterns of cognitive processes; and (3) each domain has its own problem solving representation for conceptualizing specific solutions. Specifically, mental representation is critical for creativity.

For example, the representation used to solve physics problems differs from the representation used to solve accounting problems, because accounting problems are numeric in nature while physic problems are formula oriented (Larkin et al. 1980). Needless to say, architectural problems are graphic driven (Akin and Lin 1995). Thus, special knowledge is required in each domain, requiring at least 10 years of consistent training and practicing for achieving an expert or master level (Hayes 1981). Even though the fundamental cognitive processes in problem solving are the same across fields to a certain degree (Simon 1985), creativity in design, as highlighted in Chap. 2, has its own character.

The next chapter will focus on the study of cognitive factors and structures that contribute to creative thinking, which also will explain the characteristics of a designer's individual style and creativity and their correlations. Methods used are based on the four master architectural designers' design data collected as case examples to show identified stylistic and creative features, and search from designers' writings to analyze the key factors leading to their creativity. Detailed information on causative factors of design creativity can be further collected from cognitive experiments and verified by protocol data.

References

Akin O, Lin C (1995) Design protocol data and novel design decisions. Des Stud 16:211–236

Albert RS, Runco MA (1999) A history of research on creativity. In: Sternberg RJ (ed) Handbook of creativity. Cambridge University Press, Cambridge, pp 16–34

Amabile TM (1983a) The social psychology of creativity: a componential conceptualisation. J Pers Soc Psychol 45:357–376

Amabile TM (1983b) The social psychology of creativity. Springer, New York

Amabile TM (1997) Entrepreneurial creativity through motivational synergy. J Creat Behav 31:18–26

Amabile TM (1999) How to kill creativity. In: Harvard business review on breakthrough thinking. Harvard Business School Press, Boston, pp 1–28

Annarella LA (1999) Encouraging creativity and imagination in the classroom. ERIC ED434380, Viewpoints, Chicago

Bachelor PA, Michael WB (1997) The structure-of-intellect model revisited. In: Runco MA (ed) The creativity research handbook, vol 1. Hampton Press, Cresskill, pp 155–182

Baer JM (1996) The effects of task-specific divergent-thinking training. J Creat Behav 30:183–187

Barron FX (1955) The disposition toward originality. J Abnorm Soc Psychol 51:478–485

Barron FX (1969) Creative person and creative process. Holt, Rinehart & Winston, New York

Barron FX (1988) Putting creativity to work. In: Sternberg RJ (ed) The nature of creativity. Cambridge University Press, New York, pp 76–98

Barron FX, Harrington DM (1981) Creativity, intelligence and personality. Annual review of psychology. Annual Reviews, Palo Alto, pp 439–476

Basadur MS, Graen GB, Scandura TA (1986) Training effects on attitudes toward divergent thinking among manufacturing engineers. J Appl Psychol 71:612–617

Basadur M, Runco MA, Vega LA (2000) Understanding how creative thinking skills, attitudes, and behaviors work together: a causal process model. J Creat Behav 34:77–100

Baughman WA, Mumford MD (1995) Process-analytic models of creative capacities: operations influencing the combination-and-reorganization process. Creat Res J 8:37–62

Beghetto RA, Kaufman JC (2007) Toward a broader conception of creativity: a case for mini-c creativity. Psychol Aesthet Creat Arts 1:73–79

Boorstin DJ (1992) The creators: a history of heroes of the imagination. Random House, New York

Bradshaw G, Langley PW, Simon HA (1983) Studying scientific discovery by computer simulation. Science 222:971–975

Brolin C (1992) Creativity and critical thinking. Tools for preparedness for the future. Krut 53:64–71

Burstiner I (1973) Creativity training: management tool for high school department chairmen. J Exp Educ 41:17–19

Carroll JM, Thomas JC, Malhotra A (1980) Presentation and representation in design problem-solving. Br J Psychol 71:143–153

Castillo LC (1998) The effect of analogy instruction on young children's metaphor comprehension. Roeper Rev 21:27–31

Chan CS (1990) Cognitive processes in architectural design problem solving. Des Stud 11(2):60–80

Chan CS (1997) Mental image and internal representation. J Arch Plan Res 14(1):52–77

Chan CS (2008) Design cognition: cognitive science in design. China Architecture & Building Press, Beijing

Chan CS (2011) Design representation and perception in virtual environments. In: Wang XY, Tsai J (eds) Collaborative design virtual environments. Springer, Amsterdam, pp 29–40

Clapham MM (1997) Ideational skills training: a key element in creativity training programs. Creat Res J 10:33–44

Clapham MM, Schuster DH (1992) Can engineering students be trained to think more creatively. J Creat Behav 26:165–171

Cooper R, Press M (1995) The design agenda: a guide to successful design management. Wiley, Chichester

Cox CM (1926) The early mental traits of three hundred geniuses. Stanford University Press, Stanford

Craft A (2000) Creativity across the primary curriculum. Routledge, London

Csikszentmihalyi M (1996) Creativity: flow and the psychology of discovery and invention. HarperCollins, New York

Csikszentmihalyi M (1999) Implications of a systems perspective for the study of creativity. In: Sternberg RJ (ed) Handbook of creativity. Cambridge University Press, Cambridge, pp 313–335

Dacey J (1999) Concepts of creativity: a history. In: Runco MA, Pritzer SR (eds) Encyclopedia of creativity, vol 1. Elsevier, Amsterdam, pp 309–322

Dacey J, Lennon K (2000) Understanding creativity: the interplay of biological, psychological and social factors. Creative Education Foundation, Buffalo

Daniels RR, Heath RG, Enns KS (1985) Fostering creative behavior among university women. Roeper Rev 7:164–166

Dorst K, Cross N (2001) Creativity in the design process: co-evolution of problem-solution. Des Stud 22(5):425–437

Duncker K, Lees LS (1945) On problem-solving. Psychol Monogr 58(5):i

Eastman C (2001) New directions in design cognition: studies of representation and recall. In: Eastman C, McCracken M, Newsteller W (eds) Design knowing and learning. Elsevier, Amsterdam, pp 147–198

Eisenmann R (1997) Mental illness, deviance and creativity. In: Runco MA (ed) The creative research handbook, vol 1. Hampton Press, Cresskill, pp 295–312

Ericsson KA, Simon HA (1980) Verbal reports as data. Psychol Rev 87:215–251

Ericsson KA, Simon HA (1996) Protocol analysis: verbal reports as data. MIT Press, Cambridge, MA, pp 48–62

Eysenck HJ (1993) Creativity and personality: suggestions for a theory. Psychol Inq 4(3):147–178

Eysenck H (1995) Genius: the natural history of creativity. Cambridge University Press, Cambridge

Eysenck HJ (1997) Creativity and personality. In: Runco MA (ed) The creativity research handbook, vol 1. Hampton Press, Cresskill, pp 41–66

Feist GJ (2010) The function of personality in creativity: the nature and nurture of the creative person. In: Kaufman JC, Sternberg RJ (eds) Cambridge handbook of creativity. Cambridge University Press, New York, pp 113–130

Feldman DH (1999) The development of creativity. In: Sternberg RJ (ed) Handbook of creativity. Cambridge University Press, Cambridge, pp 169–188

Feldman DH, Czikszentmihalyi M, Gardner H (1994) Changing the world, a framework for the study of creativity. Praeger, Westport

Finke RA, Ward TB, Smith SM (1992) Creative cognition: theory, research, and applications. MIT Press, Cambridge, MA

Flach F (1990) Disorders of the pathways involved in the creative process. Creat Res J 3:158–165

Fritz RL (1993) Problem solving attitude among secondary marketing education students. Mark Educ J 19:45–59

Galton F (1869) Hereditary genius: an inquiry into its laws and consequences. Macmillan, London

Gardner H (1982) Art, mind, and brain: a cognitive approach to creativity. Basic Books, New York

Gardner H (1983) Frames of mind: the theory of multiple intelligences. Basic Books, New York, pp 170–204

Gardner H (1993) Creating minds. Basic Books, New York

Getzels SW, Csikszentmihalyi M (1976) The creative vision: a longitudinal study of problem finding in art. Wiley, New York

Ghiselin B (1955) The creative process. Mentor, New York

Glover JA, Ronning RR, Reynolds CR (1989) Handbook of creativity: perspectives on individual differences. Plenum Press, New York

Goldschmidt G, Porter W (eds) (2004) Design representation. Springer, London

Gombrich EH (1960) Art and illusion: a study in the psychology of pictorial representation. Pantheon, New York

Greco A (1995) The concept of representation in psychology. Cogn Syst 4(2):247–256

Guilford JP (1950) Creativity. American Psychologist 5:444–454

Guilford JP (1959) Traits of creativity. In: Anderson HH, Anderson MS (eds) Creativity and its cultivation, addresses presented at the interdisciplinary symposia on creativity. Harper, New York, pp 142–161

Guilford JP (1967) The nature of human intelligence. McGraw-Hill, New York

Gur RC, Reyher J (1976) Enhancement of creativity via free image and hypnosis. Am J Clin Hypn 18:237–249

Hausman CS (1984) A discourse on novelty and creation. State University of New York Press, Albany

Hayes JR (1981) The complete problem solver. The Franklin Institute, Philadelphia, pp 51–69

Health T (1993) Social aspect of creativity and their impact on creativity modeling creativity. In: Gero JS, Maher ML (eds) Modeling creativity and knowledge-based creative design. Erlbaum, Hillsdale, pp 9–23

Isaksen S (1988) Educational implications of creativity research: an updated rationale for creative learning. In: Gronhaug K, Kaufmann G (eds) Innovation: a cross-disciplinary perspective. Norwegian University Press, Oslo, pp 167–203

Kaplan C, Simon HA (1990) In search of insight. Cogn Psychol 22:374–419

Kaufman JC, Beghetto RA (2009) Beyond big and little: the four C model of creativity. Rev Gen Psychol 13:1–12

Kaufman SB, Kaufman JC (2007) Ten years to expertise, many more to greatness: an investigation of modern writers. J Creat Behav 41:114–124

Korf RE (1980) Toward a model of representational changes. Artif Intell 14:41–78

Kosslyn SM (1975) Information representation in visual images. Cogn Psychol 7:341–370

Kubie LS (1958) Neurotic distortion of the creative process. University of Kansas press, Lawrence

Kulkarni D, Simon HA (1988) The processes of scientific discovery: the strategy of experimentation. Cognit Sci 12:139–175

Lakoff G (1993) The contemporary theory of metaphor. In: Ortony A (ed) Metaphor and thought. Cambridge University Press, New York, pp 202–251

Langley P, Simon HA, Bradshaw GL, Zytkow JM (1987) Scientific discovery. MIT Press, Cambridge, MA

Larkin JH (1980) Teaching problem solving in physics: the psychological laboratory and the practical classroom. In: Tuma DT, Reif F (eds) Problem solving and education: issues in teaching and research. Lawrence Erlbaum Associates, Hillsdale, pp 111–126

Larkin JH, McDermott J, Simon DP, Simon HA (1980) Expert and novice performance in solving physics problems. Science 208:1335–1342

Martindale C (1990) The clockwork muse: the predictability of artistic change. Basic Books, New York

Martindale C (1995) Creativity and connectionism. In: Smith SM, Ward TB, Finke RA (eds) The creative cognition approach. MIT Press, Cambridge, MA, pp 250–268

McCormack AJ (1974) Training creative thinking in general education science. J Coll Sci Teach 4:10–15

Meador KS (1994) The effects of synectics training on gifted and non-gifted kindergarten students. J Educ Gift 18:55–73

Mooney RL (1963) A conceptual model for integrating four approaches to the identification of creative talent. In: Taylor CW, Barron F (eds) Scientific creativity: its recognition and development. Wiley, New York, pp 331–340

Moran S, John-Steiner V (2003) Creativity in the making: Vygotsky's contemporary contribution to the dialectic of development and creativity. In: Sawyer RK, John-Steiner V, Moran S, Sternberg RJ, Feldman DH, Nakamura J, Csikszentmihalyi M (eds) Creativity and development. Oxford University Press, New York, pp 61–90

Mumford MD (2003) Where have we been, where are we going? Taking stock in creativity research. Creat Res J 15:107–120

Mumford MD, Gustafson SG (1988) Creativity syndrome: integration, application, and innovation. Psychol Bull, 103:27–43

Mumford MD, Mobley MI, Uhlman CE, Reiter-Palmon R, Doares LM (1991) Process analytic models of creative capacities. Creat Res J 4:91–122

Mumford MD, Peterson NG, Childs RA (1999) Basic and cross-functional skills: taxonomies, measures, and findings in assessing job skill requirements. In: Peterson NG, Mumford MD, Borman WC, Jeanneret PR, Fleishman EA (eds) An occupational information system for the 21st century: the development of O*NET. American Psychological Association, Washington, DC, pp 49–76

Mumford MD, Baughman WA, Sager CE (2003) Picking the right material: cognitive processing skills and their role in creative thought. In: Runco MA (ed) Critical and creative thinking. Hampton, Cresskill, pp 19–68

Newell A, Simon HA (1972) Human problem solving. Prentice-Hall, Englewood Cliffs

Olson CB (1984) Fostering critical thinking skills through writing. Educ Lead 42:28–39

Patrick C (1937) Creative thought in artists. J Psychol 4(1):35–73

Peirce CS (1997) Pragmatism as a principle and method of right thinking: the 1903 Harvard lectures on pragmatism. SUNY Press, Albany

Peirce CS (1998) On the logic of drawing history from ancient documents. In: Peirce Edition Project (ed) The essential Peirce: selected philosophical writings, 1893—1913, by Charles S. Peirce. Indiana University Press, Bloomington, p 95

Piaget J (1967) The psychology of intelligence. Routledge & Kegan Paul, London

Rand HJ (1952) Creativity – its social, economic and political significance. In: Olsen F (ed) The nature of creative thinking. Industrial Research Institute, Inc., New York, pp 12–15

Rickards T, Freedman B (1979) A re-appraisal of creativity techniques in industrial training. J Eur Ind Train 3:3–8

Rossman J (1931) The psychology of the inventor; a study of the patentee. Inventors Publishing Co., Washington, DC

Rostan SM (1994) Problem finding, problem solving, and cognitive controls: an empirical investigation of critically acclaimed productivity. Creat Res J 7:97–110

Rubinstein MF (1975) Patterns of problem solving. Prentice-Hall, Englewood Cliffs

Runco MA (1986) Flexibility and originality in children's divergent thinking. J Psychol 120:345–352

Runco MA (1991) Divergent thinking. Ablex Publishing, Westport, CT

Runco MA (1994) Creativity and its discontents. In: Shaw MP, Runco MA (eds) Creativity and affect. Ablex, Norwood, pp 102–123

Ryhammer L, Brolin C (1999) Creativity research: historical considerations and main lines of development. Scand J Educ Res 43(3):259–273

Sashin JI (1985) Affect tolerance: a model of affect-response using catastrophe theory. J Soc Biol Struct 8(2):175–202

Schoenfeld AH, Herrmann DJ (1982) Problem perception and knowledge structure in expert and novice mathematical problem solvers. J Exp Psychol Learn Mem Cogn 8(5):484–494

Scott G, Leritz L, Mumford MD (2004) The effectiveness of creativity training: a quantitative review. Creat Res J 16(4):361–388

Seltzer K, Bentley T (1999) The creative age: knowledge and skills for the new economy. Demos, London

Shaw MP, Runco MA (1994) Creativity and effect. Alex, Norwood

Shepard RN (1974) Representation of structure in similarity data: problems and prospects. Psychometrika 39:373–421

Simon HA (1985) What we know about the creative process, frontiers in creative and innovative management. Ballinger Publishing Co, Cambridge, MA, pp 3–20

Simonton DK (1977) Creative productivity, age, and stress: a biographical time-series analysis of 10 classical composers. J Pers Soc Psychol 35:791–804

Simonton DK (1994) Greatness: who makes history and why. The Guilford Press, New York

Simonton DK (1998) Fickle fashion versus immortal fame: transhistorical assessments of creative products in the opera house. J Pers Soc Psychol 75:198–210

Simonton DK (2000) Creative development as acquired expertise: theoretical issues and an empirical test. Dev Rev 20:283–318

Simonton DK (2004) Creativity in science: chance, logic, genius, and zeitgeist. Cambridge University Press, Cambridge

Sinha P (2002) Creativity in fashion. J Text Appar Tech Manage 2(IV):1–16

Spearman CE (1931) The creative mind. Appleton-Century, New York

Stein MI (1953) Creativity and culture. J Psychol 36:311–322

Sternberg RJ (1988) The nature of creativity. Cambridge University Press, Cambridge, MA

Sternberg RJ (1999) Handbook of creativity. Cambridge University Press, New York

Sternberg RJ (2006) The nature of creativity. Creat Res J 18(1):87–98

Sternberg RJ (2010) College admissions for the 21st century. Harvard University Press, Cambridge, MA

Sternberg RJ (2012) The assessment of creativity: an investment-based approach. Creat Res J 24(1):3–12

Sternberg RJ, Gastel J (1989) Coping with novelty in human intelligence: an empirical investigation. Intelligence 13:187–197

Sternberg RJ, Kaufman JC (2010) The Cambridge handbook of creativity. Cambridge University Press, Cambridge

Taylor CW (ed) (1964) Creativity: progress and potential. McGraw-Hill, New York, p 178

Taylor CW (1988) Various approaches to and definitions of creativity. In: Sternberg RJ (ed) The nature of creativity: contemporary psychological perspectives. Cambridge University Press, Cambridge, pp 99–124

Thevenot C, Oakhill J (2006) Representations and strategies for solving dynamic and static arithmetic word problems: the role of working memory capacities. Eur J Cogn Psychol 18:756–775

Thorndike EL (1911) Animal intelligence. Macmillan, New York

Torrance EP (1962) Guiding creative talent. Prentice Hall, Englewood Cliffs

Torrance EP (1966) Torrance tests of creativity. Personnel Press, Princeton

Torrance EP (1969) Creativity, vol 28, What research says to the teacher. National Education Association, Washington, DC

Torrance EP (1971) Are the Torrance tests of creative thinking biased against or in favor of "disadvantaged" groups? Gifted Child Quarterly 15:75–80

Torrance EP (1974) The Torrance tests of creative thinking: norms-technical manual. Personal Press, Princeton

Torrance EP (1988) The nature of creativity as manifest in its testing. In: Sternberg RJ (ed) The nature of creativity. Cambridge University Press, New York, pp 43–75

Tralbaut ME (1969) Van Gogh, le mal aimé. Edita, Lausanne

Vernon PE (1989) The nature-nurture problem in creativity. In: Glover JA, Ronning RR, Reynolds CR (eds) Handbook of creativity: perspectives on individual differences. Plenum, New York, pp 93–110

Visser W (2006) Designing as construction of representations: a dynamic viewpoint in cognitive design research. Hum Comput Interact 21:103–152

Wallas G (1926) The art of thought. Harcourt, Brace and Company, New York

Wertheimer M (1959) Productive thinking. Harper & Row, New York

Woodman RW, Schoenfeldt LF (1989) Individual differences in creativity: an interactionist perspective. In: Glover JA, Ronning RR, Reynolds CR (eds) Handbook of creativity, perspectives on individual differences. Plenum Press, New York, pp 3–32

Part IV
Style and Creativity

Chapter 8
Creative Processes and Style

Cognition, as introduced in Chap. 2, is the human intelligence process of organizing personal information through the use of human conscious awareness, visual perception, reasoning, and judgment to accomplish everyday tasks. In design fields, designers apply design cognition to organize design information for creating artifacts. Here, design cognition is the ability to manipulate images, utilize rationale, and create three-dimensional forms to generate a product that serves a function. This ability, usually occurring in the design process, is recognized as a phenomenon and pattern of doing things. For example, as shown in Chap. 6, designers consciously utilize some invariant knowledge, rules, mental images, and certain fixed sequences in design processes; certain constant features are also generated and distinguished as the representation of a style. Therefore, patterns of constant utilization of knowledge in design are described as the phenomena of style coming from cognitive operations, and the cognitive mechanisms applied in design provide the incentive for a style. Similarly, creativity is a phenomenon of cognitive operational results that share similar cognitive driving forces. This chapter explains the connection and correlation between the two phenomena of style and creativity from a cognitive perspective.

8.1 Degree of Style and Creativity

In design, some designers do not have a strong style expressed in their works, but some do. Similarly, some designers are more creative than others. An interesting question arises as to why there are various stylistic expressions and different types of creativity among different designers. Are there correlations between style and creativity? To answer these questions, it is necessary to explore the reasons relating to the causes of degree of individual style before exploring the nature of creativity. Three theories: (1) personal design grammar and rules; (2) geographical context and building typologies of the project; and (3) design intentions in each design, could

© Springer International Publishing Switzerland 2015

C.-S. Chan, *Style and Creativity in Design*, Studies in Applied Philosophy, Epistemology and Rational Ethics 17, DOI 10.1007/978-3-319-14017-9_8 well explain the factors that shape the phenomenon of the changes of degree of style across designers. These theories share some similarities and correlations with the degree of creativity in design. Four master architects' works are used as case studies in the following sections to examine the variation of the degree of style between them.

8.2   Case Studies

8.2.1   Personal Design Grammar and Rules—Wright

The first reason why some designers have stronger style than others is their use of specific grammar, rules, constraints, and preferred forms in design. The preferred forms, in fact, are applied directly into design and expressed powerfully as a designer's signature; whereas the applied grammar, rules, and constraints have their own clear steps and procedures embedded in their utilization that might lead to the creation of certain features. If such grammar, rules, and constraints together with the same steps, procedures, and attributes are repeated in many projects by the same designer, the results definitely yield many similar features and a stronger individual style is generated. For example, **Frank Lloyd Wright's Prairie School style** has stronger expression than **Richard Meier's New York Five** style or even **Charles Moore's Vernacular style**. This is because Wright had more rules, grammar, and features applied across projects (see Chap. 5).

8.2.2   Geographical Context of the Project and Building Typologies—Otto Wagner

The second reason that explains a stronger style for some designers is that the projects designed share the same geographical location and building typology. If projects are located in similar geographic environments, the site considerations are similar. If projects have similar building types, the same type of users and usage share similar functional considerations as well. Thus, similarities across design occur more frequently to manifest a style than when different building types are located in different geographic zones. For instance, **Wright**'s projects from 1901 to 1910 were mainly residences in the Midwestern United States. Thus, similar country sites allowed the designer to create similar patterns and thus maintain the same style across buildings.

   Otto Wagner's design is another example that explains this phenomenon well. **Wagner** (1841–1918) is regarded as the founder of the Modernist Movement in architecture at the turn of the twentieth century. His designs ranged from row houses, apartment buildings, and a subway pavilion (different building types), but mostly located in the city of Vienna (the same geographic location). The geographical and typological impacts on the formation of style can be seen in his three mixed

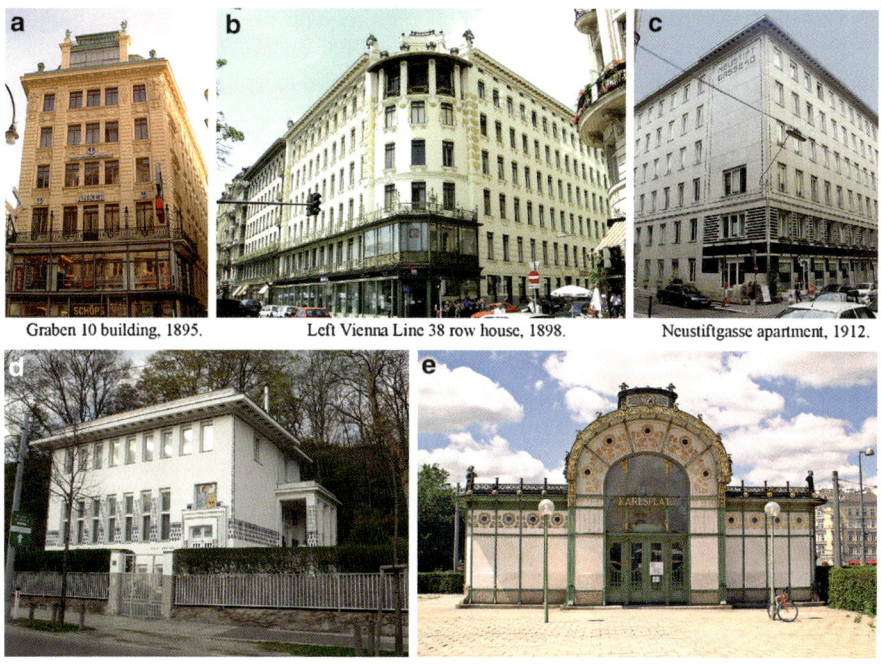

Fig. 8.1 Five projects in Vienna designed by Otto Wagner (Sources: (**a**) Graben 10 building, by author. (**b**) Row house: Gryffindor/Wikimedia Commons, http://commons.wikimedia.org/wiki/File:Otto_Wagner_Vienna_June_2006_020.jpg. Accessed 10 Oct 2013. (**c**) Neustiftgasse apartment: Heardjoin/Wikimedia Commons, http://commons.wikimedia.org/wiki/File:Otto_Wagner,_1909-1912,_A1070_Wien,_Neustiftgasse_40,_p1.jpg. Accessed 10 Oct 2013. (**d**) Zweite Villa: Welleschik/Wikimedia Commons, http://commons.wikimedia.org/wiki/File:Otto_Wagner_zweite_Villa2.JPG. Accessed 10 Oct 2013. (**e**) Karlsplatz subway pavilion, by author)

residential and commercial buildings in photos a, b, and c in Fig. 8.1. These three buildings share common features of: (1) extended continuous eaves on the roof; (2) similar proportions of the long and narrow vertical window openings; (3) horizontal bands define the second floor; and (4) continuous second floor block. Photo d in Fig. 8.1 is a residential building design in Vienna that shares the same features of extended eave, window module, and solid base. However, when typology changes to a subway pavilion, features are totally different (see photo e in Fig. 8.1).

8.2.3 Design Intentions Developed for Each Project—Otto Wagner

The big change of form in **Wagner**'s design could be associated with another cognitive factor, of design intention, utilized in different building projects. Design intention means that designers have certain intended steps to accomplish goals.

Historically, **Wagner** created a style in his architectural practice (Geretsegger and Peintner 1979, p. 31) which not only embraced but clearly manifested a distinct change in traditional sensitivity, the almost total breakdown of Romanticism and the emergence of purpose-built structures in every sphere of architecture. In his design career, he developed an architectural theory of evolution, in which structure fulfilled the role of the "original cell" of the architectural form. **Wagner** indicated that "every architectural form has arisen from a structural form and then developed by successive stages into an artistic form" (Geretsegger and Peintner 1979, p. 28). Thus, the functionalities that the buildings serve and the architect's design intentions dominate form generation, which affects feature expression. In this example, design intention dominates **Wagner**'s design creation.

As explained by architectural historians, Wagner took some influences from the **Art Nouveau Movement,** using new methods occurring at the beginning of the modern period in architectural design history. **Wagner** concentrated on structural components first and applied additional details to make the form artistic.[1] This method of utilizing structural components first was repeatedly applied in most of his designs. Pictures shown in Fig. 8.2 reflect the use of architectural forms from structural elements with artistic expressions in **Wagner**'s Church Am Steinhof Project, one that contributed to **Wagner**'s international reputation (Geretsegger and Peintner 1979, p. 206). Similar efforts, repeated in his other design projects, are considered a part of his general method of creation. However, having no cognitive data, the effort of developing design by using specialized knowledge to get a final solution in the design process case by case, cannot be clearly comprehended, observed, or identified in resulting products. But it is safe to say that Wagner's style of design was to

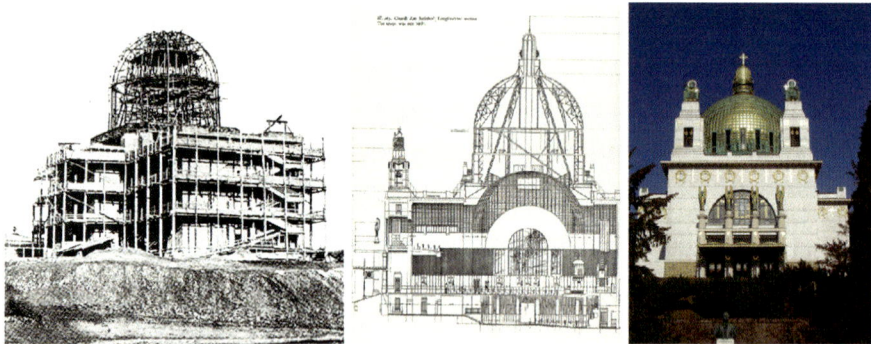

Fig. 8.2 Church Am Steinhof in Vienna, 1902 (Sources: *Left* and *center*: Drawings are from Geretsegger and Peintner (1979); *Right*: Perspective is from the following Web page: Welleschik/ Wikimedia Commons, http://it.wikipedia.org/wiki/Chiesa_di_San_Leopoldo_(Steinhof). Accessed 8 Feb 2014)

---

[1] **Art Nouveau movement** was an important bridge in architectural history between the Neoclassicism and Modernism.

incorporate the use of new materials and new forms to reflect the fact that society itself was changing. In fact, the second floor block with different form shown in the Neustiftgasse apartment building in Vienna (see photo c in Fig. 8.1) is a new design done at that time for modern apartment buildings.

8.2.4 Other Examples of Renzo Piano and Tadao Ando

Regarding the impact of geographic location, building typology, and design intentions to the formation and degree of style, **Renzo Piano**'s design is another good example. **Piano,** who was awarded the 2008 Gold Medal by the American Institute of Architects, has projects designed across a diverse array of building types, geography, and solutions to different sustainability issues plus material considerations (Piano 1998, pp. 56–59). In building types, his work ranges from a convention center (**Centre Pompidou** in Paris, 1971), airport (**Kansai Airport** in Osaka, 1994, see footnote 4), art museum (**High Museum Expansion** in Atlanta, 2005), office tower (**New York Times Tower**, 2007), science museum (California **Academy of Sciences** in San Francisco, 2008), to monumental museum (**Paul Klee Center** in Bern, Switzerland, 2005), just to name a few. Figure 8.3 shows some examples of his designs. These projects vary in design solutions applied to the generation of forms, and each solution reveals a careful concentration of particular detail. Thus different locations and different typologies of the projects require different design approaches and intentions that change the character of the design products. Forms differ and not many features repeatedly occur across all his design projects.

Explained from the cognitive perspective of design intentions, **Piano** performed systematic research to understand the constructional materials he used, and he used them to the best of their capacity. He controlled and applied technology to its extreme, but also balanced creativity and science. In other words, he studied the methodology of creation for architectural form within each design. Thus, he would start at the beginning with a special abstract developed and arrive at a solution specific to it. Because each design had a separate design program (or specification), there is no consistent and artistic development or expression appearing in **Piano**'s work. Furthermore, there are two other reasons that might cause differences in final form.

One reason is that starting from 1971, his completed famous projects were created cooperatively with three architects across three periods (Piano 1989, p. 9). The process of design decision making in these projects may have been influenced by other persons. His form generation might not have considered the same features. For instance, in 1989, he had three offices in Genoa (Italy), Paris, and Houston working on a number of projects, with fifty people speaking four languages. At that time, several, according to Piano, "two or three of these projects are still in that magical state when the idea hasn't come yet, but it is gradually taking shape. During this period, silence—meaning devoting oneself to absorbing—is fundamental. In this phase, only a superficial architect, in whom thinking and doing are separated, would

Pompidou Center in Paris, 1971.

Kansai Airport in Osaka, 1994.

High Museum Expansion in Atlanta, 2005.

California Academy of Sciences in San Francisco, 2008.

New York Times Tower, 2007.

Paul Klee Center in Bern, Switzerland, 2005.

Fig. 8.3 Pictures of six design projects done by Ranzo Piano (Sources: (**a**) Pompidou: Image is by photoeverywhere.co.uk/Creative Commons, http://photoeverywhere.co.uk/west/paris/slides/pompidou_centre2964.htm. Accessed 12 July 2014. (**b**) Kansai: Hide-sp/Wikimedia Commons, http://commons.wikimedia.org/wiki/File:Kansai_International_Airport_02.JPG. Accessed 10 Oct 2013. (**c**) High Museum: by author. (**d**) Academy of Science: by author. (**e**) NY Times: Eden, Janine & Jim/Wikimedia Commons, http://commons.wikimedia.org/wiki/File:New_York_Times_Building. jpg. Accessed 12 July 2014. (**f**) Klee Center: Florian. Arnd/Wikimedia Commons, http://commons. wikimedia.org/wiki/File:Paul-klee-zentrum-ansicht-zoom.jpg. Accessed 12 July 2014)

confidently start designing. You have to wait, tie your hands behind your back and be patient until the outlines of the new adventure have taken shape" (Piano 1989, p. 22). After the outlines were decided, a number of experiments were conducted by groups to test the materiality and constructional technology in the next conceptual design and design development stages. This example shows the phenomenon of group style and group creativity, which are covered in the later part of this chapter.

The other reason is that Piano's projects have utilized high-tech methods, which change from time to time (Piano 1989, p. 18; Pearson 2009). Because of the rapid changes and improvements of high-tech materials, the generated forms may have been revolutionized through different techniques applied. Yet, the common essential character of his designs, as shown in the buildings in Fig. 8.3, was that there were some articulated geometric structures generated through repetition of certain standard elements (or features).

A final example is **Tadao Ando**'s design. **Ando** has designed museums, religious structures, residential, and commercial buildings in Germany, Spain, Italy, France, and his native Japan. He established his design firm in 1969 in Osaka, and was mainly responsible for all the design works he built. In 1995, he was awarded the Pritzker Prize, architecture's most prestigious award. In 2002, he was awarded the AIA gold medal. His design works, mostly located in Japan, show outstanding design patterns that provide strong suggestions as to what cognitive factors trigger the phenomenon of style. From analyzing his design projects, his design cognition could be categorized into four major elements: (1) preferred graphic component and special design methodologies used to implement the composition; (2) preferred building materials; (3) design intentions of utilizing natural light; and (4) unique design knowledge learned from experience. Specifically, **Ando**'s cognition processes are explained through his writings describing his design of the **Tea House in Oyodo** in the mid-1980s (Isozaki et al. 2007, pp. 58–61).

According to his writing, there are four major steps in his design process (see Fig. 8.4). First, when he conceived a new building, he started by formulating a single schematic concept in his mind. During that mental process, he carefully examined the surrounded circumstances piece by piece to study the place, natural climate, historical, social, and cultural background of where the building would exist. At the end of this stage, he had a clear picture of the characteristics of the space and a schematic concept for the building. He said, "when conceiving a new building I begin by rendering it in my mind as a single schematic concept. This is an abstract mental process, by thoroughly stripping away the full circumstances within which the building will exist—place, climate, weather, historical and cultural background, and so on—I can clarify what the role of the space should be (p. 58)."

Second, he applied geometry to resolve the schematic concept developed in his mind. Through the geometric arrangements, he also considered spatial order implemented in the geometric forms. He wrote that "geometry is the tool the mind uses to resolve this theoretical problem, and once an idea has been distilled to its fundamental

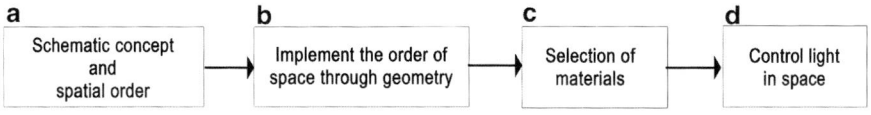

Fig. 8.4 Diagram of Ando's design processes from **a** to **d**

geometrical form, the architect has successes in expressing his concept in terms of spatial order" (p. 58). Third, he selected construction materials for the building. Concrete block, veneer, and tent canvas were his preferred materials for their easiness to obtain, writing that "concrete block, veneer, and tent canvas—all easily obtained everyday materials—were selected as construction materials" (p. 60). Fourth, during his design processes, he also applied light to create the character of the space. He wrote, "another of my methodological concerns was the need to control light, since my limited materials and the effect of minor details on light and shadow would profoundly affect the character of the space" (p. 60). These writings provide general but representative information on his design thinking. Thus, based on the descriptions written by **Ando**, design cognitions occurring in his design process can be categorized into four major items.

1. Schematic concept and spatial order development
   At the stage of developing schematic concepts and spatial order, he applied his knowledge learned from experience or from life to develop his design concepts. Specifically, he pondered his personal experience, added intelligent ideas, and incorporated these experience and ideas into his very essence (Furuyama 2006, p. 7). For example, in an interview, he said, "starting around age fifteen, I often went out to study old rural houses and other examples of traditional Japanese architecture, in order to impress upon myself the image of traditional Japanese space. This is not expressed directly in forms, but rather in the sensitivity I use when I design something" (Ando 1984, p. 131). At this stage, he applied what he had learned from experience and from what he had perceived through observations to come up with a concept on spatial order.
2. Preferred form or fundamental geometric components
   Regarding his geometric composition, he usually used simple geometry of cubes and cylinders that were inter-connected. He said, "I create architectural order on the basis of a geometry, the basic axis of which is simple forms—subdivisions of the square, the rectangle, and the circle" (Ando 1984, p. 139). Inside the form or on the site, the basic concept of the "closed/open pair" or "emptiness/solid pair" was also a fundamental characteristic of his method of articulating the site or the space (Ando 1984, p. 15). The designs for **Rokko Housing One** (1983) and **Rokko Housing Two** (1993) illustrate a range of issues in the traditional architectural vocabulary—the interplay of solid and void, the alternatives of open and closed, the contrasts of light and darkness. Other than the residence houses that were designed around a courtyard, such as the **Sumiyoshi House** (1979) and **Kidosaki House** (1986), his public building designs also have an outdoor opening in the center of the building mass, as in **Naoshima Contemporary Art Museum** (1992), **Museum of Wood** (1994), and **Benetton Communication Research Center** (2000) etc.
   He also used a long linear geometry, a long and straight passage way, to define the entrance path, which after passing through, affected people's perception. The path, for instance, could be a covered colonnade space of glazed promenade, as

Fig. 8.5   The colonnade shaped entrance path in the Chapel of Mt. Rokko (Source: http://commons. wikimedia.org/wiki/File:Rokko_Mount_Chapel_Tadao_Ando.jpg. Accessed 10 Oct 2013)

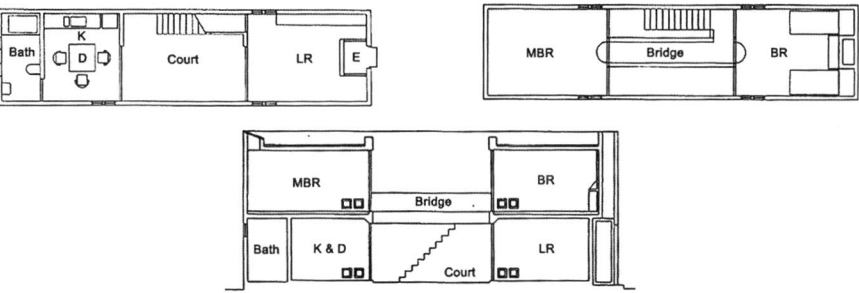

Fig. 8.6   Azuma House floor plans (*top left* and *top right*) and section drawing (Source: Mariana, http://en.wikiarquitectura.com/index.php/File:Azuma2.jpg. Accessed 10 Oct 2013)

used in **Chapel of Mt. Rokko**, 1986 (see Fig. 8.5). At the end of that colonnade, visitors are led from light infused space on the right into the interior of the chapel. The same long linear path, but open rather than covered, can be found in his **Church of the Water** (1988), where the dominant freestanding wall on one side sets up the approaching and marching direction to the church.

The idea of a path was also used to connect spaces by an independent corridor. For example, a centralized corridor that links two rooms through an open interior courtyard in Sumiyoshi row house (1976) in Osaka (see Fig. 8.6) clearly explains this notion. Other examples showing the use of paths are the walkway path that turns around the entrance wall (see photo a in Fig. 8.7), and leads to the

Fig. 8.7 The passage path (**a**) and the entrance path (**b**) in the Water Temple, and axis in Westin Hotel (**c**) (Sources: (**a**) Water Temple path: Mungo Binkie/Creative Commons, http://www.flickr.com/photos/mungobinkie/3864035067/in/set-72157622046374933/. Accessed 10 Oct 2013. (**b**) Water Temple: Mungo Binkie/Creative Commons, https://www.flickr.com/photos/mungobinkie/3864823232/in/photostream/. Accessed 13 July 2014. (**c**) Westin Awaji Island Hotel: Chris 73/Wikimedia Commons, http://zh.wikipedia.org/zh-tw/File:Westin_Awaji_Island_Hotel_06.jpg. Accessed 10 Oct 2013)

underground building entrance of the **Water Temple** (1991) (see photo b in Fig. 8.7). He also used a path as the axis to orient the building mass (see photo c in Fig. 8.7) in **Awaji-Yumebutai** (2000). To him, path defined by a corridor represents order. He said, "I prefer to use simple forms, and to focus on the mazes created by the paths of people's daily activities. Stairs and corridors fix these patterns; they are places where one cannot move irregularly" (Ando 1984, p. 132). His notion of path really defines the architectural order.

3. Preference of concrete materials

   In most of his design projects, he selected concrete as the constructional material. He explained, "I attempt to use a modern material—concrete and specifically, concrete walls—in simplified forms to realize a kind of space that is possible…It seems to me that, at present, concrete is the most suitable material for realizing spaces created by rays of sunlight. But the concrete I employ does not have plastic rigidity or weight. Instead, it must be homogeneous and light and must create surfaces" (Ando 1984, p. 142).

4. Element of lights

   Light is an essential and important design factor, which has been developed as a body of knowledge by **Ando**, utilized as a design constraint and operated together with walls as part of his design language. Therefore, the use of light in space, which has certain rules embedded in it, is also a part of his design grammar. To him, the meaning of a space can change simply by controlling the amount of light. He explains, "in the **Bansho house** (1976), light only enters from above on one side. In the morning it shines on the dining room table, and reflects into the living room as well as into the second floor bedroom. Then it gradually spreads out, and as it moves, the rooms change in character. One can make completely different spaces depending on whether light is direct or reflected, and whether or not it comes in from one direction. When light coming in from on direction, I try to create different spaces depending upon the way it is received. This method is found in traditional Japanese architecture – in tea room for example, or in houses built in the style of tea rooms" (Ando 1984, p. 132).

He also said, "in the **Koshino House** (1981), the light behaves in a sculptural way for only twenty or thirty minutes when the sun shines directly in from the west. Light can be used as an object, but a more gentle light is generally necessary for everyday life. There is the artistic problem of using light sculpturally, and then there is the problem of gently manipulating light to facilitate daily activities, to make the environment more comfortable, or to provide a pleasant atmosphere. For example, light entering from a low window is very gentle; but because it is diffuse it does not photograph well. Light entering from the top, though, can be very beautiful. Simply by manipulating light in these different ways, the space can be made sufficiently rich. This method was not used by modern architects since they were concerned more with economic problems, with building in large quantities, with building quickly" (Ando 1984, p. 132).

5. Special methods on the use of walls to implement the geometry and light

Ando's architecture is the architecture of **walls**, which plays a crucial role in composition. To him, walls were the most basic elements of architecture, and three major methods could be categorized in his handling of walls. The first method was to apply a freestanding wall for defining outdoor space. For instance, an L-shape freestanding wall was used to partially enclose the landscape, as shown in the **Chapel on Mt. Rokko**, (1986) (photo a, Fig. 8.8). The same method is used in the **Church of the Water**, (1988) (photo b, Fig. 8.8), where the L-shaped freestanding wall framed the natural surroundings as visitors approached the church, and it extended along one side of the lake, and wrapped around the back of the church. The third example of using an L-shape freestanding wall is in the **Forest of Tombs Museum** (1992) (see the lower L shape wall on photo c, Fig. 8.8). There, an L-shape wall framed the landscape on the south and west.

The second method was to use the freestanding wall to penetrate a volume or another wall, which was done in the **Forest of Tombs Museum**, where an L-shape wall pierced the center of the circular courtyard. Inside the courtyard, a ramp winding around the wall led visitors through displays of artifacts (see upper L shape on photo c, Fig. 8.8). The second example is the **Church of the Light** (1989) in Osaka, where a fifteen-degree free-standing wall pierced through a concrete triple cube dividing the space into the chapel and a triangular entrance space (see photo a, Fig. 8.9). The third example is the **Stone Hill Center** (2008), a combination of galleries and art-conservation labs, in Williamstown, Mass. Here, a free-standing wall symbolically transected the main volumes of the workspace on the north side at a diagonal (see photo b, Fig. 8.9), and the form continuously projected out to the east end of the building (see photo c, Fig. 8.9). Such a squared archway in a freestanding diagonal created a view frame for the landscape (see photo d, Fig. 8.9). Similar methods of intersecting a freestanding wall within another wall can also be found in the entrance path at the **Children's**

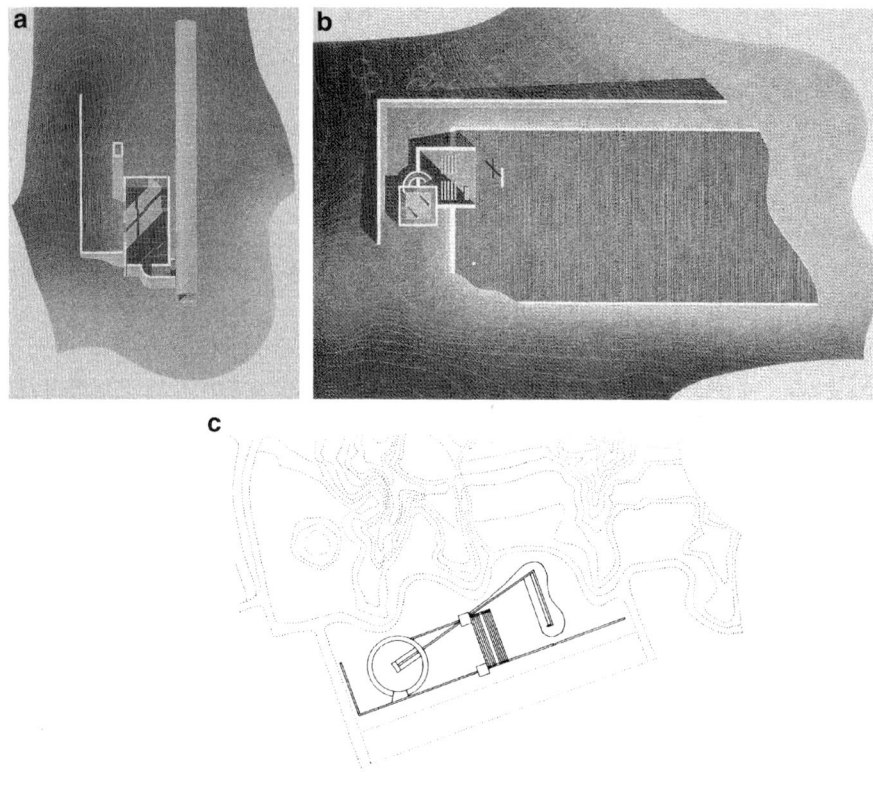

Fig. 8.8 Drawings of L-shape freestanding wall in the Chapel on Mt. Rokko (**a**) Church on the Water (**b**) and Forest of Tombs Museum (**c**) (Sources: Drawing images were from Furuyama (1993) with permission to use from Tadao Ando Architects & Associates; including (**a**) Plan of Mt. Rokko Chapel, p. 135, (**b**) Plan of Church on the Water, p. 137, and (**c**) Plan of the Forest of Tombs Museum, p. 168)

Museum (1989) in Hyogo. The intersection defines the transition of entrance passageways.

The third method he used was to narrowly cut the walls to let a strip of light shine into the space, which gave character to the space, as the 20 cm cut to show the cross in the Church of the Light (see photo a, Fig. 8.10). Other applications of this method included leaving the void edge between the ceiling and the independent wall, or cutting the back wall to let the side one through (see photo b, Fig. 8.10), or cutting the wall for visually connecting the interior and exterior (photo c and d, Fig. 8.10), such as in the **Chichu Art Museum** (2004). Other examples can be found in the Meditation Space at **UNESCO** in Paris (1995), and the **Pulitzer Foundation for the Arts** in St. Louis, Missouri (2001).

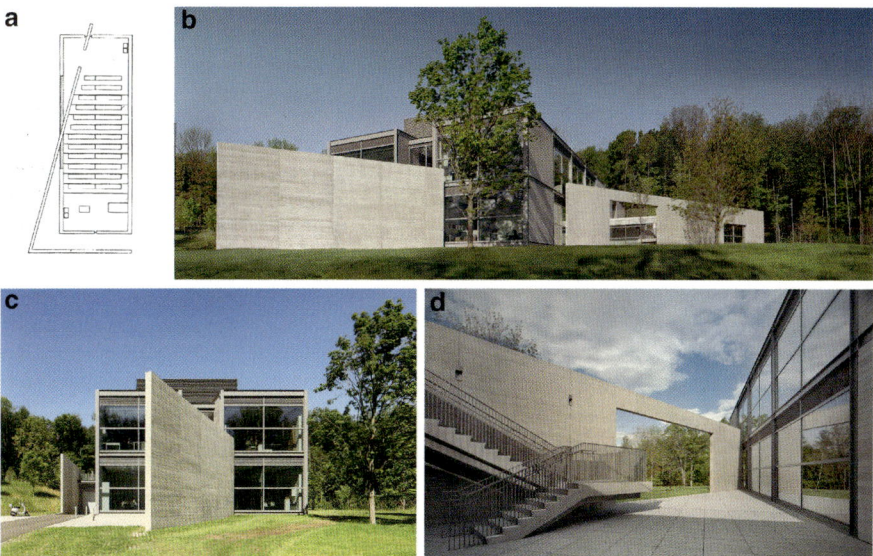

Fig. 8.9 Floor plan of the Church of the Light (**a**), North elevation of the Lunder Center at Stone Hill, Clark Art Institute, Williamstown MA, (**b**), east elevation (**c**), and the north inner court (**d**) (Sources: (**a**) Church of the Light: Furuyama (1993) with permission to use from Tadao Ando Architects & Associates. (**b**) North Elevation: © Sterling and Francine Clark Art Institute, Williamstown, MA, USA (photo by Richard Pare). (**c**) East façade: © Jeff Goldberg/Esto. (**d**) North Inner Court: © Sterling and Francine Clark Art Institute, Williamstown, MA, USA (photo by Richard Pare))

From studying the works done by **Wright**, **Wagner**, **Piano**, and **Ando**, their styles can be summarized by the degree of their expressed style. **Wright** had a very strong Prairie House style in products due to the fact that he applied rigid elevation grammar and fixed features in design. **Ando** also has a strong style in his design for he had: (1) the same design principles of using simple and minimal geometry; (2) considered the same design constraints of light (sometimes water)[2] and nature; (3) utilized the same design grammar for handling walls; (4) strong personal preferences in the use of concrete materials; and (5) unique design knowledge. Even though his works do not have repeated features showing a strong pattern of style, there are traces of methods and processes that show a similar style of doing things and similar messages across projects. **Wagner**'s designs have weak style, as his design intentions are concentrated on considerations of structural components, and he applied loose design principles across projects. **Piano** utilized various high tech

---

[2] Some of his designs have water involved. Particularly, a number of the buildings done after the Water Temple (1991) were surrounded by or visually rested on the water; for instance, the Church on the Water (1998) in Hokkaido, **Komyo Ji Temple** (2000) in Ehime, **Modern Art Museum** (2002) in Texas, and **Langen Foundation** (2004) in German etc.

Fig. 8.10 Wall cuts in the Church of the Light (**a**, **b**), and the Chichu Art Museum (**c**, **d**) (Sources: (**a**) Cross in the Church of Light: Attila Bujdoso/Wikimedia Commons, http://commons.wikime-dia.org/wiki/File:Church_of_Light.JPG. Accessed 13 July 2014. (**b**) Top wall cut in the Church of the Light: http://www.paulamontessketchbook.com/2010/06/church-of-light-by-tadao-ando/. Accessed 10 Oct 2013. (**c**) Chichu Art Museum wall: CTG/SF/Creative Commons, https://www.flickr.com/photos/27966213@N08/3728709510/in/gallery-43355952@N06-721576228 22787568/. Accessed 10 Oct 2013. (**d**) Chichu Art Museum wall cut: Todd Lappin/Creative Commons, https://www.flickr.com/photos/telstar/204531148/. Accessed 10 Oct 2013)

construction methods and different materials, which decreases the number of common features and causes a weak style. Also, **Piano**'s projects had a group of designers involved who explored many different solutions to accomplish the major task. As a result, **Piano**'s explorations of digital architecture in design yielded different products that diversified his firm's stylistic expression.

These four master architects show different styles, yet they are all recognized and considered as creative designers. What are the factors that define creativity and what are the factors that make these masters so creative? A convincing and persuasive method to explain the factors attributing to creativity is the design cognition approach, for it scientifically explores design thinking. In the following sections, the contributing factors that cause creativity are explained.

8.3 Factors of Creativity in Design

Creativity implies freedom of action and is a result of cognitive operations. Notions of human cognition and natures of creativity have been covered in detail in Chaps. 2 and 7, respectively. As summarized in Chap. 7, the fundamental human cognition processes in problem solving are the same to some degree, but design cognition has its own character, and design creativity has reasons for its unique character. Here are the reasons.

Unlike other areas, design is a special problem solving endeavor of shaping objects to meet purposes or constructing a structure adapted to objectives; it requires professional consideration of aesthetic beauty, constructional technologies, functional uses, social symbols, market demands and supplies. All man-made design has as its fundamental essence that it is consciously driven by certain intentions and achieved by a series of actions. Approached from design cognition theory (Chan 2001, 2008; Cross 2001; Eastman 2001), design can be scientifically explained as the mental process of recognizing problem contexts, choosing design issues to be considered, setting up goals or constraints to address the chosen issues, finding suitable courses of action to implement the goals, and evaluating and choosing among alternative actions to achieve a satisfactory solution. Thus, the processes of solving a design problem consist of a series of selecting operators through a number of routines (or paths) that lead to the final solution. But, within a huge **problem space** with an immense number of possible paths, which path leads to creativity and the creation of a novel product? This big question can be answered by considering the cognitive factors that guide a designer to a creative path, as the nine cognitive items proposed in the summary of Chap. 2. From the studies of **Wright**, **Wagner, Piano**, and **Ando**, creative design masters, four cognitive factors can be identified that answer most of this question.

8.3.1 Problem Representation

Representation symbolizes knowledge that represents, in a certain way, an external task being confronted. *Internal representation*, established in the mind to mentally visualize the potential solution, is a key factor in design cognition. Otherwise, it is impossible to find any solutions, which remains true under the situation of having inappropriate internal representations. The mutilated checkerboard (MC) problem, explained in Chap. 2, is a good example. To successfully solve this problem, one must switch the internal representation from dominos, the numbers of squares and their geometrical arrangement to the parity of the squares—the unbalanced number between black and white. Without changing the internal domino representation to solve the checkerboard external representation, one would not be able to solve it (Newell 1965; Wickelgren 1974; Anderson 1980; Korf 1980). However, the external representation in the MC problem, one of the well-defined problems, is already given as a solid physical object.

In solving ill-defined problems, no external representation is provided. Designers must construct both internal and external ones. For example, if a designer is required to design a façade of a residence, he has to generate an internal representation of a mental image to process the design dialog. Then, an external representation through drawing, physical modeling, or a digital modeling will be constructed to externally display the design concept developed in the mind. Since the external and internal representations are both constructed by the designer, who might have enormous diversified categories of representations to choose from, ill-defined problems provide more opportunities for designers to develop creative thinking.

The notion of using different representations to generate various solutions can be explained first in the following well-defined problem. The problem is: "how to put ten coins in five straight rows with four coins in each row?" It is impossible to put four coins in five rows to come up with the total of ten. We can use five sticks and ten real coins to demonstrate the solution. But using dots and lines in drawing as different external representations easily and effectively solves the problem. The same problem was given to a graduate class of "Design Thinking and Cognition" at the School of Architecture, Tianjin University, in fall 2012. Most students in the class of 24 participants submitted two or three solutions. Among them, one student drew six solutions (see Fig. 8.11). Thus, this student is recognized as a creative thinker.

In architectural design, representation can be seen from the design methodology perspective as the technique used to create a ***design scenario*** in the early design stage. In an experimental study on design problem solving (Chan 1990), it was found that in the very beginning of a design, the designer first concentrated on understanding the design problem, program, and requirements, which is the first stage of *problem task understanding*. Then, the designer used sketches to study the character of the site and its influence on the building design, which was the second

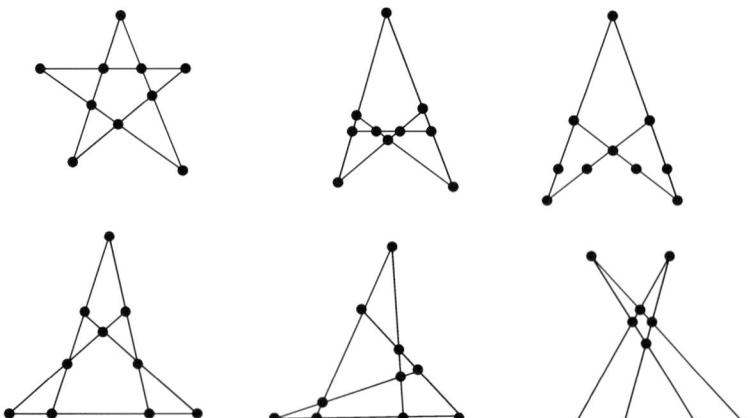

Fig. 8.11   The ten coins puzzle problem

stage of *site organization*. After this, the designer used mental images, diagrams, and sketches together with certain social-cultural, aesthetic, structural, or sustainability considerations to create a scenario that produced the overall graphic outline of a final product. Such a design scenario can also be seen as a higher level of design representation termed by Wright (Wright 1928; Scully 1960) and Piano as ***design abstract***.

But, how can an internal representation be constructed mentally? Usually, **heuristics** are applied. An ***heuristic*** is a specific **rule-of-thumb**, or an argument derived from experience or convention. Heuristics stand for experience-based techniques that let a problem solver simplify a decision task or problem to make it more tractable. In the learning process, heuristics relate to the process of gaining knowledge from an educated guess or trial-and-error rather than by following some pre-established formula. This, in fact, is abductive reasoning, basing on limited information and unrelated components to make connections in design. Other than using heuristics, any internal representation could also be constructed from searching through memory for a previously used one, or even quickly creating one by following given instructions. In lab experiments, when subjects were given a problem, they did not initially choose deliberately among problem representations, but usually adopted the representation suggested by the given problem statement (Hayes and Simon 1974; Simon and Hayes 1976).

Quantitatively speaking, the representation available in well-defined problems might have more limited applications than in ill-defined problems. Studies in comparing different representations of the same problem found that it is rare for two people, acting independently, to define an ill-defined problem in the same way (Kotovsky et al. 1985). Thus, ill-defined problems have different representations to choose from, and a creative person who solves the problem when others cannot, might be the one who chose the best representation of the problem. It is also found that in well-defined problems, an appropriate representation will solve the problem quickly, but sometimes the problem solver needs to change the internal representation to successfully solve the problem. The MC problem is, again, a good example, by changing the representation from domino to the parity of the squares. When the problem solvers have a particular representation that they believe will allow them to solve the problem, they will keep searching within the corresponding problem space unless they find the urgency to change (Kaplan and Simon 1990). Therefore, personality characteristics of flexibility and openness for change are needed for creating new solutions or for creating new representations for solving old problems.

A similar phenomenon in ill-defined problems of changing a problem representation to creatively solve the problem was also reported in a protocol analysis study conducted in the field of industrial design. It was found that when designers changed the representation or problem structure, a new and amazing solution was created immediately. The collected protocol data had also showed that when designers found that the problem structure could be changed, modified, or adjusted by representation, there was an opportunity to create new solutions (Cross 2001). In everyday thinking, it is also true that the change or restructure of problem representation plays an important role in bringing up a sudden comprehension, and a solution is discovered at the same time (Duncker and Lees 1945, p. 29). This change of representation in design thinking is similar to the notion of "**divergent thinking**." Divergent thinking uses knowledge and experience to multi-directionally process existing information to generate diversified information. It is arriving at many different solutions with innovative results, which also is one factor used to measure creativity (Guilford 1950, 1988). But divergent thinking is not part of the discussion here, for it does not fully explain the cognitive reasons for the causes of creativity. The causes of creativity, however, could be explained through the notion of cognitive activities and representation utilized in the **problem space**. A simple design problem is given below. The problem is a façade design for a single family residence.

If a façade facing south in North America is given as the design project, then the designers' design cognition in the process could be identified hypothetically through constructing its problem space, as shown in Table 8.1. This problem space represents the entire mental process of solving an elevation design, and elements inside the space are the cognitive mechanisms occurring in the process, which could be identified by associated cognitive activities. For instance, at the starting point of solving a design problem, the designer knows his **design problem**, understands his current position of the **design elements**, develops intended **design constraints** (including design issues), and generates a schematic image/diagram as a **representation**, then applies the known design rules as **operators** to create a design result. Thus, all these elements could be listed systematically. Table 8.1 is a generic example of solving a façade design. Any changes in the problem space would change the problem structure. However, the change of the internal or external design representation would dominantly alter the mental image/diagram and their corresponding operators, which would ultimately generate a different design result that could solve the problem by accident. This elaborates the situation that triggers one of the creative processes. In most cases, designers should have the expert knowledge on

Table 8.1 A generic example of a problem space in a façade design

Problem	Design the facade of a single family residence
Elements	A wall with entrance door and windows
Design intentions	Some generated design constraints and issues
Internal representation	Some schematic image/diagram of the facade/wall
Operators	Apply principles of geometrical composition to put the entrance door and windows on appropriate positions, that match the structural layouts and satisfy the imposed constraints
Initial state of knowledge	Know the geographical location of the project, the orientation of the walls, and the specifications of the door and window, the materials to be used
Knowledge available	Information about constructing wood structure
	Information about materials to be chosen
	Heuristics of design constraints and issues
	Principles of facade design

tectonic information on wall, door, and window, correlations of these elements, and heuristics of implementing the imposed design constraints (and issues) to generate an acceptable design.

But, were the results generated from the changes made on internal or external representation considered to be creative acts? This could be determined by the creative choices made in selecting their representations. For instance, in the façade design problem, designers could have a wood wall representation, or other isomorphic images of sold concrete, transparent curtain wall, or concrete and masonry unit (CMU) among others to choose from; needless to say that they could also apply any of past design cases to serve as one. Even within the wood structural representation, designers could also heuristically choose from different selections to compose siding, insulation, stud, vapor membranes, and interior layer of finishing, or even impose different sets of design constraints to generate a special design for texture consideration. If there are combinations of special design intentions and unique constraints imposed to the specially arranged representation, then the results should be creative. **Tadao Ando's** wall design in Azuma House (in Sumiyoshi) is a good example.

According to **Ando**, the small Azuma House was the point of origin for his subsequent work (Ando 1987). In this house, the façade was a standalone element with an entrance opening without door panel axially and symmetrically arranged in the center (see Fig. 8.12). The simplicity of the design makes it a very unique image that had not been done by other designers. This explains the notion of DC2, proposed in Chap. 2, that the structure of unique mental images causes creativity. As such, the

Fig. 8.12 The façade of the Azuma House designed by Tadao Ando (Source: Wikimedia commons, http:// en.wikipedia.org/wiki/ File:Azuma_house.JPG. Accessed 10 Oct 2013)

Table 8.2 An example of Ando's façade design problem space

Problem	Design the facade of a single family residence in Osaka
Elements	The facade wall and entrance door
Design intentions	Insert a concrete box into the wooden row houses site, and create a microcosm inside the box (global intention)
	Symmetry, simplification and closure as global constraints
	The indentation on the front wall is the door and it is the only hint of life (local intention)
Internal representation	Simple concrete enclosed box w/lights, the wall stands alone
Operators	Put the entrance door in the middle of the concrete wall to satisfy the simplification and constraints
Initial state of knowledge	Knowing the geographical location of Osaka, orientation of the walls, and the specifications of the door opening
Knowledge available	Information of concrete construction
	Heuristics of simplification
	Principles of symmetry

cognitive elements occurring in his problem space and its representation applied in designing the façade, based on the data obtained from his writings (Ando 1984), is theoretically organized in Table 8.2. For instance, he explains that his intention of this design "was to insert a concrete box and to create within a microcosm. A simple composition but with diverse spaces, closed but dramatized by light – such was the image I sought to develop (Ando 1987, p. 42)." In this example, he developed a design intention, which served as his global internal representation throughout the design. While he worked on the front wall, he says that "the house completely closes itself from the street. The life inside cannot be sensed from the outside. The indentation on the front wall, serving as the entry is the only hint of life" (Ando 1987, p. 42). He applied simplification and symmetry as local constraints to create the façade. Thus, the entire unique problem space and its internal representation coming from the unique design intentions created a novel design, which defines his creativity. This verifies the notion of DC1, DC6, and DC7 proposed in Chap. 2, that expertise knowledge, problem structure, and internal and external representations, respectively, generate creativity.

8.3.2 Expert Knowledge and Unique Combination of Knowledge

As explained earlier, it takes time for a creative person to acquire knowledge and skills relevant to the creative act. Chess players require about 10 years of preparation before they reach the grand master level (Simon and Chase 1973). The required 10 years of preparation before a person produces work for which they are famous is consistent across fields of music, composing, and painting (Hayes 1989). However, the use of existing knowledge in creative thinking is a very controversial issue. Some argue the importance of utilizing expert knowledge to avoid reinventing the wheel, particularly the problem solver with expert knowledge would know when and how to use the knowledge. Others argue that there are times to suspend expertise to discover truly novel ideals and insights. However, in design, it is necessary to have design knowledge and training established to know what appropriate information to search for and utilize.

For example, designs by **Renzo Piano** applied new high tech methods and skills based on extensive research performed by his design team. Thus, it is necessary to acquire expert knowledge to accomplish high level work, especially if the task is new or the designer is working on a new design approach. **Tadao Ando** is another extreme. He never attended a university and never apprenticed to any architectural firm. As a singular architect, he taught himself almost everything. He has profound knowledge but it was drawn primarily from his direct experience. One area of specialized knowledge he developed early was on the wall design using light to define the character of the space in his **Bansho House** project in 1976. Approximately 10 years later, his knowledge of light and concrete wall allowed him to design the famous Church of Light (1987), whose construction was completed in 1989 (see Fig. 8.10). This explains the item of DC1 proposed in Chap. 2, that the development of expertise knowledge causes creativity.

Other than the development of expert knowledge, it is also key to be able to combine knowledge into a novel form that has not been done before by others. **Frank Lloyd Wright** could combine his knowledge on horizontality and Mother Nature to generate the style of Prairie Houses. Such a style had not been done at the turn of the twentieth century, and demonstrates his genius design ability. This explains the item DC3 proposed in Chap. 2, that the making of associations between diversified knowledge causes creativity.

8.3.3   The Uniqueness of Design Constraints Utilized

Design constraints can be seen as placing limitations on the conditions under which a design is developed, or on the requirement of the design to be met. If the design constraints are uniquely defined by combining some theories or hypotheses, and imposed into the design, then the results could be novel. Take **Tadao Ando's** design as an example again. There are two major constraints together with ideas that he applied in several creative designs:

- **Simplification of form**: Simplification is a unique constraint. He thought, "simplification through the elimination of all surface decorations, the employment of minimal, symmetrical compositions, and limited materials constitutes a challenge to contemporary architecture" (Ando 1984). Thus, he used simple geometry with concrete walls to generate some amazing designs.

- **Interaction with nature**: The Azuma House design was **Tadao Ando's** early design in which a few special considerations and generalizations were his original and creative works. A courtyard was used to bring in nature but separate the living room and dining room, and was a unique arrangement incorporating nature as one of his design constraints. He said, "introducing nature into a building has been an important theme in my work since the design of the Azuma House in 1976" (Ando 1984, p. 25). True nature, as opposed to artificial and domesticated nature, to **Ando**, was capable of confronting the individual and it had to be enclosed in the building in order to be experienced (Ando 1987). This idea of nature was a part of his design intention, which serves as one of his design constraints. He indicated that, "of course, bringing nature into the house tends to make life more severe…However, it was precisely in this way that traditional Japanese townhouses were enriched despite their physically cramped form and spatial impoverishment…but a house into which nature has been introduced is more suitable for man and is more true to the basic character of the house. The courtyard is an important place where seasonal changes can be perceived directly through the senses" (Ando 1984, p. 25). Thus, the courtyard inside the geometry was totally open (see Fig. 8.6), which again, was a novel design.

These two examples of using unique constraints in design explains the cognitive notion of DC4 proposed in Chap. 2, the unique applications of external and internal constraints causing creativity.

8.3.4   The Unique Design Methods or Grammar Applied

Frank Lloyd Wright constantly used grid system and elevation grammar in his Prairie Houses designs. Similarly, **Tadao Ando** had his design rules and methods as well, which can be seen as part of his design grammar. Wall design is one example. He indicated, "by positioning a number of walls at certain intervals, I create openings. Walls are freed from the simple role of closure and are given a new objective… The relationship between inside and outside is based on the act of cutting on walls" (Ando 1984, p. 24). Collected from his writings and writings about him, plus repeated similar methods used in several projects, his wall design grammar can be summarized in the following.

- If two walls are independent units, then cut wall A vertically to let wall B pass through. The vertical opening on wall A is not closed to define each own identity.
- If the wall is an exterior wall that needs to give some character of its interior space, then cut this wall horizontally nearby the ceiling to directly let sunshine/light in.

Table 8.3 Justified cognitive items causing creativity and style

ID:	Cognition causing design creativity (DC) and individual style (IS)
DC1:	Knowledge repertoire.
DC2:	Structuring of mental image.
DC3:	Association of diversified knowledge information.
DC4:	Unique application of external and internal constraints.
DC5:	Quantity of environmental constraints.
DC6:	Problem structuring and restructuring.
DC7:	Internal and external representations.
DC8:	Design strategy used in design.
DC9:	Reasoning used to strategically activate design strategies.
IS1:	Repetition defines style.

- If the wall is an exterior wall that needs to connect interior and exterior spaces, then cut this wall horizontally 3 ft above the ground to let view go through.
- If walls are used to define pathway, then use a L-shape wall with ninety degrees turn to delineate the passage way and to generate a visual transition.

These design methods were special and creative, for they had not been done by others. This demonstrates the item DC8 proposed in Chap. 2, that unique design strategies used in design cause creativity.

From summarizing the case study data, it is clear that the ability of structuring a unique mental image is so critical for generating a novel form. Therefore, structuring mental image ability is added as item DC2 to the updated summary table in Table 8.3. Among the nine proposed cognitive factors causing creativity, most have been explained by the four case studies. Three of them, the highlighted items in Table 8.3, need to be justified. The first one in DC5, the quantity of environmental constraints causing creativity, indicates that fewer external constraints imposed by the environment would improve creativity. For instance, museum design typology, which has fewer external constraints imposed by the clients than commercial design typology, would cause more creativity. **Tadao Ando**'s Azuma house is another example, shown in Fig. 8.6. In that design, the open courtyard in the middle with a bridge connecting the two bedrooms on two sides is an extreme case, but that was accepted by the client without objections.

Item DC6, unique problem structuring, is formulated in the beginning of the problem solving process and can be seen by the uniqueness of the project program or project documents. Problem restructuring in the design process would provide chances for creativity, as has been reported by Cross (2011). Regarding item DC9, the application of reasoning to strategically activate unique design strategies to cause creativity, protocol analysis must be employed to further prove this point.

8.4 Group Big Creativity and Little Group Style

With scientific building technology continually improving, special knowledge and skills are needed in the design process. It is especially true for complex design projects. Big design firms greatly rely on teams with various knowledge skills to handle sophisticated information processing on complex designs. For instance, analyzing vertical elevator circulation patterns in a skyscraper, finding out health facilities used in a hospital, or figuring out complicated airplane, luggage, logistic, and passenger circulations in airport terminal designs usually involve groups of consultant companies. In the current commercial globalization and digital media era, the study of group style and team creativity is a new research direction on how to improve a company's creative productivity. In the following, group creativity and style are discussed.

First, if a style is the combination of a number of similar works done by a group of individual creators sharing a similar style as an assembly, it is called **Big-G** *group style*, for instance, the **Prairie House** group style. Style created from group projects representing a teamwork style is called *little-g group style*, for instance, the **Renzo Piano**'s group style. The little-g style refers to the ways that the group operates and the expression that the group has shown. Similarly, in studying creativity, some scientists may develop new techniques or knowledge that may have profound impact on society. Some architects may generate new designs that influence the design professions. These types of creativity have been called **Big-C** *creativity* and *little-c creativity* in everyday life when solving problems at home (Gardner 1993). In this chapter, the factors that influence team thinking efforts while working on the same project are the focus in studying a group of Big-C creativity, which also is defined from the product's perspective as *the team effort of generating products that are thought to be interesting, novel, and to have social value.*

Team effort in generating a design can be seen as a group of cognitions interacting. Individual creativity inside the eminent creativity (Big-C) relates to how a designer cognitively processes his design information in a creative way to generate creative productivity for the group, whereas group creativity relates to how the information used by the group is effectively shared and applied for generating a novel and socially valuable product. Thus, the information used by the group has to be openly shared and has to flow across the members fluently to accomplish creative acts. Factors of popular beliefs and preferences shared by group members that would affect the proportional use of group information have been discussed (Nemeth and Nemeth-Brown 2003). For group decision making processes, it also has been suggested that the group must develop group norms emphasizing high standards, open exchange of ideas, and attention to contrary points of view for the group to be successful (Stasser and Birchmeier 2003).

Other components that promote creative and innovative outcomes in group projects include diversity (Austin 1997) and expertise. In group thinking, particularly when the group members are from the same field or sharing the same professional background, it may be difficult to think of novel ideas when previously expressed ideas are very salient. Thus, diversity of group members is needed, and differences in expertise among group members can also be beneficial for the creative process. Even though it has been argued that the groups with members who differ from each other on one or more salient characteristics may experience higher levels of conflict (Jehn et al. 1997), a carefully managed group process could eliminate or reduce negative impact. Take **Renzo Piano**'s designs for example. Most of his famous projects were team works, and he would strategically have a major architect develop the major design scenario first, and then have different teams with different expertise work out details (Piano 1989). This top down driven method could explain why Piano's team worked successfully. Of course, it is important to further explore the interaction, cooperation, and transformation of cognitive messages occurring within the group and across groups to study group creativity.

8.5   Creativity and Style

In studies conducted in this chapter, creativity and style are treated as two readable entities, explored from the macro level, approached from design cognition. Methods used are to recognize both entities appearing in the design products, then to analyze the potential cognitive factors that drive the emergence of style and creativity. After analyzing the overall studies developed in these several chapters, it is sufficient to draw conclusions about the relationship between creativity and style. For example, creativity is identified by the original generation of a signature feature that is novel and has significant value. If a number of such original features appear in a number of designs, they signify the appearance of a style. More of such original features created by the same designer indicate his depth of creativity. Similarly, the more these features appear in his designs, the more stylistic the designer is with stronger and longer style. Yet, the repetition of the same features in designs happened in a long period of time would block the opportunities for creative generation, because the characteristic of creativity is that a designer must create new things from time to time. Works done by **Wright, Ando, Piano,** and **Wagner** explain this phenomenon.

   Wright's Prairie Houses style has been identified as having 10–11 representative features appearing across his projects from 1901 to 1910. Most features were creatively generated and combined by Wright himself, which does signify his creativity and strong style within this period of time. **Ando**'s designs used similar principles of geometric composition and similar methods in handling wall cuts and paths, plus repeatedly applied natural light to define spaces. His design works are shown to have consistent design principles utilized with different constraints implemented, but no identical features were reapplied. As such, the design results generated were driven by similar principles causing some degree of similarity in forms, and **Ando** is very creative with some style. **Piano** has been categorized as (little-g) group style and group creativity. **Wagner**'s design has shown a totally different approach of invisible style on structure driven process, but his designs, at his age, were very avant-garde and creative.

From the cognitive processes standpoint on exploring the factors causing the creation of the signature features (creativity) and the causes making the features reappear (style), findings could be summarized by the cognitive elements applied, developed, or nurtured in the design processes, including constraints, mental images, grammar, and representations, into the following three major points.

1. **Design intentions, constraints, or design methods**: If a designer utilizes the same design intention to accomplish intended goals, implements the same design constraints with similar heuristics, or the same design methods in a number of designs, then results might have similar features occurring to declare a style. If these intentions, constraints, or methods are newly developed in the original design process and yield a novel feature, then creativity emerges from the first creation of these features.

2. **Images, or design grammars**: If the designer uses the same mental image and design grammar, then results could "directly" reflect the appearance of the same features. If a mental image is applied in design products, its image pattern is very easy to be perceived and recognized through vision. Regarding design grammar, the embedded rules in the grammar are more logical and precise for the same or similar form generation, which would have more impact to sustain a style. However, creativity does occur under changing situations. Thus, in order to have creative acts, the same style cannot be held too long. New mental images and new grammar must be invented for new form generation.

3. **Design representation**: Theoretically speaking, repetitious application of the same design representation would generate the same style. Yet, no evidence has been found to explain that representation has any impact on the formation of an individual style, for it is an indirect factor. However, representations have impact on creativity. For instance, if a designer develops a novel representation in a process of creating a new design result, then it has the chance of generating a creative product.

8.6 Conclusions

Style and creativity do share some similarities and dissimilarities in their essence. In the fields of literature, fine arts, and performing arts, any artistic object or performance has its process of creation that follows certain unique ways and methods of generation. If such methods are novel and generate novel products, then they are valued as creative acts. If such methods are repeated over time to reproduce similar forms or actions, a pattern of characteristics emerges and a style is formed. Thus, style could be utilized as symbols or identities for recognizing an individual, a group, a period, or a culture, or for identifying the differences between individual works, group works, periods, or fashion culture. By the same token, creativity could also be used as a measurement to recognize a person's working performance. This explains the similar cognitive aspects of style and creativity and their relationship studied from the human thinking approach.

If a style is seen as an expression that relates to the quality of art work, then it should refer to how the expression would affect observers and how observers would perceive the work. In other words, style could be interpreted as how a work of art can serve to stimulate appreciation. Appreciation is a sense perception, involving a comparatively large amount of interpretation and understanding of meanings (Munro 1967, p. 49). In Western art of the fourteenth century, a new iconography of the life of Christ and Mary was created to express suffering. New patterns of line and color, which possess a lyrical and pathetic aspect than other styles, were emerging. Such new patterns were commonly accepted by audiences when they were created. Modern art expression, having the taste for the constructive and rational in industrial society, used mechanical forms that showed precision, coolness, and power. Designers also applied such fashions of form in their designs. As such, new generations of any changes define creative evolution and such creative actions cause social-cultural changes to educate human beings.

A style, as explained in this book, is the way of doing things or the way of performing things, in which choices are purposefully made to satisfy certain constraints imposed through a series of mental searches to fulfill design intentions. The factors influencing the choices made from alternatives in design have multiple dimensions. The possible factors involved in making choices include personal knowledge, mental image, rules and procedures of handling domain-specific tasks, and individual preferences. Individual preference, analyzed from the perspective of social science, could be affected by cultural forces in society. The cultural forces of fashion, value, and custom, symbolized conventions could affect the characteristics of art forms, and artists ought to select among given conventions while they are designing. Thus, the decisions made among choices determine the generation of a style, and the relationships among the made choices indicate the history, evolution, and changes of style. On the other side, options generated as the cases for making choices could be used by an individual designer to achieve a creative product or by a group to achieve group creativity. During the choice generating process, more or diversified options searched from various resources by the individual or provided by the group will definitely enhance the chances for creativity. But, this dimension of diversity efforts will increase creativity but decrease the degree of style, because it will add more chances for change and reduce the possibilities to keep the common features recycling in different designs.

In conclusion, a creative person may be a person who has cognitive and motivational factors involved in creative performance. The person would work very hard, be very flexible, always looking for originality. On the cognition side, this person would set up goals to accomplish with many years of preparation. On the problem solving side, while working on problems, this creative person would choose good problem representation, apply good constraints, and effectively evaluate the outcome and take actions to revise shortcoming. Such personality, characteristics and cognitions are also parts of the attributes of a stylistic person. As long as the same ways of doing things appear across tasks, a style is automatically manifested through the visible outcomes.

References

Anderson JR (1980) Cognitive psychology and its implications. W. H. Freeman, San Francisco

Ando T (1984) Tadao Ando: buildings, projects, writings. Rizzoli International, New York

Ando T (1987) Tadao Ando. GA Architect, 8. ADA Edita, Tokyo

Austin JR (1997) A cognitive framework for understanding demographic influences in groups. Int J Organ Anal 5:342–359

Chan CS (1990) Cognitive processes in architectural design problem solving. Des Stud 11(2):60–80

Chan CS (2001) An examination of the forces that generate a style. Des Stud 22(4):319–346

Chan CS (2008) Design cognition: cognitive science in design. China Architecture & Building Press, Beijing

Cross N (2001) Design cognition: results from protocol and other empirical studies of deign activity. In: Eastman E, McCracken M, Newstetter W (eds) Design knowing and learning: cognition in design education. Elsevier, Amsterdam, pp 79–103

Cross N (2011) Design thinking: understanding how designers think and work. Berg, Oxford

Duncker K, Lees LS (1945) On problem-solving. Psychol Monogr 58(5):i

Eastman C (2001) New directions in design cognition: studies of representation and recall. In: Eastman C, McCracken M, Newsteller W (eds) Design knowing and learning. Elsevier, Amsterdam, pp 147–198

Furuyama M (1993) Tadao Ando. Artemis Verlags-AG, Zurich

Furuyama M (2006) Tadao Ando, *1941: the geometry of human space. Taschen, Hong Kong, pp 7–20

Gardner H (1993) Seven creators of the modern era. In: Brockman J (ed) Creativity. Simon & Schuster, New York, pp 28–47

Geretsegger H, Peintner M (1979) Otto Wagner, 1841–1918, the expanding City, the beginning of modern architecture. Rizzoli, New York, pp 26–31

Guilford JP (1950) Creativity. American Psychologist 5:444–454

Guilford JP (1988) Some changes in the structure-of-intellect model. Educ Psychol Meas 48:1–4

Hayes JR (1989) Cognitive processes in creativity. In: Torrance EP, Glover JA, Ronning RR, Reynolds CR (eds) Handbook of creativity. Plenum Press, New York, pp 135–146

Hayes JR, Simon HA (1974) Understanding written problem instructions. In: Gregg L (ed) Knowledge and cognition. Lawrence Erlbaum Associates, Hillsdale, pp 167–200

Isozaki A, Ando T, Fujimori T (2007) The contemporary tea house: Japan's top architects redefine a tradition. Kodansha International, Tokyo, pp 58–61

Jehn KA, Chadwick C, Thatcher SM (1997) To agree or not to agree: the effects of value congruence, individual demographic, dissimilarity, and conflict on workgroup outcomes. Int J Confl Manag 8:287–305

Kaplan C, Simon HA (1990) In search of insight. Cogn Psychol 22:374–419

Korf RE (1980) Toward a model of representational changes. Artif Intell 14:41–78

Kotovsky K, Hayes JR, Simon HA (1985) Why are some problems hard: evidence from Tower of Hanoi. Cogn Psychol 17:248–294

Munro T (1967) The morphology of art as a branch of aesthetic. In: Beardsley M, Schueller H (eds) Aesthetic inquiry: essays on art criticism and the philosophy of art. Dickenson Publishing, Belmont, pp 43–53

Nemeth CJ, Nemeth-Brown B (2003) Better than individuals? The potential benefits of dissent and diversity for group creativity. In: Paulus PB, Nijstad BA (eds) Group creativity: innovation through collaboration. Oxford University Press, New York, pp 63–84

Newell A (1965) Limitations of the current stock of ideas about problem solving. In: Kent A, Taulbee OE (eds) Electronic information handling. Spartan, Washington, DC, pp 195–208

Pearson C (2009) California academy of science. Archit Rec 2009–01:58–69

Piano R (1989) Renzo Piano: building workshop: 1964–1988. A+U Publishing Co., Tokyo, pp 14–22

Piano R (1998) Renzo Piano: sustainable architectures: arquitecturas sostenibles. Gingko Press, Corte Madera, pp 56–59

Scully VJ (1960) Frank Lloyd wright. George Braziller, New York

Simon HA, Chase W (1973) Skill in Chess. Am Sci 61:394–403

Simon HA, Hayes JR (1976) The understanding process: problem isomorphs. Cogn Psychol 8:165–190

Stasser G, Birchmeier Z (2003) Group creativity and collective choice. In: Paulus PB, Nijstad BA (eds) Group creativity: innovation through collaboration. Oxford University Press, New York, pp 85–109

Wickelgren WA (1974) How to solve problems. Freeman, San Francisco

Wright FL (1928) In the cause of architecture. Archit Rec 63:49–57

Chapter 9
Cognitive Theory of Style and Creativity

Life and beauty both play essential roles in our lives. Good designs will have it all—aesthetic pleasure, art, creativity—and at the same time be usable, workable, and enjoyable. There is no need to sacrifice budget costs, function, manufacturing, or sales market to achieve the maximum value of a product. For skillful craftsmen, it is their passion, motivation, and ability to create things that are usable and creative, pleasurable and completely workable. That is why style of beauty and creativity of products are so critical in design, manufacturing, engineering, fashion, and everyday lives.

Many studies in different fields have explored the meaning and type of style, but these studies have concentrated mainly on the interpretations of the stylistic expressions rather than on the substance of style. On the other hand, studies in creativity have also focused on exploring intelligence of genius performance; techniques of divergent thinking for education, methods of assessing creativity, and possible reasons that initiate creation in problem solving rather than on the in depth exploration of the cognitive causes of creativity. In fact, there are correlations between style of beauty and creativity of products, but they have not been discussed so far.

As commonly recognized by scholars, style and creativity are products of thinking operated by human cognition. Studies on both must be approached from the cognitive perspective for gaining scientific understanding. By the same token, the cognitive correlations between the two must also be explained from the cognitive perspective. This chapter will sequentially summarize the operations of design cognition that cause style and creativity, the cognitive relations between them, the validity of the theory proposed, the possibility of improving style and creativity, and future studies in this area.

© Springer International Publishing Switzerland 2015
C.-S. Chan, *Style and Creativity in Design*, Studies in Applied Philosophy,
Epistemology and Rational Ethics 17, DOI 10.1007/978-3-319-14017-9_9

9.1 Thinking and Cognition

Human cognition relates to the acquisition and utilization of information for accomplishing certain things. Raw information data is obtained through sensory input from seeing, hearing, smelling, tasting, and touching. After these input data are received through the sensory register (Atkinson and Shiffrin 1971), the senses transfer the information into short-term memory, convert the information through digestion or interpretation, transform the semantic results into knowledge chunks, and selectively store the chunks permanently in long-term memory. The related phenomena of encoding and decoding the verbal, audio and visual codes are the fundamental cognition that occurs in every human's thinking across fields. It has been explained by cognitive scientists that all memory is unconscious before it becomes conscious. Only one percent of the information can reach conscious awareness by the mind. Therefore, the unconscious mind is intellectually richer and has more data to draw upon than the conscious part of the mind (Goleman et al. 1992). But, it is the conscious part that operates the process of information rationally while solving problems.

When thinking involves reasoning for decision making, the level of cognition is more complicated, and can be illustrated by knowledge processing in reading. For instance, words are images. Reading begins with seeing the image of a word. While perceiving the word, human eyes will bring the image through the cornea and lens to create an image on the retina, which triggers optic nerve impulses. These impulses are put together sequentially and passed along to the brain. In the human brain, neural models of words are always developed after words are learned and are used in future reading as patterns for instant recognition. Usually, after the image of a written word is sent by the retina to the visual cortex, visual information is sent to other parts of the brain for processing. It has been reported that a perceived image is translated from image to its individual sounds, and analyzed to articulate and understand its meaning (Shaywitz 2003). For experienced readers, words are saved as symbols and their corresponding pronunciations and meanings are recalled instantly or automatically for pattern recognition. These brain activities are attributed to fundamental cognition. However, in order to combine meanings of a group of words for catching messages, making judgments, and determining selections, the mind and brain must work together to accomplish these advanced tasks and higher levels of **reading cognition**. This relates to the phenomena of perceiving and interpreting information for learning in reading.

In fine arts, techniques of sketching, similar to writing, are the drawing language used by painters for visual communication. After practicing and exercising, drawing motions become automatic skills using brushes or pens, and are essential cognitive activities relating to the actions for representation. Expert painters can easily portray their mental image or sketch out a picture to imitate the perceived image. However, utilizing strokes to complete a representation involves: (1) interpreting the perceived image and turning the perception into abstraction; (2) determining the strength and dimensions of the strokes to be put on drawing for expressing the appearances of the shapes; (3) considering the composition and proportion of shapes in a picture to show beauty; and (4) applying colors and textures to represent abstraction or realism. These activities are the cognitive processes of combining knowledge of aesthetic and procedural knowledge of painting to implement the internal representation in the mind. Broadly speaking, in the making of a picture, certain rules and patterns are considered by expert painters. Their drawing processes also involve the use of visual language (a set of arbitrary signs and symbols to communicate thoughts and emotion) and certain grammar (a body of rules and principles used for visual communication). All these cognitive operations in thinking compose the higher level of **drawing cognition**. Similarly, in design thinking, after basic cognition is utilized, advanced level of design language and grammars would be further developed to achieve the professional design level. Such activities are termed "**design cognition**" for design thinking.

9.2 Design Thinking and Processes

Design thinking is a general term describing the mental activities on consciously considering a design, or intending to produce a design product. In architecture design, design thinking relates to how architects mentally process the information needed to construct a building and make it habitable with aesthetic values. In graphic design, it relates to how graphic designers use the information to construct a composition for visually communicating messages. In fashion design, it relates to how fashion designers apply related information to design clothing and accessories that provide comfort, beauty, and style to the clients. Regardless the various natures of different fields and outcomes, the major mechanism of thinking activities in design processes is design cognition, which is very similar cognitively across design fields.

Design cognition is the term used to describe the mental activities that occur in design thinking. It explains specifically how certain mental faculties are applied to process thinking. For instance, it illustrates how designers accumulate knowledge, manipulate (gather, analyze, accumulate, and recall) information, utilize reasoning, and operate strategies for solving design (ill-defined) problems. From the information processing point of view, design cognition is the processes of making moves on processing design. Yet, the most difficult part of the thinking in solving design problems is the beginning stage of formulating the design problem from understanding the design requirements, developing a goal, finding representations, and organizing steps to accomplish the goals. Only after the problem is structured and sequences of sub-goals are clear, the processes are initiated step by step with inductive, deductive, and abductive reasoning.[1] Through the use of such reasoning, certain strategies

---

[1] Some scholars have argued that no logic of creative processes exists and that a rational model of discovering is impossible. Recent studies had shown that it is possible to provide and implement rational models of creative reasoning and scientific discovery through computational modeling. In this intellectual framework, abduction is used to unify the different perspectives of human inferences (Magnani 2009, p. 60). Details of the computational modeling for abduction could be found in the Chaps. 3 and 4 in his book.

or methods would be developed in the processes. There also is a **spatial reasoning** commonly used in everyday design tasks, which is the operation of visual codes through certain special compositions for making a product.

In complex design problems, however, certain design strategies and methods, such as higher levels of **design cognition,** are needed to manage massive information. Sometimes it is helpful to use operation methods to filter out unnecessary information to solve functionally related problems. Sometimes it is helpful to use certain strategies for form generation, for instance, the uses of metaphor, analogy, morphosis, or past cases. Sometimes certain methodologies and strategies are developed by designers over time. These uniquely developed strategies and methodologies would be used by designers repeatedly across designs. Such a cognitive phenomenon of repetition defines the designers' cognitive style, which differs from the design style explained in Chaps. 4, 5, and 6. The cognitive style is a personal way of utilizing special sets of cognition on processing information.

In some cases, designers make "what-if" conjectures to make moves. This "what-if" process is driven by inductive or deductive reasoning until sometimes abductive reasoning is needed and occurs.[2] The "what-if" process may create new problems to be considered and resolve through making hypotheses. This is a possible, second type of cognitive style in thinking. The possible third cognitive style is to think divergently, rather than convergently, to generate a variety of responses to a given problem. The way of processing divergent thinking is to make more associations between knowledge, or make connections to different aspects of the problem for creating more diversified alternative solutions. At this time, designers must make choices to select the final solution from the alternatives. Of course, convergent thinking is necessary for concentrating on one direction to get more in depth level of consideration. But, it happens only after the conceptual framework of the problem has been considered through divergent thinking.

Another aspect in the processes is the problem context. The problem needs to be structured into a framework and the problem situation needs to be consciously checked, and sometimes restructured to resolve or avoid conflicts. While looking for solutions, some sub-problems could be easy to resolve for an experienced designer who has been practicing for a decade. In some cases, certain knowledge accumulated under years of practice becomes procedural knowledge that has automatic skills embedded and is effortlessly used in new designs. Most of the time, the generated sub-solutions would be evaluated immediately to guarantee their feasibility. After all the sub-problems are solved stepwise with satisfactory sub-solutions achieved, a final product is created. If it meets all required constraints and is marketable, then the design processes are complete. This summarizes the possible design cognition that would occur in the design processes. But, as explained, cognition is the

---

[2] This could be explained operationally, in the design processes, as to get specific facts from general data (deduction), or to generalize from a sample data (induction), until up to the point of making hypotheses to infer causes from effect (abduction).

mechanisms applied in certain ways for solving problems. What basic mechanisms are used generically and what are the specific mechanisms used in certain fields? This needs more research effort.

9.3   Design Cognition, Style, and Creativity

The basic cognitive mechanisms are attention, recognition, chunking, perception, association, representation, repetition, goal seeking, problem structuring, and reasoning. Advanced cognitive skills are the combinational utilization of the basic mechanisms in multiple ways. More layers of cognitive mechanisms applied in a design course, i.e., applying reasoning to select association for the making of a representation, might generate a higher stylistic and creative result. Of course, the quality of application is more important than the quantity of the combinational applications of mechanism. There is a prevailing belief that a basic level of intelligence is a necessary requirement for creativity in the generation of novel ideas (Sternberg 1997). Therefore, basic professional knowledge and intelligence skills are the prerequisites for generating creative products in the considered field of design.

Style has also been defined as the acts of choices made between alternatives (Gombrich 1960). The acts of making choices are decision making activities. For instance, when a number of alternative solutions are generated in the process, designers must make a choice to select the final solution from the alternative. Then, style comes from making this choice. This notion of style is to define style from the history of taste and fashion point of view. However, seeing style from the design cognition perspective, it should be defined by the common set of features that appears in design products and manifests the designer's way of doing things. Research data, collected from the case study on **Frank Lloyd Wright** (Chap. 5) and from the 12 protocol analysis experiments (Chaps. 4 and 6), could be summarized into the following items to scientifically describe the substance of style.

- **Repetition:** a cognitive mechanism, is the major force generating a design style or a cognitive style. For cognitive style, it indicates that the problem solver would repeatedly apply the same thinking pattern to solve a number of problems. For design style, it indicates that the preferred features or pre-solutions are repeated, or the same design processes are reapplied in many projects to create similar design products. These repeatedly appearing features consist of a common feature set, which defines an individual style. Repetition will also generate **rhythm**, which is another type of stylistic creation (Chaps. 2, 4, and 5).
- **Across products:** If there are at least four common features appearing in at least three products designed by the same designer, then an individual style is manifested and it could be identified as the designer's individual style (Chap. 4). Features of companies' logos are different, for the logo legally represents a company under the copyright.

- **Within products:** If more than three features from the same common set appear in one product, the product has the designer's style, or the style exists in that product. If the stylistic feature appearing in a product is less than three, then the style of this product is not recognizable or it is a weak style (Chaps. 4 and 6).
- **Threshold of feature modification:** Each feature has its own compositional character. Any of its topological change will impact the visual recognition of the style more significantly than its geometrical change. The threshold of geometrical change is 40 % of proportion. If it is distorted more than 40 % in any dimension, it would not be recognized as the same feature style (Chap. 4).
- **Design constraints and methods:** Constant application of the same internal constraints and the algorithms developed by the designer would cause similar features that signify the designer's style. For instance, an architect would have his own mathematic method generated from experience for determining the unit size and use philosophy to decide a design scenario (or design parti) at the conceptual design stage (Chap. 6).

Creativity is identified by the features appearing in products, which are visible and recognizable. As long as the processes that trigger the creation of novel features can be identified, then the cognitive factors generating creativity are able to be detected and explained. Following the data collected from the case studies done on Chaps. 5 and 8, the cognitive mechanisms that cause creativity, labeled as DCx on Tables 2.1, 2.2, 2.3, and 2.4 in Chap. 2, are itemized here to describe the substance of design creativity. The corresponding variables are included in the beginning of the items for reference.[3]

- **DC2, association of diversified knowledge information:** Creativity can happen when the designer makes unconventional associations with different knowledge chunks, and apply different knowledge to solve the current problem. For example, **Wright** used grid system and modules of units based on the default casement window system for the floor plan, Froebel system of geometric blocks for the solid form, elevation grammar for the facades, and the combination of knowledge on horizontality and nature in his Prairie Houses design (Chap. 5).
- **DC5, problem structuring and restructuring from what-if reflection:** Creativity can happen in the very beginning of generating the parti, schema, abstract, or scenario stage. For example, **Tadao Ando** applied some philosophical thoughts in the beginning of the design and implemented the abstract notion to solid form (Chap. 8). Another example is **Santiago Calatrava**, who constantly used drawings on developing analogy and applied analysis methods to create an overall conceptual scenario in the beginning of the design. Because the problem definitions are set earlier, the unique program shapes the problem structure that may guide the design results into a new and creative form.

---

[3] Table 8.3 shown in Chap. 8 has the item of DC2 added to explain the significance of mental image to creativity. The ability of constructing a unique mental image is a critical factor to creativity. Other variables are the same in sequence on both tables.

Creativity can also happen when the designer asks a good question to himself that changes the problem structure (problem-restructuring) which leads to a creative solution generation (Cross 2011).

- **DC6, internal and external representations:** Creativity can happen when a designer changes the representation or uses a very different representation for the design solution. For instance, **Tadao Ando** used a simple box with only one opening to represent the entrance in his **Azuma House** (Chap. 8). The result is very unique.
- **DC6, structuring of mental image (internal representation):** Creativity can happen at the stage of developing a visual image as a sub-solution for a sub-problem. For example, a special feature might be developed right in the middle of the design processes with a new image, and designers would integrate the new feature image into the entire design context to create a new form (Chan 1990).
- **DC7, design strategy used in design:** Creativity can happen when a designer uses a particular method to solve the problem, for example, a particular method other than metaphor, analogy, or morphosis. The cut on the wall to let sunshine in for giving the space a character done by **Tadao Ando** is a good example (Chap. 8).

9.4   Cognitive Theory of Style and Creativity

As shown on the itemized list of summaries, it is clear that creativity may occur anytime in the entire design processes and is defined by the new and novel creation of any forms in the end result. By combining the notions of style and creativity, the **cognitive theory** of style and creativity can explain how features define both style and creativity. But, the fundamental differences between them are that the feature in style must be repeated and features in creativity must be new and novel. Only after more than three new features are created and recognized as novel forms, and repeated more than three times, a style is generated.

Regarding the production of creativity, basic design knowledge is required, then operations of the same cognitive mechanism (or the same intelligence of the designer) generates creativity and style. It is possible that the qualitative aspect of knowledge (DC1), would be a factor for the degree of style and creativity. However, it is more likely that because of the special uses (or changes) of the cognitive mechanisms at a particular moment, creativity occurs. Yet, different mechanisms have different impacts on style and creativity. For instance, repetition (IS1) and representation (DC6) are the most important ones that affect style directly. Without repeated action in learning (rehearsal), procedural knowledge would not be established. Without repeated actions in applying the same procedures to generate the same features, a style cannot be identified either. Without utilizing the same representation, the created features might not be the same, and style would be altered. Without good representation, features would not be that stylistically generated in the design processes, and the designers' artistic expression shown in the product might not be that creative. Other mechanisms of association (DC2), problem structuring and restructuring (DC5), goal sequences and constraint (DC3) are indirect factors on creativity, which influence the generation of creativity more than style. The remaining design creativity variables of structuring of mental image (DC6), the number of extrinsic given constraints (DC4), design goal and constraints (DC3) developed, strategies used (DC7), and reasoning applied (DC8) for decision making are intangible variables that happen in the design processes. More studies are needed for detailed explanations.

On the other hand, the number of novel features created by designers in their career is an index of their creativity. Likewise, the number of common features created by the same designers showing in their designs is another index of the level of individual style. Yet, the style should not last for too long or the same set of features repeated too many times. It is because that too many repetitions intensively visualized in a short period of time would cause visual fatigue with the sense of boredom in consequence, particularly in fashion designs. Similarly, without new creation, creativity is not convincing. More creative design does not mean having stronger style. But, having stronger style would definitely show a high level of creativity. Among the case studies done so far, **Frank Lloyd Wright** had his famous early **Prairie House** style and his late **Usonian House** style. It is rare in modern design history that an architect would have created two famous styles in his/her career. But Wright accomplished them and both of the styles have more of stylistic common features apparent in the buildings. Thus, among the four studied architects here, Wright is the most creative.

In fashion design, as we all recognize, dress designs must be creative and stylistic to lead cultural trends for every season. Thus, fashion design is the most challenging profession, with quick fashion changes in style and creativity, and thus is the most stylistic and creative profession among all design fields.

In short, creativity happens at the moment when the generated solution (or product) has the following characteristics: (1) it changes the conventional solution; (2) it is different to the conventional one; (3) it is beautiful and functional; and (4) it is accepted by users. Creativity does not just play a role in the arts, invention, and innovation; it also is a part of our everyday routines (Runco and Richards 1997). Thus, creativity is definitely useful, and it can be applied each day to many aspects of our lives.

9.5 Could Style and Creativity Go Together?

As proposed, and supported by the data given on Sect. 8.5 in Chap. 8, style and creativity are different things attributing to cognitive phenomena. Creativity is all about creating something that is innovative, novel, or not created before. Style is all about elements that are "created first with special values" and then repeated over and over. The first created element with special values in this definition of style signifies a creative action that attributes creativity. If the element is not created by the designer but copied from other sources and repeated, then it should not be recognized as creative with authentic stylistic expression. Therefore creative designers or individuals would have the tendency and opportunity to be stylistic, but stylistic designers might not be creative under the condition that either the style is not original or the style has not been changed over some period of time.

Exploring the notion from the other perspective, style is the ways of doing things that are revealed in products. The elements of the repetitious ways of doing things could be defined as "process style", and the repetitious features in products generated by repeating such ways of doing things in the processes are defined as "product style". The elements shown in product style might be the features selected from designers' preferred forms, applied directly to the product, and repeated. In this regard, such a pattern has nothing to do with creativity but merely replication. However, in process style, a designer would reapply the same process again, which might show innovation occurring in products through variations of some cognitive mechanism applied. Therefore, creativity may come from process style rather than product style. But, this does not mean that studies of the driving forces of creativity should be approached from the process style. Instead, creativity is an independent variable by its own right, and style is a dependent variable of creativity with another variable of repetition added to the formula, as shown below, to conceptually define the correlation between style and creativity.

$$S\left(X,\ F\right) = C\left(X_i,\ F_j\right) + \left(k\ *\ R\left(F_j\right)\right), \quad i \geq 3, j \geq 4, k \geq 3.$$

In this conceptual formula, X represents a set of objects or buildings. F is a set of common features appearing in all members of the object set X. X_i and F_j are member in sets X and F. Thus, a given style S(X, F), with a set of common features F shown in all objects in X, is the sum of X_i, an original creation with the feature F_j identified as a novel feature, plus k times of repetition of feature F_j. Features are created by a series of cognitive processes on managing design information, or by implementing a set of design rules. From the collected experimental data (see Chap. 4), three is the threshold for style recognition, which also depends upon the contents of features. Stronger features would be easily perceived than weaker features. Thus, the number of repetitions, represented by k, must be three or greater than three. In other words, the number of features appearing in X_i to represent its style must be four; and when the number is down to 3 or fewer, the represented style is not recognizable ($j \geq 4$, see p. 100). If there are three features repetitiously appearing in X three times, then the style of X can barely be recognized with low probability ($i \geq 3$ and $k \geq 3$, see p. 105). To be clear, C (X_i, F_j) in this formula is not applied prescriptively to measure creativity, but, to signify the new and novel generation of F_j in X_i.

In this concept, the cognitive mechanism of repetition could generate similar, repeated features across design products, which would cause the phenomenon of rhythm to highlight the value of a product style, as described in Chap. 2. Similarly, the repetition mechanism, if applied appropriately in design processes, could also make the processes more efficient. This is the nature of design automation, which is the aspect of having parts of the design procedures been familiar enough to be utilized directly without any further reflection. Such repeated automatic processes across designs, characterized as a "process style," could possibly generate similar features as end results for each design and particularize a specific product style.

On the other hand, inappropriately utilizing repetition will lead to certain undesirable side effects. For instance, in design, some designers might keep repeating the same thing over and over again, unable to break free of one idea. This phenomenon of "**design fixation**"[4] which is the result of repetition (Adamson 1952) is considered to block creativity and is not what a good designer aspires to be. Therefore, in doing things, designers must apply different methods with variations to meet the problem context for new creations. After all, once being exposed to the same product style for some time, the style must be changed to promote creativity. Otherwise, obsession over the same style will either be boring to users or beholders, or would block creative thinking for the designer.

9.6   Improving Cognition, Creativity and Style

9.6.1   Can Cognition Be Improved?

There is an interesting and arguable experiment on brain training through computer games done on the Internet to help people improve their mental ability. The experiment recruits volunteers to practice a series of online tasks for a minimum of 10 min a day, three times a week, for 6 weeks. The experiments have three groups. One group focuses on general intelligence—reasoning, planning, and problem-solving abilities. The second group is trained on attention, short-term memory, visual-spatial abilities and mathematics. The third group is on the use of the Internet to answer obscure questions.[5] Results reported in the journal *Nature* indicated that "people who practice a certain mental task—for instance, remembering a series of numbers in sequence, a popular brainteaser used by many video games—improve dramatically on that task, but the improvement does not carry over to cognitive function in general." The phenomenon of "practice makes perfect" did explain that practice of a particular task would improve the performance on that task, but improvement did not translate beyond anything other than that specific task. The test, with 11,430

---

[4] In design, observed in students and practitioners alike, designers would have a premature commitment to a particular problem solution. It is found that it is difficult to move away from an idea or a problem solution that has been developed or an existing precedent solution. This effect is called *functional fixedness* (Jansson and Smith 1991) or fixation in psychology (Chysikou and Weisberg 2005). Functional fixedness is a type of cognitive bias that involves a tendency to see objects as only working in a particular way or in the way it is traditionally used. It can also impair our ability to think of novel solutions to problems. Such a fixation refers to situations where innovation is blocked.

[5] This article can be found on line at: http://www.nature.com/news/2010/100420/full/4641111a.html

volunteers aged from 18 to 60, has not shown their general cognition abilities of memory, reasoning, and learning (Katsnelson 2010). It is not clear whether training in creative design would help everyday life creativity, or if computer games enhance anything beyond task-specific performance (Weinberg 1989; Smith et al. 1999; Green and Bavelier 2003).

On the contrary, other studies (Jaeggi et al. 2008) in neural science did demonstrate that intelligence is not purely genetic or determined at birth. Studies show that with appropriate training by having tasks exercised in the working memory, cognitive skills and intelligence can be improved. In one study, the trained task was entirely different from the intelligence test itself. Furthermore, the studies also demonstrate that the extent of gain in intelligence critically depends on the amount of training: the more training, the more improvement in **fluid intelligence**, which is the ability to reason and to solve new problems independently of previously acquired knowledge. Similarly, with more training in design, design cognition can be improved as well. As long as the designers' cognition has been improved, will their style and creativity be improved as well?

9.6.2 Can Style Be Improved?

Style is defined in this book as the repetition of features. Any of such features should be new creations of novel forms to achieve a significant level of style—to invent a signature feature representing a new style. Therefore, style could be improved by inventing more new, pleasant, attractive, and expressive features in design. The distinctive way of expression in a product is another notion of style. For instance, New York Five style in architectural design has its own assigned color (purely white), material used (brick or concrete painted in white), and compositional grammars (full height of window openings and corridors connecting spaces, semi-circular staircase landing) applied. These features had not been seen when they were developed, and they are avant-garde. The combinations of these features, indeed, manifest the style that is different from others in the same period. Thus, for improving a person's design style, a designer should look for (or generate new) unique material, special combinational principles, and techniques that reflect (or lead) the cultural trends. Only after new features are created (creativity) and applied across at least three projects, then a style emerges. Thus, creativity always comes first, and style emerges second.

9.6.3 Can Creativity Be Improved?

Everyday lives (little-c) creativity refers to how people think imaginatively and approach problems intuitively. In order to achieve higher levels of creativity, they have to have strong intrinsic motivation, adequate expertise in the field, and

Table 9.1 Summaries of improving cognitive skills to improve style and creativity

	Cognitive items	Possible improvements for style and creativity
1	Expert knowledge	Enhance field knowledge from exercise and experimentation
2	Personality	Keep the intrinsic motivation and be consistent with it
3	Association	Make associations (or links) between different notions in different fields
4	Representation	Apply new representation; or change, modify, improve the used one, if necessary
5	Design constraint	Apply unique and different constraints to generate unique design scenarios
6	Problem structuring	Constantly check the problem situation. Modify the problem structure, if needed, to create a new situation
7	Repetition	Skillfully utilize repetition to create unconventional rhythm and style
8	Design strategy	Use metaphor, analog, morphosis or other techniques through different representations to create new features
9	Feature creation	Make at least 3 signature features. Repeat them in 4 designs to create a style. Then change and invent new features to generate another style
10	Environment	Advocate an encouraging environment to promote creativity
11	Expressive intention	Make the composition and features to be expressive and stylitic

divergent thinking skills. In handling ill-defined design problems, creativity must also have the same three essential components of motivation, expertise, and divergent thinking; plus five others of association, representation, constraint, problem structuring, and strategies. These eight factors are critical for the generation of design creativity. Among these factors, representation and strategy do apply to the generation of style. However, the factor of repetition is not an essential component of creativity; it is an essential and critical component of style.

As the causes of creativity have been explained in the last section, it is clear that the improvement of creativity depends upon the improvement of each of the cognitive skills. The more practice, the better creativity can be achieved. The possible improvements on creativity and style are summarized in Table 9.1. Lastly, the key to creativity is a willingness to put up with non-conformism above all, and that the more conformist an environment is, by and large, the less creative it will be in all dimensions of human activity, but particularly in technological invention. Therefore, an open environment is an encouraging one that promotes creativity.

9.7 Validity of the Theory

How do we know the theories developed so far in this book are valid? Particularly, parts of the experiments of the research on style presented here were conducted in laboratories rather than in real settings. Also, the laboratory settings and the experimental tasks are not exactly the same as the professional setting in real design projects. Are the findings sound?

In psychology, scholars use the term **"ecological validity"** to explain the degree to which the findings of research studies can be generalized to other settings in reality (Brewer 2000). Ecological validity is the degree to which the behaviors observed and recorded in a study reflect the behaviors that actually occur in natural settings. It basically is the extent to which findings from a study can be generalized to the real world.

It has been assumed that if an experiment is realistic or true-to-life, then there is a greater likelihood that the results of the findings can be generalized. If both the laboratory setting and the tasks are artificial, then there is less likelihood that the findings can be generalized (Brewer 2000). In the studies of design thinking conducted in this book, the lab and the design tasks are very much the same as what happen in any design firm and therefore would have similar thinking procedures applied in real life. Thus, the findings can be generalized to real life with high ecological validity; and the findings discovered and theory established in this book are, to some extent, valid enough to explain the psychological phenomenon of style and creativity in design thinking. Of course, the validation of the theory should be left to the reader to decide.

9.8 Future Studies

Thinking involves many different mental activities and patterns due to individual differences. Thinking patterns and behaviors are also different across disciplines because: (1) the nature and character of problems are different; (2) corresponding knowledge, methodologies, and strategies utilized are different; and (3) the problem-solving techniques required are different as well. In order to get insightful understanding of human thinking for problem solving, more studies on different levels of performance in different fields are needed. Future studies are outlined in this final section.

After the literature on design studies published from 1950 to the present time and the theories developed by significant scholars in the field are reviewed, it is fair to say that thinking is cognition, and cognition relates to the operation of cognitive mechanisms. It is also fair to say that design is dependent upon a set of circumstances of problem context that creates possibilities for potential solutions. The relation between the problem context and the possibilities for actions defines the design situations. As such, reasoning used to recognize the situation in order to strategically determine actions is what categorizes design cognition.

Taking the brain-as-computer metaphor, the hardware is human brain, whereas software is the consciousness in the mind. The hardware is the place containing knowledge. The consciousness, or the software, arranges strategies, methods, and representation to manipulate external data and internal knowledge for creating a design. Therefore, the study of human hardware, or the brain, should be on: (1) where the design knowledge is stored; (2) how knowledge would be inspired; and

(3) what cognitive results occur right after knowledge is inspired. The body of knowledge utilized by individual designers is a critical variable in thinking, which has its own favored contents or specialties shaped by personal preferences and individual mindsets, and is built up gradually through long periods of professional education and training. Because of this specialized body of knowledge, designers have unique thinking routines that define their individual style (Chan 2001). Thus, further study in this regard could help to understand the hardware component of knowledge.

In order to see what might happen in the brain, the concept of **neurocognition** (or cognitive neuroscience) is a potential approach method. It is the combination of neuroscience and cognitive psychology. Neuroscience concentrates on the analysis of mechanisms of "neurons," which are the functional units of the nervous system. The human brain is home to 100 billion neurons, each of which stores information. A study analyzing brain activity in relation to the processes of imagery (O'Craven and Kanwisher 2000) showed an interesting result. In that study, colorful brain scans performed through **Magnetic Resonance Imaging** (**MRI**) revealed different hot spots (indicating the blood flow through the brain) when people saw a place and when they saw a face. In the experiment, subjects would first see pictures of a face and a place, and were then asked to recall the same face and place later (Chan 2008, pp. 154–155). Interestingly, the same brain areas lit up when people merely imagined the same faces and places. This indicates that similar information is processed by the same brain area and stored in the same area, and different sets of information are handled by different areas. However, design thinking relates to reasoning and the manipulation of information. If information processing could be transparently visualized, then the "mind is like a black box" metaphor would no longer hold true. Thus, through the MRI scan aided with protocol analysis, patterns of design thinking could be accurately recorded, and data could be used to better explore the real processes of information acquisition, storage and retrieval. From studying the changes in neurons on the MRI scanned images, the process of knowledge could be identified, and the nature of knowledge could be explained.

On the other hand, studies of consciousness, the software side, could also be done by collecting protocol data through psychological experiments and MRI scanned images to analyze the reasoning applied on making decisions, and on the cognitive processes while developing design strategies. This would improve the understanding of design thinking processes, reasoning used, the formation of stylistic features, and the resources of creation. Particularly, if a robust and reliable method of studying design thinking is established by means of combining MRI methods used in cognitive neuroscience and protocol analysis in psychology, then empirical research with experiments could be conducted to achieve a better understanding of creative problem-solving patterns. In fact, protocol analysis could be aided by MRI brain scan images to compensate for the criticized deficiencies and verbal gaps in protocol data; the transparent internal processes of neurons reflected in the brain scans would reflect the fundamental driving forces of human thinking. Particularly, by combining methods of MRI scan and protocol analysis, it would be easy to identify the types of knowledge used in design, the relationships between knowledge, how related knowledge is recalled and applied in design, and how master designers could methodologically use their knowledge at critical moments to generate a creative solution. Of course, the effectiveness in revealing the nuances of design thinking is yet to be proved.

Regarding other research directions, design reasoning and cognitive design skills are the future major research items to be considered. Reasoning has been studied recently, in how it is applied in design to figure out the character of design thinking. But, how reasoning is used in geometric modeling to generate creative forms, how people reason in virtual environments, what types of reasoning in the design processes would be more likely to trigger the creative moment, and what kind of problem situations are more likely to enhance creativity, do need further discussions. For the cognitive design skills, digital design resulting from technological invention would bring new cultural influence to representation in thinking. In the digital media dominated environment, designers might use methods to process digital information that differs from the conventional way of designing. Thus, what cognitive strategies used, while working in digital design, is another question to ask.

The third important research item is design representation. More research is needed to explore how internal representation is constructed to match with the digital external representation. This line of research could be explained from comparing the four different design models used historically.

1. If designers use sketch through pencil and paper mode, the design would be concentrated more on functionality than on the form consideration. Drawing and sketching are part of professional designers' automatic skills after long practice. They could put ideas directly to drawings and spend more time focusing on solving functional requirements.
2. If design is done through study model mode, then designers must think three-dimensionally while they are facing three-dimensional models. In other words, there are three-dimensional internal representations existing in designers' minds, and their thinking turns naturally toward three dimensions. The generated form would be richer than the use of sketch mode, but less effort would be paid to resolve functional requirements within the same time constraints.
3. If a digital model or a virtual reality model mode is applied, then designers must develop some design rationale for formulating possible three-dimensional forms before they start designing, and search for appropriate system commands to accomplish the creation of the expected form. But, in the design processes, especially with novice users, we can expect to have lengthy search for software commands, system functions, and appropriate sequences for form generation. In result, the architectural design thinking in this case would be interrupted from time to time, which would prevent design continuation.
4. If a designer applies scripting mode for design representation, designers must know the questions they are facing, the strategies, variables, and algorithms they will apply, procedures of encoding to be processed, software to be utilized, and system commands to be included in the script codes. After all the knowledge is well prepared, then designers need to execute the codes for form generation and

evaluate the results to test whether functions are satisfied, or problems resolved. These complicated and lengthy procedures in script modeling—used particularly in parametric modeling and rapid prototyping—are more complicated than the conventional design representations, because designers are simultaneously working on a geometric form design plus a software engineering design. Their internal representations are in multiple formats of mathematic equations, computer program codes, and mental images. Designers have to find the balance among them.

Due to the different nature of these internal representations used in the mind, the strategies applied for handling the transformation between the internal and external would be different among designers. It will be interesting to find out how external digital representation affects creative thinking due to its multi-task nature.

Regarding the future research on style, a design style can be seen as consisting of certain elements and ways to compose these elements, as shown in the pointed arch for Gothic and Islamic architecture, the round arch for Roman and Byzantine. As such, formal properties of shape, color, arrangement, texture, size, and orientation are visual elements of a style. In this regard, a design style relates to not just the features of an object, but also the relationships between the components of the object that can be recognized explicitly. Thus, a style can be described not only in terms of the common components of a design, but also on how they are arranged methodologically. For instance, the Familia Sagrada Cathedral in Barcelona designed by **Antoni Gaudi** is a good example on using organic features on facades with special connections among features. Such arrangements might have certain personal considerations attached, which are the designer's intended expressions or the results of certain Gestalt consideration. Further studies should be on the topological context of features and the relationships between the type of knowledge and type of features created.

Regarding future study on creativity, it would be appropriate to find out how creativity can be cognitively measured. One possible way would be to compare the cognitive performance of three groups of eminent, good, and average architects by their designs on the same design task. Cognitive performance could be compared by concentrating on the major factors triggering creativity, including problem structuring, representation used, reasoning utilized for decision making, and knowledge level accomplished among other factors. Such studies could be run in laboratory settings with methods of video recording or computer simulation. Findings from the comparison of cognitive data could provide clues for the practitioners to improve their creativity, and for students to learn how they could achieve outstanding levels of design creativity.

All these possible future research directions, as explained in this section, are classified by groups and outlined on Table 9.2. Hopefully, the overall plan of study will provide enough further information to rightfully justify the indirect cognitive factors to the generation of creativity and the expressive nature of style in design.

Table 9.2 Future studies itemized by categories

	Research category	Research contents
1	Reasoning	How reasoning is utilized in virtual environment and in geometric modeling?
2	Digital design cognition	What are the differences of information processing between traditional design cognition and the digital design cognition?
3	Cognitive skills	What cognitive mechanisms that effect design thinking the most?
4	Internal representation	How visual information of design knowledge is represented in memory?
5	Digital representation	Impact of digital design representation to design thinking?
6	Design strategies	How design strategies are utilized to generate a form for solution?
7	Style	What the expressive features are and how do they define a style?
8	Creativity	How to measure design creativity?
9	Virtual environment	Definition of privacy and virtual thinking in virtual space?

References

Adamson RE (1952) Functional fixedness as related to problem solving: a repetition of three experiments. J Exp Psychol 44(4):288–291

Atkinson RC, Shiffrin RM (1971) The control of short-term memory. Sci Am 225(2):82–90

Brewer MB (2000) Research design and issues of validity. In: Reis H, Judd C (eds) Handbook of research methods in social and personality psychology. Cambridge University Press, Cambridge, pp 3–16

Chan CS (1990) Cognitive processes in architectural design problem solving. Des Stud 11(2):60–80

Chan CS (2001) An examination of the forces that generate a style. Des Stud 22(4):319–346

Chan CS (2008) Design cognition: cognitive science in design. China Architecture and Building Press, Beijing

Chysikou EG, Weisberg RW (2005) Following the wrong footsteps: fixation effects of pictorial examples in design problem-solving task. J Exp Psychol Lear Mem Cogn 31(5):1134–1148

Cross N (2011) Design thinking: understanding how designers think and work. Berg, Oxford

Goleman D, Kaufman P, Ray M (1992) The creative spirit. Dutton, New York

Gombrich EH (1960) Art and illusion: a study in the psychology of pictorial representation. Pantheon, New York

Green CS, Bavelier D (2003) Action video game modifies visual selective attention. Nature 423:534–537

Jaeggi SM, Buschkuehl M, Jonides J, Perrig WJ (2008) Improving fluid intelligence with training on working memory. Proc Natl Acad Sci 105(19):6829–6833

Jansson DG, Smith SM (1991) Design fixation. Des Stud 12(1):3–11

Katsnelson A (2010) No gain from brain training. Nature 464:1111. doi:10.1038/4641111a

Magnani L (2009) Abductive cognition. The epistemological and eco-cognitive dimensions of hypothetical reasoning. Springer, Berlin/Heidelberg

O'Craven KM, Kanwisher NK (2000) Mental imagery of faces and places activates corresponding stimulus-specific brain regions. J Cogn Neurosci 12(6):1013–1023

Runco MA, Richards R (eds) (1997) Eminent creativity, everyday creativity, and health. Ablex, Greenwich, Conn

Shaywitz S (2003) Overcoming dyslexia: a new and complete science-based program for reading problems at any level. Alfred Knopf Publisher, New York

Smith ME, McEvoy LK, Gevins A (1999) Neurophysiological indices of strategy development and
    skill acquisition. Cogn Brain Res 7(3):389–404
Sternberg RJ (1997) Successful intelligence. Plume, New York
Weinberg R (1989) Intelligence and IQ: landmark issues and great debates. Am Psychol
    44(2):98–104

Name Index

A

Aalto, A., 3, 54, 55, 56, 65–66
Ackerman, J., 96, 98
Akin, O., 19, 22, 24, 26, 34, 44, 66, 67, 159, 266
Alessi, 1
Alexander, C., 13–17, 22, 111, 223
Ando, T., 5, 263, 279–289, 293–297, 299, 310, 311
  Awaji Yumebutai, 2000, 284
  Azuma House, 1976, 283, 293, 296, 297, 311
  Bansho house, 1976, 284, 295
  Benetton Communication Research Center, 2000, 282
  Chapel of Mt. Rokko, 1986, 283
  Chichu Art Museum, 2004, 286, 288
  Children's Museum, 1989, 285–286
  Church of the Light, 1989, 285–288
  Church on the Water, 1988, 286, 287
  Forest of Tombs Museum, 1992, 285, 286
  Kidosaki House, 1986, 282
  Komyo Ji Temple, 2000, 287
  Koshino House, 1981, 285
  Langen Foundation (2004), 287
  Modern Art Museum, 2002, 287
  Museum of Wood, 1994, 282
  Naoshima Contemporary Art Museum, 1992, 282
  Pulitzer Foundation for the Arts, 2001, 286
  Rokko Housing One, 1983, 282
  Rokko Housing Two, 1993, 282
  Stone Hill Center, 2008, 285
  Sumiyoshi House, 1976, 283
  Tea House in Oyodo, 1981, 281

  UNESCO, 1995, 286
  Water Temple, 1991, 283–284, 287
Archer, B., 16, 17
Association of Computer Machinery (ACM), 14, 20

B

Bramante, D., 108
Braque, G., 88, 89, 91
Breton, A., 90

C

Cezanne, P., 112, 113
Cox, C., 250, 251, 263

D

Dali, S., 27, 90
Darke, J., 26, 63, 64
Darwin, C., 97, 251
De Quincey, T., 96
Dearle, J.H., 87
Delacroix, E., 113
di Bondone, G., 96

E

Eastman, C., 10, 24, 26, 42, 45, 66, 67, 198, 255, 258, 289
Ebbinghaus, H., 28, 29
Eisenman, P., 48, 50
Ericsson, K.A., 64, 67, 68, 198, 264

© Springer International Publishing Switzerland 2015
C.-S. Chan, *Style and Creativity in Design*, Studies in Applied Philosophy,
Epistemology and Rational Ethics 17, DOI 10.1007/978-3-319-14017-9

F
Foz, A.T.K., 26, 45, 66, 186, 226
Frith, C., 28, 30, 91

G
Gaudi, A., 320
Gombrich, E.H., 30, 98, 107, 111, 159, 203,
        258, 309
Graves, M., 1, 116, 118, 119, 228, 229
Gropius, W., 6

H
Hitchcock, H.-R., 117, 129, 130, 185
Howe, J., 177–179

J
Jobs, S., 2
Johnson, M., 45, 122
Jones, J.C., 13, 17, 22

K
Kepler, J., 265
Klee, P., 12, 279
Klimt, G., 87, 88
Kubler, G., 98, 103, 158

L
Lakoff, G., 45, 122, 255
LaRue, J., 159
Leroy, L., 86
Li, J., 97

M
MacCormac, R., 173, 182
Manet, E., 86
Manson, G., 117, 129, 153, 170, 173, 182,
        186–189, 203
March, L., 26
Meier, R., 116, 118–122, 126, 127, 128,
        148, 149
Meyer, L., 159
Monet, C., 112, 113
Moore, C., 3, 116, 119–122, 124, 125, 126,
        128, 135, 148
Morris, W., 87

N
Newell, A., vii–viii, 21–24, 38, 41, 66, 201,
        214, 256, 289
Newton, E., 110, 157, 158

P
Pablo, P., 88, 89
Paivio, A., 29
Peirce, C.S., 51–52, 257, 258
Piano, R., 5, 46–47, 279–289, 291,
        295, 298, 299
    Academy of Sciences, 2008, 279
    Centre Pompidou, 1971, 279
    High Museum Expansion,
        2005, 279
    Kansai Airport, 1994, 279
    New York Times Tower, 2007, 279
    Paul Klee Center, 2005, 279
Plato, 83–85, 157, 246

R
Rapoport, A., 160
Renoi, P.-A., 113, 149
Reynolds, J., 85
Riegl, A., 97–98
Rorick, H., 174
Rosch, E., 32, 122, 123, 148
Rowe, P., 10

S
Saarinen, E., 46
Schapiro, M., 83, 98, 110, 111,
        158, 200
Scully, V., 170, 173, 177–179, 181,
        185, 291
Simon, H., ix, 11, 18, 21–24, 26, 30, 34,
        38, 41, 43, 64, 66, 67, 68, 111,
        159, 160, 198, 201–203, 214,
        245, 250, 256, 258, 264–266,
        291, 294
Sparshott, F., 103, 111, 158
Stiny, G., 26, 99, 101
Sullivan, L., 153, 162, 187

T
Target, 1
Turing, A., 27

U

Ullman House, 175
Unity Temple, 162, 163, 166, 167, 171, 173,
    174, 176, 182, 184–186, 191, 192

V

van der Rohe, M., 48, 88
van Doesburg, T., 11
van Gogh, V.W., 107, 108, 112–113, 253
Vasari, G., 85, 96
Vauxcelles, H., 88
Venturi, R., 116, 119, 120, 126
Vermeer, J., 151–153

W

Wagner, O., 5, 276–279, 287, 289, 299
Wallas, G., 245, 251, 255
Watson, V., 116, 117, 119
White, C.E. Jr., 168, 169, 172, 174, 176–177,
    178, 179
Willitts House, 170–171, 185
Winslow House, 170, 171, 185
Wright, F.L., viii, ix, 3, 4, 5, 26, 58, 65, 101,
    102, 107, 113, 114, 116, 117, 119, 120,
    124, 126–130, 136–138, 140, 143, 144,
    146, 148, 149, 153, 161, 162–165,
    167 –193, 200, 203, 215, 234, 236,
    263, 276, 287, 289, 291, 295, 296, 299,
    309, 310, 312

Bagley House, 187, 188
Barton House, 117, 130,
    182, 186
Blossom House, 187, 188
Bradley House, 188
Coonley House, 185
Dana House, 117, 129, 188
Davenport House, 186, 188
Evans House, 130, 182
Foster House, 188
Hickox House, 129, 188
Horner House, 186
Ingalls House, 186
Johnson Wax Building, 178, 181,
    185, 186
Kaufmann House, 185
Larkin Building, 162, 163, 166,
    173, 182, 184–186,
    191–192
Little House, 117, 130,
    136–138, 143, 145,
    186, 236
Martin House, 117, 129,
    130, 186
Ross House, 170–171,
    182, 183
A Small House with Lots of
    Room in It, 188
Walser House, 186
Yahara Boat Club, 162, 163, 166, 185,
    191–192

Subject Index

A

Abstract-expressionism, 32, 90, 219
Aesthetics, 2, 3, 9–10, 52, 81, 84–86, 92,
    158, 172, 236, 246, 261–263, 289,
    291, 305, 307
Affect tolerance, 252
Art Nouveau, 87, 278
Arts, 1–3, 11, 12, 17, 19, 20, 23, 46, 54, 55,
    65, 71, 81–98, 103, 107, 108, 110–113,
    115, 118, 122, 123, 128, 150, 153,
    157–159, 161, 175, 199, 245–250, 253,
    255, 256, 258, 259, 278, 279, 282,
    285–288, 300, 301, 305, 306, 312
Arts and crafts movement, 12, 87
Associationism, vii, 14, 16–18, 21, 22, 24,
    26–33, 36, 49, 53, 60, 65, 82, 87, 107,
    108, 111, 118, 131, 158, 162, 168, 179,
    181, 201, 202, 207, 214, 221, 223, 247,
    251, 255, 256–258, 262, 277, 286, 287,
    292, 295, 308–310, 312, 316

B

Bauhaus, school of art and craft, 12
Building Information Modeling (BIM), 61

C

Categorization, 1, 10, 13, 17, 21, 23, 25, 30, 36,
    42–43, 62, 81, 84, 99, 101, 103, 110, 113,
    115, 117, 19, 122–126, 136, 148, 163,
    168, 172, 204, 251, 252, 254, 255, 257,
    259, 281, 282, 285, 290, 299, 317, 321
Chunk, 29, 30, 32, 33, 41, 42, 60, 202, 256,
    306, 309, 310

Cognitive strategy, 2, 20, 21, 43–51, 54, 57,
    60, 319
Cognitive theory of style and creativity,
    5, 305–321
Column order
    Corinthian column, 108
    Doric column, 111
    Ionic column, 108
Conceptual-driven process, 61
Connectionism, viii, 18, 20, 26, 28, 30,
    32, 60, 61, 70, 90, 92, 93, 103,
    140, 158, 250, 254, 256, 275,
    291, 308, 320
Consciousness, 2, 4, 10, 27, 29, 30, 53, 57, 91,
    159, 254, 255, 262, 275, 289, 306–308,
    317, 318
Constraints, vii, 4, 21–23, 33–38, 41, 60,
    63, 71, 94, 95, 97, 159–161,
    167–171, 179, 180–182, 185–186,
    189–190, 192, 193, 197, 201–205,
    211, 214, 216, 219–225, 227,
    231–235, 237, 238, 254, 276, 284,
    287, 289, 292–297, 299–301 308, 310,
    312, 316, 319
    budget constraint, 179, 216, 221,
        232, 233
    context constraint, 220,
        221, 232
    design constraint, 4, 38, 159, 160, 161,
        169, 170, 179, 181, 185–186, 192, 197,
        201–205, 211, 214, 216, 219, 223,
        231–233, 235, 237, 254, 284, 287, 292,
        293, 295–296, 300, 310
    global constraint, 36, 63, 179, 201, 219,
        223, 234

© Springer International Publishing Switzerland 2015
C.-S. Chan, *Style and Creativity in Design*, Studies in Applied Philosophy,
Epistemology and Rational Ethics 17, DOI 10.1007/978-3-319-14017-9

Constraints (*cont.*)
   local constraint, 171, 202, 221, 294
   sunshine/light constraint, 219, 220,
      222–224, 233, 296
   view constraint, 220, 224, 225
Creativity
   assess of creativity, 4, 245, 258–259, 263,
      265–266, 305
   Big-C creativity, 261, 298
   cognitive processes, vii, viii, 22, 26, 34,
      35–36, 49, 62, 66, 67, 91, 254–258,
      263, 264, 265, 266–300, 307,
      313, 318
   creative environment, 252, 254
   creativity data collection, 262–266
   history, xviii, 56, 245–252
   innovation, 36, 37, 43, 245, 263, 298
   invention, 9, 96, 113, 243
   little-C creativity, 251, 262, 266,
      298, 315
   mini-c creativity, 251, 256
   operational definition, viii, 160, 245, 251,
      254, 260–263, 305, 311
   pro-c creativity, 251
   train, 40
   traits, 250, 252–253
Cubism, 27, 88, 91

D
Dada, 90
Declarative knowledge, 27, 28, 201,
   214, 223
Declarative memory, 30, 31
Design
   abstract, 18, 44, 168, 178, 192, 234,
      291, 310
   activity, 10, 18–22, 26, 34, 52, 60, 62, 63,
      66, 67, 102, 103, 200, 204, 209
   brief, 5, 65, 204, 207, 214
   cognition, 1–3, 9–72, 161, 168, 197, 262,
      275, 281, 282, 288, 289, 292, 299, 305,
      307–311, 315, 317
   constraint, 4, 33–38, 71, 159–161, 169,
      170, 179, 181, 185–186, 192, 197,
      201–205, 211, 214, 216, 219, 223,
      231–233, 235, 237, 254, 284, 287, 292,
      293, 295–296, 300, 310
   global constraint, 36, 63, 179, 201, 219,
      223, 234
   goal driven, 17, 33, 214
   parti, 26, 44, 45, 66, 226, 310
   scenario, 34–35, 49, 63, 219, 224, 226,
      290, 291, 299, 310

   strategy, 33, 36, 43–51, 53, 61, 66, 71, 200,
      297, 308, 311, 318
Design cognition, 1–3, 9–72, 161, 168,
   197, 262, 275, 281, 282, 288,
   289, 292, 299, 305, 307–311,
   315, 317
   conceptual driven, 61
   operational definition, 59–61, 115–128,
      260–263
   study methods, 19, 62–72
   top-down driven, 44, 61
Design fixation, 314
   functional fixedness, 314
Design Method Group, 17
Design method movement,
   12, 20
Design methodology, vii, 2, 12, 17, 18, 45,
   181–190, 281, 290
Design research, 11, 19, 20
Design Research Society, 17
Design studies, 2, 5, 11–21, 67,
   231, 317
Design thinking, vii, viii, 1–3, 10–11, 18, 19,
   21, 25, 26, 30, 32, 34, 37, 40, 53,
   59–62, 67, 68, 72, 99, 102–103, 163,
   199, 254, 263, 282, 288, 290, 292,
   307–309, 317–319
De Stijl, 11, 88, 89
Divergent thinking, 60, 257–260, 292, 305,
   308, 315–316
Drawing cognition, 307

E
Ecological validity, 5, 317
Emotionalist, 54, 85, 91, 168,
   246, 252
Empiricism, 18
Environmental Design Research
   Association, 17
Episodic memory, 30, 31
Ergonomics, 13
Expert system, 23

F
Feature
   categorization, 103, 117, 122,
      136, 163
   common feature, 113–117, 119, 120,
      122–130, 133, 135, 138, 142, 148, 149,
      151, 153, 154, 161, 162, 167, 182,
      190–191, 197, 214, 215, 220, 277, 288,
      301, 309, 312, 313

constant feature, 153, 162, 186, 189, 225, 275
critical common feature, 114, 115, 117, 128, 135, 149, 153, 154
effectiveness, 139–141
feature matching, 122
frequency, 122, 136–142
identification, 112–114, 119–120, 126
interaction, 3, 141, 142
prominent feature, 107, 113, 189
recognizability, 1, 3, 4, 107, 119, 128, 135, 136, 142–149, 153, 197, 227
signature feature, 113, 299, 300, 315
similarity, 3, 117, 122, 123
Fine arts, 81, 84, 91, 107, 112, 150, 158, 199, 300, 306
Fluid intelligence, 315
Formalism, 11, 65, 88
Froebel Kindergarten education, 173

G
Generality, 148
General linear model, 131, 134
Gestalt psychologist, 22
Gestalt psychology, 12, 92, 159, 256
Graph theory, 16
Grid system, 61, 65, 172–174, 177, 179, 181–183, 192, 193, 296, 310
Group style
  Big-G group style, 298
  little-G group style, 298, 299

H
Heuristics, 43, 60, 256, 258, 261, 265, 291, 293, 300
Hexadecimal, 111, 112
Hippocampus, 30–31

I
Ill-defined problem, 18, 23–25, 41, 42, 43, 62, 99, 254, 260, 290, 291, 307, 316
Ill-structured problem, 23, 24
Impressionism, 86–87, 110, 112, 149
Individual style, viii, 1–4, 34, 58, 68, 71, 72, 103, 107, 111, 113–115, 124, 135, 142, 146, 153, 154, 160–162, 190, 191, 193, 197, 199, 203, 204, 232–234, 236–238, 245, 264, 266–267, 275, 276, 300, 309, 312, 318
Information, vii, 1, 3, 10, 12, 19, 20, 22–24, 26, 27, 29–33, 35–37, 40, 41, 44, 51,

53, 59–62, 64–67, 72, 82, 91, 92, 103, 111, 123, 124, 160, 161, 163, 169, 176, 187, 197–200, 202, 207, 208, 213, 216, 219, 228, 237, 245, 254, 256–258, 260, 261, 264, 267, 275, 282, 291–295, 298, 306–308, 310, 313, 318–320
Information processing (IP) theory, vii, 10, 12, 22, 26, 27, 30, 61, 92, 160, 198, 199, 298, 307, 318
Innovation, 36, 37, 43, 64, 65, 112, 170, 175, 188, 243–245, 253, 257, 260, 263, 266, 292, 298, 312–314
Inside-out strategy, 44
Interview method, 63–65, 263
Intuition, 13, 28, 60, 70, 91, 99, 199, 315
Invention, 113, 183–184, 243–246, 262, 264–266, 312, 315, 316, 319
IP theory. *See* Information processing (IP) theory

K
Knotted-String records, 82

M
Magnetic resonance imaging (MRI), 318–319
Mental image, viii, 1, 13, 29, 32, 42, 43, 67, 153, 197, 199, 202, 203, 214, 227–231, 258, 275, 290–293, 297, 300, 301, 306, 310–312, 320
Method of studying cognition, viii, 62–72
  case study type, 64–66
  controlled experiment type, 62–63
  interview type, 63–64
  think aloud type, 66–68
  Virtual Architectural Design Tool (VADeT), 68–70
Methodological approach to design period, 17
MRI. *See* Magnetic resonance imaging (MRI)

N
Network theory, 29, 31, 32
Neurocognition, 318
Neuron, 32, 33, 92, 319

O
Object-oriented programming (OOP), 16
Operations research (OR), 12, 17, 19, 40
Outside-in strategy, 44

P

Pattern language, 13–16, 223
Perception,1–3, 10, 12, 20, 22, 29, 39, 40, 53,
    54, 57, 68–70, 91, 92, 112, 115, 120,
    142, 148, 149, 154, 159, 175, 227, 228,
    260, 275, 282, 296, 300, 301, 306,
    309, 313
Positivism, 18
Pre-solution model, 30, 45, 186–187, 189,
    203, 226, 227, 231, 232, 237, 238
Primary auditory cortex, 30
Primitive art, 82
Priori concept, 226
Problem behavior graph, viii, 38
Problem solving theory, 2, 17, 18, 22, 25, 26
    critical problem situation, 39–40
    definition of problem, 22
    problem behavior graph, viii, 38
    problem space, 23, 24, 66, 201, 289, 291,
        292, 294
    problem structuring, 21, 37–40, 44, 60,
        71, 256, 291, 292, 294, 297, 309–312,
        316, 320
    solution structuring, 38–40, 44, 60
Procedural knowledge, 27, 28, 54, 201, 202,
    214, 223, 307, 308, 311
Procedural memory, 30
Production system, 100, 101, 201, 214, 223
Protocol analysis, viii, 4, 19, 66–68, 70, 72,
    163, 198, 263–265, 291, 297, 309, 318
Prototype model, 122, 148

R

Random error, 131, 134, 136
Random selection, 204
Reading cognition, 306
Reasoning, 20, 21, 35, 44, 45, 51–53, 59–61,
    65, 66, 70–72, 94, 199, 226, 257–258,
    262, 266, 275, 291, 297, 306–309, 312,
    314, 315, 317–320
    abductive reasoning, 51, 52, 257–258, 291,
        307, 308
    case-based reasoning, 44, 45, 226
    deductive reasoning, 51, 257, 308
    inductive reasoning, 51
    spatial reasoning, 53, 308
Reflection-in-action, 28, 39
Rehearsal, 54, 311
Repertoire, viii, 30, 41, 201, 202, 255
Repetition, 2, 21, 29, 53–60, 66, 139, 163,
    187, 193, 198, 200, 203, 214, 227, 231,

232, 237, 281, 299, 300, 308, 309,
    311–316
Representation, viii, 1, 9, 81, 107, 158, 199,
    248, 275, 306
    conceptual representation, 26
    external representation, 41–43, 61, 258,
        289, 290, 293, 294, 311, 319
    internal representation, viii, 41–43, 53,
        62, 91, 258, 289–291, 294, 307, 311,
        319, 320
Rule-of-thumb, 291

S

Schema theory, 29, 33
Scripting, 61, 319
Seasoned knowledge, 200, 231, 234–236
Semantics
    information, 23
    memory, 30
    network theory, 29, 31, 32
    solution, 219
Sensory register, 306
Shape grammar, 26, 99–103
Spatial cognition, 20, 53
Style
    Baroque, 96, 110
    Big-G style, 298
    Carolingian style, 96
    choice, 111, 159, 160, 201, 234, 262,
        301, 309
    common features, 113–117, 119, 120,
        122–126, 128, 133, 135, 138, 142, 148,
        149, 151, 153, 154, 161, 162, 167, 190,
        191, 197, 214, 215, 220 277, 288, 301,
        309, 312, 313
    critical common features, 114, 115, 117,
        128, 135, 149, 153, 154
    degree of style, 115, 128–135, 137, 138,
        148–150, 153, 154, 162, 197, 275–276,
        279, 301, 311
    direct factor of expressing style, 231, 232
    expressiveness, 135, 154, 158
    feature matching, 122–123
    feudalism, 95
    group style, 5, 101, 107, 114, 135, 150,
        233, 281, 298–299
    humanism, 95
    identification, 103, 114–115, 134, 135,
        142, 150
    indirect factor of expressing style, 232,
        300, 312

individual style, viii, 2–4, 34, 58, 68, 71,
    72, 103, 107, 111, 114, 115, 124, 135,
    142, 146, 153, 154, 160–162, 190–191,
    193, 197, 199, 203, 204, 232–234,
    236–238, 245, 264, 267, 275, 276, 300,
    309, 312, 318
little-g style, 298, 299
measurement of style, 115, 135–142
Middle Ages, 94–96, 246
modern style, 2, 3, 116, 118–120
New York Five style, 3, 276, 315
operational definition, viii, 3,
    115–128
perceptibility, 115, 149–150, 154
period style, 97, 107, 135
Prairie House style, 3, 4, 26, 107, 113,
    161–167, 287, 295, 298, 299, 312
process style, 313, 314
product style, 153, 313, 314
recognizability, 128, 135–137, 142–148,
    149, 153
regional style, 107, 135, 233
Rococo, 92, 96, 149
seasoned knowledge, 200, 231,
    234–236
signature feature, 113, 299, 300, 315
similarity, 122–125
stilus, 92, 93
between style, 5, 115, 128, 149, 150, 154,
    157, 275, 305, 313
within style, 115, 128, 135,
    148–150, 157
Taliesin, 162, 177
Usonian House style, 153, 162,
    182, 312
vernacular style, 3, 116, 119–121, 124,
    126, 276
Surrealism, 90
Syntax, 16, 111, 112, 135,
    140, 153

T
Temporal neocortex, 30

Theory of style, viii, 3, 4, 98, 107
Theory of style and creativity,
    305–321
Troubleshooting, 23–24
True art, 83, 92
Types of problem, 23–25
    crypto-arithmetic, 22
    insight problem, 23
    missionaries and cannibals
        problem, 22
    mutilated checkerboard (MC) problem,
        41–43, 258, 289, 291
    open-ended problem, 23
    procedural problem, 23
    routine problem, 23, 258
    social policy problem, 23–25
    tame problem, 24
    Tower of Hanoi, 22
    wicked problem, 24, 25

U
Unit system, 4, 170, 172–173, 176, 177, 179,
    182, 234

V
Verbalization
    concurrent verbalization, 264
    introspective verbalization, 264
    retrospective verbalization, 264
Vienna Secession, 87, 88
Virtual Architectural Design Tool (VADeT),
    68–70

W
Well-structured (well-defined) problem,
    23–25, 37, 41–43, 62, 289–291
Will to art, 98

Y
YingZao FaShi, 97

Printed by Printforce, the Netherlands